PROCEEDINGS OF THE 30TH INTERNATIONAL GEOLOGICAL CONGRESS
VOLUME 18 PART A

GEOLOGY OF FOSSIL FUELS - OIL AND GAS

T0187746

Proceedings of the 30th International Geological Congress

PROCEEDINGS OF THE

30TH INTERNATIONAL GEOLOGICAL CONGRESS

BEIJING, CHINA, 4 - 14 AUGUST 1996

VOLUME 18 PART A

GEOLOGY OF FOSSIL FUELS - OIL AND GAS

EDITORS:

SUN ZHAOCAI
CENTRAL LABORATORY OF PETROLEUM GEOLOGY, MGMR, WUXI, CHINA
WANG TINBIN
INSTITUTE OF PETROLEUM GEOLOGY, MGMR, BEIJING, CHINA
YE DELIAO
CENTRAL LABORATORY OF PETROLEUM GEOLOGY, MGMR, WUXI, CHINA
SONG GUOJUN
CENTRAL LABORATORY OF PETROLEUM GEOLOGY, MGMR, WUXI, CHINA

CRC Press
Taylor & Francis Group
Boca Raton London New York

CRC Press is an imprint of the
Taylor & Francis Group, an **informa** business

First published 1997 by VSP BV

Published 2019 by CRC Press
Taylor & Francis Group
6000 Broken Sound Parkway NW, Suite 300
Boca Raton, FL 33487-2742

First issued in paperback 2020

© 1997 by Taylor & Francis Group, LLC
CRC Press is an imprint of Taylor & Francis Group, an Informa business

No claim to original U.S. Government works

ISBN-13: 978-0-367-57947-0 (pbk)
ISBN-13: 978-90-6764-234-7 (hbk)

Visit the Taylor & Francis Web site at
http://www.taylorandfrancis.com

and the CRC Press Web site at
http://www.crcpress.com

CONTENTS

Part I
THEORY OF OIL AND GAS GENERATION IN NONMARINE BASINS AND THEIR CONDITIONS OF ACCUMULATION

Part II
NEW PROSPECTS AND TARGETS OF HYDROCARBON IN THE 1990s

Part III
RESERVOIR CHARACTERIZATION

Part IV
MARINE CARBONATE SOURCE BEDS AND RESERVOIR

Part V
OIL/GAS BASIN SYSTEM AND OTHERS

PART I

THEORY OF OIL AND GAS GENERATION IN NONMARINE BASINS AND THEIR CONDITIONS OF ACCUMULATION

Proc. 30th Int' l. Geol. Congr., Vol. 18, pp. 3~15
Sun Z. C. *et al.* (Eds)

Advances in Hydrocarbon Generation Theory (I)— Immature Oils and Generating Hydrocarbon and Evolutionary Model

HUANG DIFAN

Research Institute of Petroleum Exploration and Development, *CNPC*, *Beijing*

Abstract

The wide existence of immature oils throughout the world especially in China broke through the hydrocarbon generation theory of kerogen thermal degradation. Studies have shown that soluble and insoluble organic matter are organically connected in the sedimentary rock, both contributing to hydrocarbon generation. In the diagenesis, kerogen can not produce oils, so immature oils derive directly from the degradation of the soluble lipids. Hydrocarbon generation process by kerogen thermal degradation takes place mainly in the early catagenesis $R^o = 0.5\%$-1. 2%, and the oils generated are normal. In the late catagenesis, hydrocarbon generation comes into wet gas stage ($R^o = 1.2\%$-2. 0%), with pyrrobitumen and oils cracking into high mature light oils and wet gases respectively. In this paper, the author presented a novel hydrocarbon generation and evolutionary model with soluble and insoluble organic matter both contributing to hydrocarbon generation.

Keywords: Immature oils, Kerogen, Soluble organic matter, Generating hydrocarbon and evolution.

HYDROCARBON GENERATION THEORY OF KEROGEN THERMAL DEGRADATION AND IMMATURE OIL

Hydrocarbon generation theory of kerogen thermal degradation was gradually developed in the late 1960' s[25,19,18,22,1,34]. Tissot & Welte and Hunt successively published two excellent work — *Petroleum Formation and Occurrence and Petroleum Geochemistry and Geology*[32,12], in which this theory was disussed systematically and scientifically, thus forming a quite integrate theory system. During the latest fifteen years, studies on petroleum origin stepped out of basic studies and developed toward applied technique field, becoming a very effective and highly applicable technique in petroleum exploration with the foundation of the hydrocarbon generation theory of kerogen thermal degradation.

The elementary thoughts of this theory are, 'At the end of diagenesis, the organic matter consists mainly of kerogen. Thermal degradation of kerogen is responsible for the generation of most hydrocarbons, i. e. , oil and gas. Kerogen is the main precursor of petroleum compounds. ' (Tissot and Welte, 1978); 'A sedimentary rock must pass through the time-temperature threshold of intense hydrocarbon generation in order to form enough hydrocarbons to yield commercial petroleum accumulations. The rate of hydrocarbon generation from kerogen increases exponentially with a linear increase in temperature. The quantity of oils and gases formed depends primarily on the kerogen type, temperature and time. ' (Hunt,1979). From this theory, geochemists established a series of highly applicable geochemical parameters and research methods dealing with evaluating hydrocarbon potential, such as division of hydrocarbon generation and evolution stages, determination of upper and lower threshold values, judgment of hydrocarbon origin and origin analysis by biomarker assemblage, determination of three factors affecting hydrocarbon generation potential (sedimentary organic matter abundance, type and maturity), establishment of hydrocarbon generation model of kerogen thermal degradation and the comprehensive analysis and numerical modeling of sedimentary history, thermal history and hydrocarbon generation and evolution history[16,17,33,4], etc.. These parameters and methods have achieved remarkable results and become the principles for petroleum exploration.

In the early 1980' s, according to hydrocarbon generation theory of kerogen thermal degradation, we made use of abundant nonmarine hydrocarbon and source rocks' geochemistry materials systematically, summarized the principles of hydrocarbon generation and evolution, defined a series of corresponding geochemistry indexes, developed and perfected nonmarine hydrocarbon generation theory, pointing out, 'The formation of terrestrical petroleum is an unavoidable outcome of geohistory development, lake evolution and chemical evolution of biological kingdom'[7,9].

Marked successes of this theory have been demonstrated in oil and gas exploration. Most of the oil and gas resources have a close relationship with this theory. However, this theory obviously ignores the contribution of soluble organic matter in the sedimentary rock to hydrocarbon generation in the diagenesis and early catagenesis. Hence, the theory can not give a reasonable and scientific explanation to the formation of immature and low mature oils.

On the basis of hydrocarbon generation theory of kerogen thermal degradation, kerogen has a loose structure, rich in O, N, S and othe heteroatoms, generating little petroleum in the diagenesis before the formation of fossil kerogen, i.e. the developing period for young kerogen. Besides, in accordance with Van Krevelen curve which reflects kerogen element composition evolution, kerogen mainly gets rid of heteroatomic groups, releasing CO_2, H_2O and H_2O and CH_4 in the process of maturation above the threshold of oil generation. Therefore, the theory can not explain the formation of immature oils. Tissot (1984) pointed out 'These immature heavy oils contain abundant heterocompounds (resins and asphaltenes) resulting from an early breakdown of kerogen with the associated high sulfur and nitrogen content'. This means that the origin of immature oils still deals with kerogen degeneration. On the contrary, we think that in the diagenesis, part of the organic matter containing biochemically synthesized hydrocarbons is decomposed and the rest of it will be condensed into kerogen or directly transform into hydrocarbons, contributing to the formation of immature oils together with the residual lipids. Therefore, there is an evident dominance of biological configuration in the immature oils.

To solve the discrepancy between hydrocarbon generation theory of kerogen thermal degradation and the wide existence of immature oils, Snowdon and Powell proposed to modify hydrocarbon generation model for terrestrial organic matter [29]. In 1983, at the Eleventh World Petroleum Congress, Palacus et al, studied immature-marginal mature oils from Cretaceous carbonate-evaporate of lagoon facies in the Southern Florida Basin, the United States, pointing out, 'These carbonate-related reservoirs form more in the late diagenesis than in the catagenesis' [24], Fu studied the immature oils in salt rock sediments and pointed out, 'Environment for salt rock deposition is strongly reductive, facilitating the early reservation and transformation of organic matter; The existence of immature oils is a challenge to hydrocarbon generation theory of kerogen thermal degradation' [3]. Huang and Li pointed out, 'Immature oils directly come from the degradation of soluble organic matter in the source rocks at low temperature' [6].

DISTRIBUTION OF IMMATURE OILS

Immature oils should be defined as the oils formed in the late diagenesis before the evolution of organic matter in the sedimentary rocks reaches the oil threshold. Their source rocks' vitrinite reflectances are 0.3%-0.7%. And their oil quality belongs to heavy oils and some condensates.

Being confined to hydrocarbon generation theory of kerogen thermal degradation, in spite of the wide existence of immature oils all over the world, people always ignore the study on the exploration and development of immature oils and their forming and accumulating conditions. Tissot estimated, 'Although some accumulations of immature non-degraded heavy oils are known, they possibly amount to less than 1% of the world reserves of this category' [32]. But if estimated according to the relative quantity of hydrocarbons and soluble heavy composition in the source rocks at different hydrocarbon generating stage (Figure 1) [12], the immature heavy oil resources should take up more than 10% of the world's total resources.

Immature and low mature oils distribute very widely, including both China and other countries from terrestrial facies to marine facies. Martin gave an example of North American immature oils in the study of normal alkanes' distribution in crude oils[21]. Besides those examples mentioned above, immature asphalt from organic-rich carbonate rocks in Upper Cretaceous Senonian in Israel[31,30], and the immature condensates from southeastern Mediterranean coastal plain in Israel are also noticeable examples[23].

Tissot and Welte (1978) mentioned some immature oil examples in the world in *Petroleum Formation and Occurence*, but they did not locate the immature oils in the general scheme of

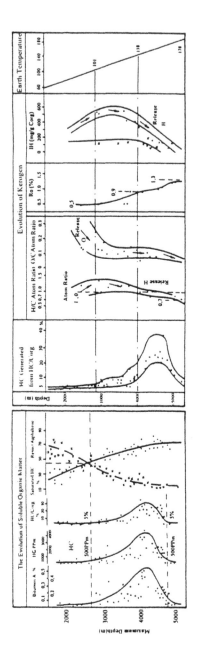

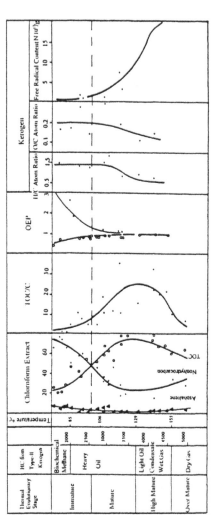

Figure 1   Generating hydrocarbon and evolutionary section of Tertiary oil source rock in Raoyang sag and Qianjiang sag (The upper is after Liang, D. G., 1982; The lower is after Jiang, J. G., 1982. The evolutionary stage is slightly modified).

the organic matter evolution. Hun (1979) introduced the typical immature oils from the oil pools in the caprocks of salt dome and knoll existing in the Gulf of Mexico and from deltas in California and Orinoco (Brazil). He also located immature oils in organic matter maturation and evolution figure. However, immature oils were located in the early diagenesis.

From above, Tissot and Hun both noticed the existence of immature oils while elaborating and summarizing hydrocarbon generation theory of kerogen thermal degradation, but they did not carry out further researches. These examples of immature oils do not seem to be in the tune with the theory itself.

Immature oils were first reported in China in 1982 (called low mature oils at that time). Shi et al [27] and Jiang et al[3] reported the immature oils from Y1-18 well in Yihezhuang Oilfield, Shandong Province and from Guanghua Temple Oilfield, Jianghan depression, respectively. Soon after, immature oils stimulate great interests of Chinese geochemists. In several years, immature and low mature oils were reported in nearly every terrestrial Mesozoic-Cenozoic hydrocaarbon basins in China [2,5,10,15,20,28,35,37] . At the same time the studies on the hydrocarbon generation precursor, generation conditions, distribution characteristics, and formation mechanism of immature oils are being further carried out[3,6,10,26,36]. Studies demonstrate that such oils exist mainly in Tertiary and next in Cretaceous. Moreover, salt lake facies and brackish water facies are especially favorable for the formation of immature oils.

CONTRIBUTION OF DISPERSED SOLUBLE ORGANIC MATTER TO EARLY HYDRO-CARBON GENERATION IN SEDIMENTARY ROCK

For hydrocarbon generation and evolution of kerogen, one of the important mark is the decrease of its H/C atom ratio or S_2 peak in Rock-Eval analysis, while the evolution of young kerogen in diagenesis before reaching the threshold of hydrocarbon generation generally shows the increase of the above two indexes [9,14] . It means that the young kerogen not only contributes little to hydrocarbon generation but also is accumulating its 'raw material' for hydrocarbon generation and preparing conditions for the following hydrocarbon generation of thermal degradation.

The soluble organic matter in sedimentary rock in diagenetic stage is extremely active and at an enormously changing state, which provide us with important information on its hydrocarbon generation. All profiles of hydrocarbon generation and evolution in Chinese oil-gas bearing basins reflect a kind of unified model in essence, with the content of non-hydrocarbon deceasing and that of hydrocarbon increasing in diagenetic stage as shown in Figure 1. In group composition, non-hydrocarbon in rocks generally decreases from 70% to 50%, whereas hydrocarbon increases from 25% to 50% or so. The cross point generally represents the threshold of hydrocarbon generation. It is apparent that the increase of hydrocarbon is mainly derived directly form decarboxylation of resins. At the same time, part of non-hydrocarbon together with humic acid is polycondensed into kerogen. Although young kerogen also undergoes dehydrocarboxylation and releases other heteroatomic radicals in diagenetic stage, it can not generate hydrocarbon. This is proved by the fact that hydrogen content of the young kerogen either maintains constant or increases slightly. All the above shows that the change of molecular geochemical parameters in rock extracts, such as the decrease of OEP and CPI, and the increases of EOP (tends to 1) all represents the evolution of soluble organic matter and has nothing to do with kerogen evolution. After the evolution of organic matter passes through threshold of oil generation in early catagenetic stage hydrocarbon generation by kerogen thermal degradation continues growing and large amount of hydrocarbon generated from kerogen thermal degradation covers or dilutes the immature-low mature characteristics of hydrocarbon derived from early soluble organic matter.

Ternary diagram (Fig. 2) of oil-rock group composition in Xialiaohe Basin also demonstrates that marginal-low mature crude oil from Gaosheng oil field can only be derived from the soluble organic matter in immature source rock, with non-hydrocarbon and asphaltenes accounting for 50%.

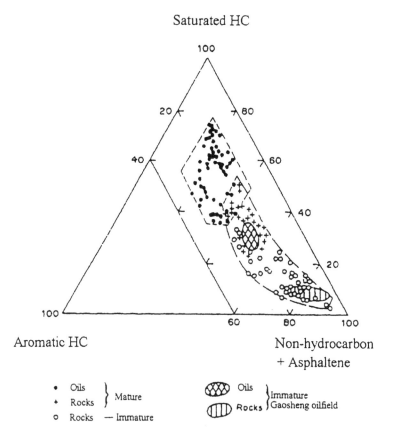

Figure 2   Triangular diagram of Tertiary oil and rock extract composition in Liaohe Basin (Modified after Piao, M. Z. , 1984).

Some evidence may be numerated, strengthening the above analysis on the contribution of kerogen and soluble organic matter to hydrocarbon generation in different stages in the natural profile of hydrocarbon generation and evolution.

Zhang et al. (1992) studied the interconversion of kerogen, humic acid and extracts in brown coal and their contribution to hydrocarbon generation at different temperatures by hydrous pyrolysis [11]. As shown in Figure 3, the humic acid and extractable matter are the major contibutors to the formation of kerogen and humic acid respectively; However, as to the formation of chloroform bitumen, especially for hydrocarbons, extractable matter is the major contributor in early period (270℃), while kerogen is the major contributor in medium to late stage (350℃ and 400℃).

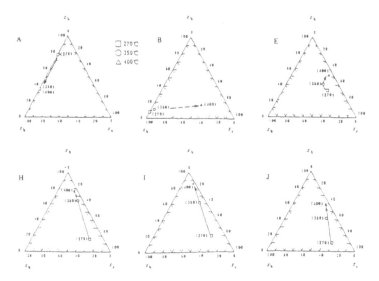

Figure 3 The relative contribution of three kinds of organic matter in brown coal at different temperatures to the generation of kerogen, humic acid and hydrocarbons (Xk, Xh and Xe represent the relative contribution to the generation of kerogen, humic acid and chloroform bitumen respectively).

Xia and Luo (1995) in Lanzhou Institute of Geology, Chinese Academy of Sciences, made a successful experiment to simulate the generation of alkane with Tertiary brown coal form Nanning. The experiment temperature ranges from 250℃ to 550℃, the interval of which is 50℃. In every interval, samples are heated isothermally for 72hr. The experiment is classified into two groups: Y are Z. The samples of group Y are heated continuously, whereas those of group Z are heated periodically, with the products at different temperatures extracted and heated at a 50℃ higher temperature. Therefore, group Y experiment shows the process of thermal evolution of all organic matter in brown coal, and group Z shows the thermal evolution of kerogen. As shown in Figure 4, the experiment clearly distinguishes the contribution of kerogen and soluble organic matter to hydrocarbon generation. At the same time, the differences between the geochemical characteristics of their products are shown in

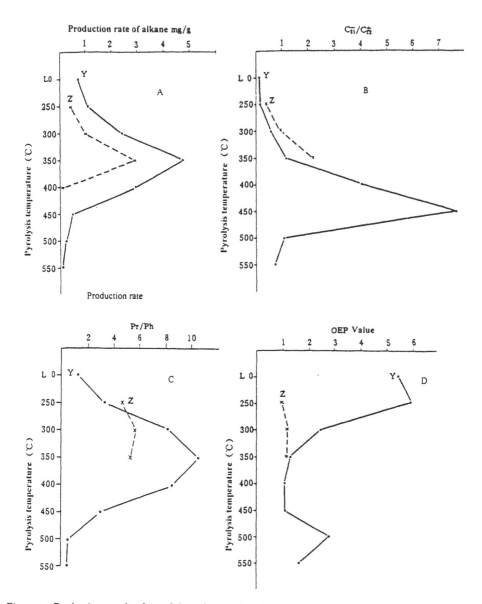

Figure 4   Production rate by thermal degradation, C_{21}^-/C_{22}^+, OEP and Pr/Ph as a function of temperature for brown coal from Nanning (Y: heat continuously; Z: heat by stages) (after Xia, Y. Q. and Luo,
B. J. , 1993).

Figure 4 clearly. Figure 4 shows that: (1) The highest peaks for hydrocarbon generation of
kerogen and soluble orbanic matter occur at 350 ℃ and 400℃ respectively. (2) At 400℃,
hydrocarbon generating potential for kerogen is nearly exhausted. Therefore, nC_{21}^-/nC_{22}^+
reaches its maximum at 450℃, showing the composition feature of normal alkane in bitumen

from kerogen thermal degradation. (3) Normal alkanes from kerogen thermal degradation have no odd-even predominance. The odd even predominance is the feature of immature soluble organic matter. It gradually decreases with the adding and blending of products from kerogen thermal evolution and disappears corresponding to the peak of hydrocarbon generation of kerogen. It also means the end of hydrocarbon generation of soluble organic matter in diagenesis. (4) Pr/Ph in the products from kerogen thermal degradation is relatively higher and doesn't vary with the temperature, which affects the ratio of Pr/Ph in soluble organic matter in the same way. Its blending and thermal degradation cause the original low Pr/Ph in soluble organic matter to increase gradually and reach the maximum value at 350°C. (5) When the temperature is above 400°C, the lower part of every hydrocarbon generation curve shows that Pr/Ph decreases, and C_{21}^-/C_{21}^+ reaches its maximum and OEP increases slightly. All these geochemical characteristics are detemined by the evolution of asphalt or pyrrobitumen from kerogen degradation.

After all, as the experiment shows, we should put emphasis on the contribution of soluble organic matter to hydrocarbon generation, especially on the important role of geochemical parameters played by the products of soluble organic matter in immature and relatively higher mature stages.

A NOVEL MODEL OF HYDROCARBON GENERATION AND EVOLUTION

Soluble and insoluble organic matter in sedimentary rock are a whole in organic relation and keep in a state of dynamic equilibrium of interconversion in the process of hydrocarbon generation with the changing of physico-chemical condition. In diagenetic stage, part of soluble organic matter in rock will be converted to immature oil, and the other part will be condensed into kerogen. In catagenetic stage, kerogen thermal degradation will inevitably undergo the intermediate stage in which bitumen is formed. Even at high level of maturity, pyrrobitumen derived from kerogen become the major contributor to high mature light oil after the hydrocarbon generating potential for kerogen getting exhausted. Hence, to establish an objective model, organic matter in rock must be regarded as a unity.

In order to understand the process of oil and gas generation comprehensively, we propose the model shown in Figure 5[8].

The model shows different paths of evolution in gas-prone and oil-prone environments (humification for gas and saprofication for oil), It also emphases the contribution of soluble organic matter (including humin) to hydrocarbon generation. The model shows not only the matrix types and their different potential for hydrocarbon generation, whether oil or gas is predominant, but also phasic characteristics of hydrocarbon generation and the maturity of corresponding products in different stages which include immature oil and gas. The imaturation sequences of crude oil in non-marine hydrocarbon basin are also reflected in the model clearly. Finally, as shown in figure 5, the three hydrocarbon generation and evolution stages

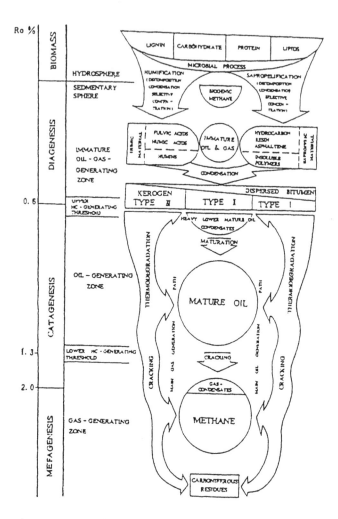

Figure 5   Hydrocarbon generation and evolutionary model of organic matter (Huang, D. F. et al. 1991).

have their own characteristics: heavy oil is formed at diagenetic stage characterized by the degradation of soluble lipid to generate hydrocarbon directly and the formation of fossil kerogen; the earlier phase of catagenesis ($R^\circ < 1.2\%$) is characterized by the formation of normal oil predominantly derived from kerogen thermal degradation, whose active energy is less than 160kJ; at the end of the earlier phase, the potential of kerogen for hydrocarbon generation is exhausted; the late phase of catagenesis is characterized by the formation of light oil and wet gas by cracking of pyrobitumen and normal oil respectively, whose active energy is 160-200kJ/mol; dry gas is formed in metagenesis stage. The relationships of hydrocarbon generation, source of original matter and nature of hydrocarbon in the evolution process of organic matter are illustrated in figure 6 clearly.

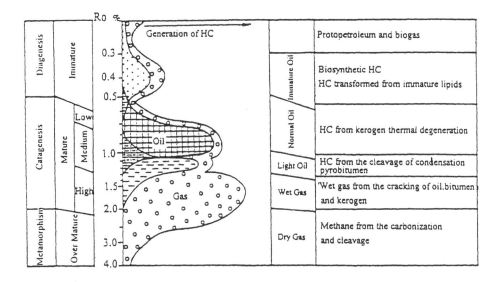

Figure 6   Hydrocarbon generation during the evolution of organic matter and its relationship with the origin of parent material and the property of oil and gas.

CONCLUSIONS

1. Immature oils are generated directly from soluble lipids which contain biosynthetic hydrocarbons, different from the "protopetroleum" in Recent and Quaternary sediments. Their wide existence and economic value break through the limits of hydrocarbon generation theory of kerogen thermal degeneration.

2. The hydrocarbon generating potential of kerogen is nearly exhausted when vitrinite reflectance reaches 1. 2%. Light oils of high maturity are formed from condensed pyrobitumen during its higher evolutionary stage.

3. On the basis of the idea that soluble and insoluble organic matter both take part in generating hydrocarbon as an integral, a new generating hydrocarbon evolutionary model of source rocks is presented in this paper.

REFERENCES

1. Albrecht P. and Ourission G.. Diagenesis des hydrocarbures satures dans une serie sedimentaire epaisse (Douala, Cameroum). Geochem. Cosmochim. Acta 33,138-142 (1969).

2. Cheng J. Y. , He B. J.. Types and genesis of the petroleum in the north of Liaodong bay. Preceedings of the 4th National Sympsium on Organic Geochemistry of China. Wuhan, China University of Geosciences Press, 89-103 (1990).

3. Fu J. M. , Sheng G. Y. , Jiang J. G.. Immature oils in gypsolith sedimentary basin. Geology of Oil and

Gas 6 : 2,150-158 (1985).

4. A. Hood, C. C. M. Gutjaha and R. L. Heacock. Organic metamorphism and the generation of petroleum. AAPG Bull. 59,986-996(1975).

5. Huang D. F. et al. The maturation series of Tertiary crude oil in China and its significance. Advances in Evaluation and Research of Oil and Gas Resources No. 2, Beijing, Press of Petroleum Industry, 104-115(1989).

6. Huang D. F. , Li J. C.. Immature oil in terrestrial sediments and its significance. Oil Acta 8 : 1,1-9 (1987).

7. Huang D. F. , Li J. C. (Editors). Terrestrial oil and gas formation in China. Beijing, Press of Petroleum Industry(1982).

8. Huang D. F. , Li J.C. , Zhang D. J.. Immature oil in Chinese nonmarine sediments and significance. In:Organic Geochemistry, Advances and Application in Energy and the Natural Environment, Manchester University Press, 39-47(1991).

9. Huang D. F. , Li J. C. , Zhou Z. H. , Gu X. Z. ,Zhang D. J.. Evolution and generating hydrocarbon mechanism of terrestrial organic matter. Beijing, Press of Petroleum Industry(1984).

10. Huang D. F. , Liao Q. J. , Xu Y. C.. Preliminary study on the genesis of immature petroleum. Annal Research Report of Biogeochemistry and Gasgeochemistry Laboratory, Lanzhou Institute of Geology, Academia Sinica. Lanzhou, Ganshu Publishing House of Science and Techology, 1-19(1987).

11. Huang D. F. , Qin K. Z. , Wang T. G. , Zhao X. G. et al.. Oil from coal:Formation and mechanism. Beijing, Press of Petroleum Industry(1995).

12. Hunt J. M.. Petroleum Geochemistry and Geology. San Francisco: W. H. Freeman and Company (1979).

13. Jiang J. G. , Zhang Q.. The generation and evolution of petroleum of the saline Qianjiang formation in Jianghan basin. Geology of Oil and Gas 3 : 1, 1-14(1982).

14. R. E. Laplante. Hydrocarbon generation in Gulf Coast Tertiary sediments. AAPG Bull, 58,1281-1289 (1974).

15. Li M. F. , Chen Z. Y. , Wang Y. S.. Distribution and characteristics of immature oil in Liaohe faulted basin. Sympsium of Proceedings of the 5th Organic Geochemistry(1992).

16. Lopatin N. V.. The main stage of petroleum formation (in Russian). Izv. Akad. Nauk. Uzb. SSR, Ser. Geol. 1,69-76(1969).

17. Lopatin N. V.. Temperature and geologic time factors in coalification. Izv. Akad. Nauk. Uzb. SSR, Ser. Geol. 3,95-106.

18. Louis M. and Tissot B. P.. Influence de la temperature et de la pression sur la formation des hydrocarbures dans les argiles a kerogen. Proc, Seventh World Petr. Cong (Mexico) 2, 47-60(1967).

19. M. Louis. Etudes geochimiques sur les "Schisters cartons" du Toarcian du Basin de Paris, In: Advances in Organic Geochemistry (Eds: G. D. Hosbson and M. C. Louis). New York: Pergamon Press 84-95(1964).

20. Ma W. Y.. Immature oil from dolostone of Biyang sag. Sympsium of Proceedings of the 3rd Organic Geochemistry(1986).

21. Martin R. L. et al.. Distribution of N-paraffins in crude oils and their implication to origin of petroleum. Nature 199,1190-1193(1963).

22. Mclver R. D.. Composition of kerogen-clue to its role in the origin of petroleum, Proc. Seventh World Petr. Cong. (Mexico) 2, 25-36(1967).

23. Nissenbaum A. et al.. Immature condensate from Southeastern Mediterranean coastal plain, Israel. AAPG Bull. 69 : 6,946-949(1985).

24. Palacas J. G. et al.. Carbonate rocks as sources of petroleum, geochemical/ chemical characteristics and oil-source correlations. Sympium of Proceedings of The Eleventh World Petroleum Congress, Beijing, Press of Petroleum Industry, 1 (Petroleum Geology), 20-34(1983).

25. Philppi G. T.. On the depth, time and mechanism of petroleum generation. Geochem. Cosmochim Acta 29, 1021-1049(1965).

26. Qin K. Z.. The thermal degradation of kerogen and its relationship with formation of immature oil. Proceedings of the 3rd National Sympsium on Organic Geochemistry of China. Beijing, Geological Publishing House, 159-168(1987).

27. Shi J. Y. , A. S. Mackenzie, G. Eglinton, A. P. Kovat, J. R. Maxwell. Biomarkers in crude oil and source rock and their application in Shengli Qilfield. Geochemistry 1,4-20(1982).

28. Shi J. Y. , Wang B. S.. Geochemical characteristics of sterane and terpane of source rock in Shubei basin and low mature source rock and crude oil in the Eastern China. Geochemistry 1,80-89(1985).

29. Snowdon L. R. and Powell T. G.. Immature oil and condensate — modification of hydrocarbon generation model for terrestrial organic matter. AAPG Bull. 66,775-788(1982).

30. E. Tannenbaum et al.. Formation of immature asphalt from organic-rich carbonate rocks- I , correlation of maturation indicates. Org. Geochem. 6,503-511(1984).

31. E. Tannenbaum et al.. Formation of immature asphalt from organic-rich carbonate rocks- I , correlation of maturation indicates. Org. Geochem. 8,191-192(1985).

32. Tissot B. P. and Welte D. H.. Petroleum Formation and Occurrence. Berlin Heidelberg New York Tokyo;Spring-Verlag(1978;1984).

33. Tissot B. P.. Primieres donnees sur le mecanismes etla cinetique de la formation fu petrole dans les sediments; Simulation d' un schema reactionnel sur ordinaleur. 1' Inst. Francais Petrol. , 24,470-501 (1969).

34. Vassoevich N. B. ,Akramkhodzhaev A. M. ,Geodekyan A. A.. Principal zone of oil formation, In;Advances in Organic Geochemistry 1973 (Eds; Tissot, B. P. , B. P. Bienner, F.). Paris; Technip, 309-314(1974).

35. Wang G. S. , Shi J. Y.. Characteristics of generating hydrocarbon in Shubei basin and its oil exploration prospect. Geochemical Symposium (1966-1986). Beijing, Science Press, 172-185(1986).

36. Wang T. G. , Zhong N. N. , Hou D. J. , Huang G. H. ,Bao J. P. ,Li X. O.. Formation mechanism and occurrence of low mature oil. Beijing, Press of Petroleum Industry(1995).

37. Zhou G. J.. Low mature oils in terrestrial faulted basin. Proceedings of 3rd National Symposium on Organic Geochemistry of China. Beijing, Geological Publishing House, 27-37(1987).

Proc. 30th Int'l. Geol. Congr., Vol. 18, pp. 17~32
Sun Z. C. *et al.* (Eds)
© VSP 1997

Characteristics and Formation Conditions of Oil/Gas Fields (Pools) in Abnormal Formation Pressure Environments in China

WANG TINGBIN

Institute of Petroleum Geology. MGMR, Beijing 100083, *CHINA*

LIU BIN

General Office of National Committee on Mineral Resources, Beijing 100812, *CHINA*

Abstract

Abnormal pressure oil/gas fields (pools) are widely distributed in China, which with the less than 0. 9 and greater than 1. 1 pore fluid pressure coefficient account for 42% of total oil/gas fields (pools), 35% of which are overpressured ones and 7%, subpressured ones. The abnormally low pressured gas fields (pools) are relatively less in numbers and each has less than $5 \times 10^9 m^3$ in reserves; the number of super-pressured gas fields (pools) is more than that of the former, but decrease with the increasing of pore fluid pressure coefficient, and reserves range from $2 \times 10^9 m^3$ to $15 \times 10^9 m^3$.

The spatial and temporal distribution, the abnormally pressured oil/gas fields (pools) are chiefly located at the formations with composite lithological-structural traps and lithological traps from the Paleozoic to the Cenozoic. Their stages of accumulation were generally in the Mesozoic and the Tertiary. The subpressured oil/gas (pools) were buried at a depth of less than 2000m, and the overpressured ones at a depth of more than 3500m. The more the depth is, the more the number of overpressured oil/gas fields (pools) becomes obviously.

The sedimentary-type overpressured oil/gas fields (pools) are mainly located in the eastern area of China, and controlled by the under-compaction, aquathermal, hydrocarbon generation and dehydration of clay minerals which all are the contributors to the pore fluid pressures. But in the central area of China, the structural-type overpressured oil/gas fields (pools) were mainly influenced by the hydrocarbon generation and the tectonic compression in the late stages. The subpressured oil/gas fields (pools) in northeast area of China are chiefly controlled by the tectonic elevation and erosion in the late stages.

Our investigation shows that the very favorable conditions to form abnormally high pressure oil/gas fields (pools) in China include the abundant soure rocks, the matching of hydrocarbon generation with the timing of traps, the effective regional seals and the pore fluid pressure coefficient between 1. 1~1. 3.

Keywords: Abnormal formation pressure, Pore fluid pressure coefficient, China

DISTRIBUTION OF ABNORMALLY PRESSURED OIL/GAS FIELDS (POOLS) IN CHINA

Abnormal pressure oil/gas fields (pools) are widely distributed in China. Fig. 1 shows the distribution of oil/gas fields (pools) with various formation pressure in China. According to Fig. 1, 42% of total oil/gas pools (formation pressure coeficient of larger than 1. 1 or less than 0. 9), 35% of them are abnormally high pressure ones and 7% of them are subnormal pressure ones. In China, abnormal pressure oil/gas fields (pools) are widely found in various petroliferous basins. The distribution, horizon, depth, pressure strength of abnormal pressure oil/gas pools vary significantly due to different geological condition in different basin types.

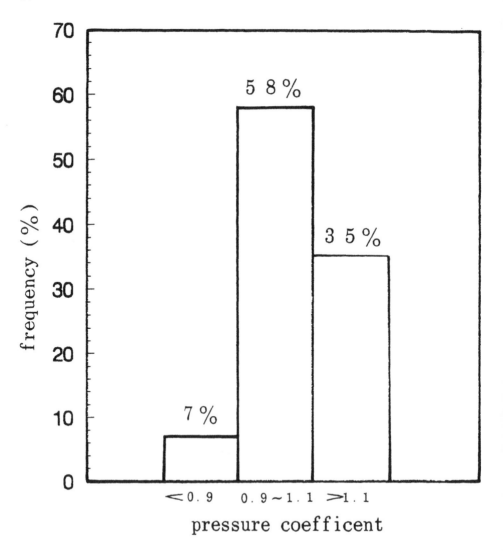

Fig. 1   Histogram of oil/gas pools with various formation pressure in China

In Sichuan basin, for example, most of abnormal pressure oil/gas fields (pools) are abnormally high pressure ones, few abnormally low pressure fields (pools) have been found. Fig. 2 presents oil/gas fields distribution in Sichuan basin, except for a few fields, most of these fields are abnormally high pressure ones which consist of abnormally high pressure oil/gas pools.

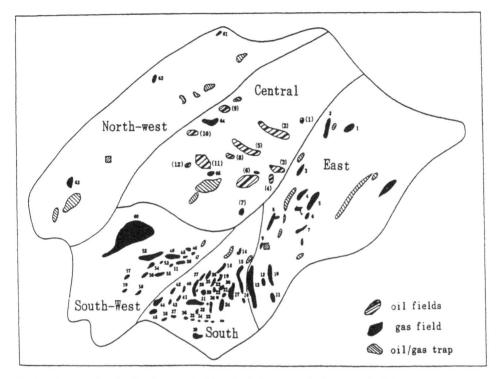

Fig. 2   Map showing the distribution of oil/gas fields in Sichuan basin, central, China.

The relationship between formation pressure coefficient and burial depth of abnormally high pressure pools in Sichuan Basin is that abnormally high formation pressure occur from 500 to 5000m in depth. However, oil/gas fields (pools) are mostly distributed from 1000 to 3000m. For most oil/gas fields (pools), their formation pressure coefficient range from 1. 05 to 1. 30 (Fig. 3).

The relationship between formation pressure coefficient and field number in Sichuan Basin is that oil/gas field with a formation pressure coefficient of more than 2. 2 are less than 1% in number. The distribution of abnormally high pressure oil/gas fields is characterized by higher pressure in the Northwest part of the basin and lower pressure in the south and southwest parts (Fig. 4).

The gas pools of Xu-4 and Xu-2 member of Xujiahe formation in upper Triassic in Sichuan basin have the highest pressure coefficient, as high as more than 2. 2, in the northwest part (Fig. 5,6). The Permian abnormally high pressure environment has larger areal extent, that is, Permian gas pools are abnormally high pressure both in the west and east parts (Fig. 7).

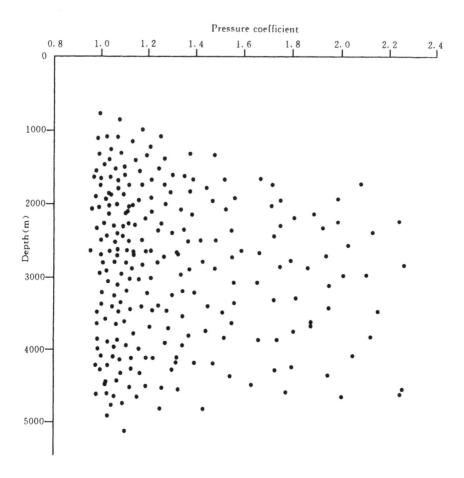

Fig. 3   Formation pressure coefficient-burial depth relationship of abnormally high pressure fields (pools) in Sichuan Basin, Central, China.

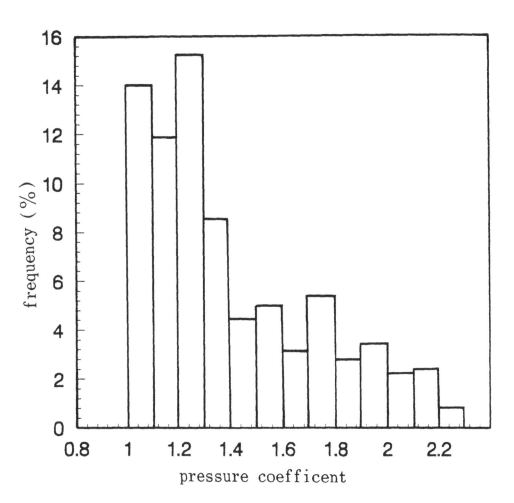

Fig. 4   Map showing the relationship between formation pressure coefficient and field number in Sichuan Basin.

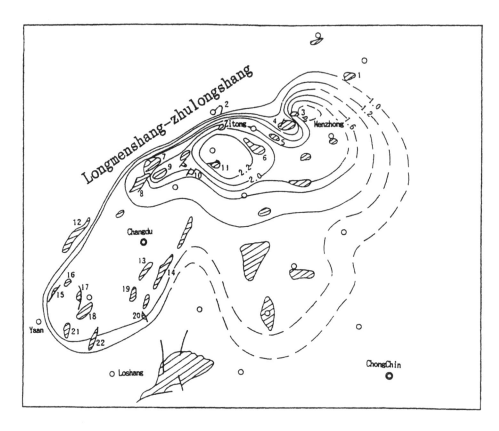

Fig. 5   Formation pressure coefficient isogram of Member Ⅳ of upper Triassic Xujiahe formation, in Sichuan basin, central China.

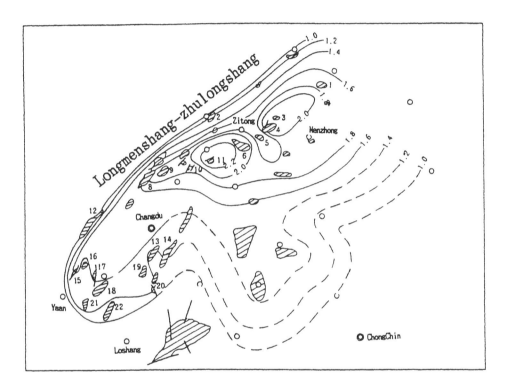

Fig. 6   Formation pressure coefficient isogram of Xu-2 Member in upper Triassic in Sichuan Basin.

Fig. 7 formation pressure coefficient isogram of Permian stratum in Sichuan basin.

As mentioned above, abnormal pressure oil/gas fields (pools) vary signifcantly in various basin. For example, abnormal pressure fields (pools) in thermal rift basins in the east part of China, are largely different from those in Sichuan basin. Rift basins in the east part contain not only abnormally high pressure fields (pools) but also subnormal pressure fields (pools). In Songliao basin, abnormally high and low pressure fields (pools) have been found. The basin is a large-sized Meso-Cenozoic continental petroliferous basin in which the largest amount of oil/ gas has been produced now in China. The basin contain Hedimiao, Saertu, Putiaohua, Gaotaizi, Fuyu and Yangdachengzi pay zones from lower to upper.

The formation pressure of Hedimiao oil-bear formation is characterized by subnormal pressure, subnormal pressure mainly occurs in the southern Changjiaweizi. The lowest formaion pressure gradient is 0. 75 in Songliao basin (Fig. 8). The formation pressure of Seartu oil-bear formation mainly is normal-abnormally high pressure. Abnormally high pressure is largely distributed along the Xingzhan-Maoxing zone (In Fig. 9). The distribution of formation pressure in Putiaohua oil-bearing formtion is very different from those of the two above oil-bear formation. Putiaohua oil-bear formation has normal, abnormally high and subnormal pressure. Abnormally high pressure enviroment mainly occurs in Gulong area in the western Songliao Basin and subnormal pressure enviroment in Xingzhan area in the eastern Songliao Basin (Fig. 10). Abnormal formation pressure of Fuyu oil-bear formation in Songliao is largely distributed along the Zhaozhou-Sizhan zone. Abnormally high pressure enviroment exists in Zhaozhou and abnormally low pressure enviroment in Sizhan (Fig. 11). The distribution Characteristics of the formation pressure in Yangdachengzi oil-bear formation in Songliao Basin is similar to that of Fuyu and the subnormal pressurte is mostly distributed in Sizhan (Fig. 12).

In general, based on the formation pressure distribution Characteristics in oil-bear formation mentioned above, oil/gas fields (pools) with abnormally high pressure are mainly distributed in the western part of Songliao Basin and oil/gas fields (pools) with subnormal pressure in the eastern part of Songliao.

Based on statistic data, the reserve of a subnormal pressure fields (pools) usually is small in China. The reserve of a abnormally high pressure fields (pools) generally is more than that of a subnormal pressure fields (pools). However, the number of oil/gas fields (pools) and reserve magnitude of each fields (pools) decrease with increasing formation pressure coefficient. In Fig. 13, we can find the another characteristic of abnormal pressure fields (pools) in China, that is, the depth of subnormal pressure pools is less than 2000m and the depth of abnormally high pressure pools is more than 2000m. With increasing burial depth, formation pressure coefficient increases but the percentage of abnormally high pressure oil/gas fields (pools) obviously decreases. In China, abnormal pressure pools are mostly distributed in Paleo-Cenozoic lithologic and lithologic-structural traps. The timing of these pools with abnormal pressure generally is from Mesozoic to Tertiary.

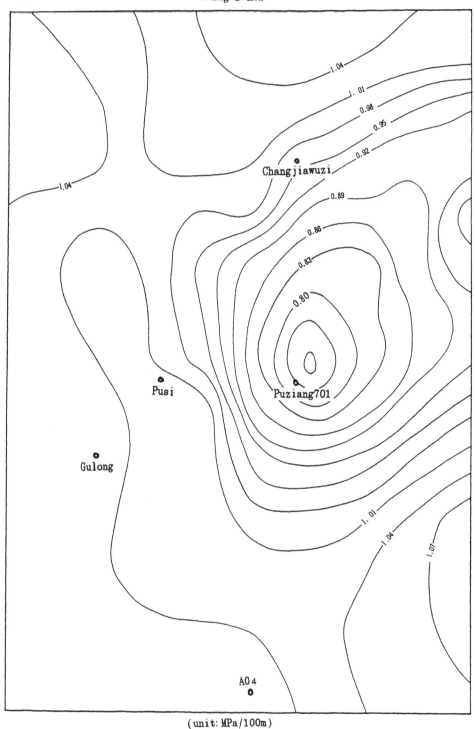

(unit: MPa/100m)

Fig. 8   Map showing formation pressure distribution in Hedimiao oil-bear formation in Songliao Basin, Northeast China.

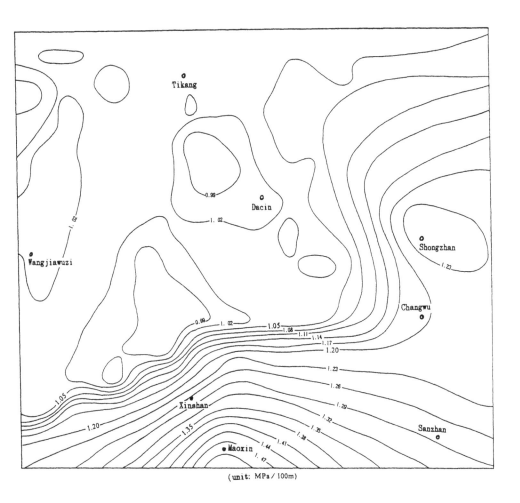

(unit: MPa / 100m)

Fig. 9　Map showing the distribution of formation pressure in Saertu oil formation in Songliao Basin.

(unit: MPa / 100m)

Fig. 10   Map showing the distribution formation pressure of Putiaohua oil formation in Songliao Basin,
Northeast China.

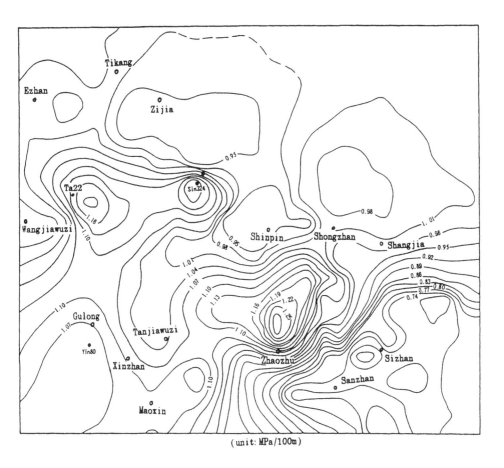

(unit: MPa/100m)

Fig. 11   Map showing the formation pressure distribution of Fuyu oil formation in Songliao Basin, Northeast China.

(unit: MPa/100m)

Fig. 12   Map showing the formation pressure distribution of Yangdachengzi oil-bear formation in Songliao Basin, Northeast China.

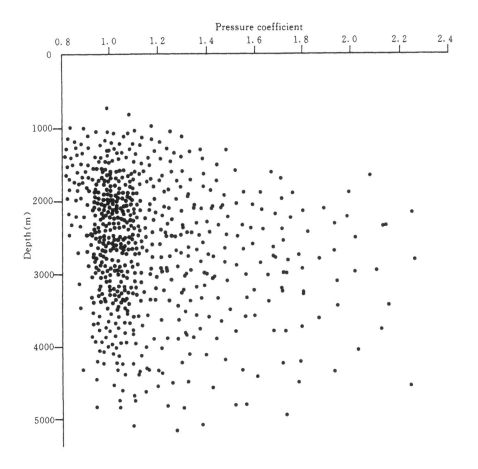

Fig. 13 Diagram showing the relationship between the formation pressure coefficient and depth of abnormal pressure oil/gas pools in China.

ORIGIN OF ABNORMAL PRESSURE OIL/GAS FIELDS (POOLS) IN CHINA

A large number of petroliferous basins with complex types exist in China. There are many geological factors influencing abnormal pressures and these factors vary significantly in various basins. In the eastern part of China, for example, the formation of sedimentary type abnormally high pressure fields (pools) is mainly controlled by dehydration of clay minerals and pressure increases caused by undercompaction, water heating and hydrocarbon generation. However, in the central part, the formation of structural-type abnormally high pressure fields (pools) is largely influenced by late hydrocarbon generation and structure compression, in the Northeast part, the formation of subnormal pressure fields (pools) is principally affected by late structural uplifting and eroding. In summary, main factors influencing abnormal pressure oil/gas fields (pools) include:

1) under compaction;

2) water heating;

3) dehydration of clay minerals;

4) hydrocarbon generation and structural processes.

In the forming process of abnormal pressure fields (pools), the importance of these geological factors is not equal to each other, usually one or two factors play a dominant role in the process. The dominant factors, associated with other factors, result in the formation of abnormal pressure fields (pools) with similar and different characteristics, in various petroliferous basins in China.

FORMATION CONDITION OF ABNORMAL PRESSURE OIL/GAS FIELDS (POOLS) IN CHINA.

The formation of abnormal pressure oil/gas fields (pools) is determined by depositional and burial history, diagenesis and porosity evolution history, and structural movement history in sedimentary basin. Less compacted sedimentary-type abnormally high pressure fields (pools) usually developed in hot basins and regions with high deposition rate, abundant organic matter, thick source rocks, late hydrocarbon-generating peak, and high geothermal gradient. In the basins and regions structural movement is dominated by extensional one, and subsidence and burial process developed continuously. Structural type subnormal pressure fields (pools) occur generally in organic matter-rich basins and regions which were buried early and eroded lately due to structural uplifting. Many subnormal pressure oil/gas fields (pools) result from the destroy and depressurization of abnormally high pressure fields (pools). Generally speaking, subnormal pressure fields (pools) have worse reservoir rocks. but are closed heavily, surrounding these pools. Summarily, abundant organic matter is a requirement to forming abnormal pressure oil/gas fields (pools). Well-sealed reservoir rock is critical to the formation of abnormal pressure oil/gas fields (pools). Fast deposition, large scale fluid generation and structural movement are important factors influencing the formation of abnormal pressure fields (pools). The most favourable conditions for abnormal pressure fields (pools) in China are abundant hydrocarbon sources, good match of hydrocarbon generation peak and trap timing, and efficient regional seal. Efficient regional seal is the most important condition of the formation of abnormal pressure fields (pools). Without sealing, no abnormal pressure fields (pools) develops. Under this condition, only normal pressure fields (pools) can be formed.

Proc. 30th Int' l. Geol. Congr. , Vol. 18, pp. 33~42
Sun Z. C. *et al.* (Eds)
© VSP 1997

Structural Style and Petroleum Systems of the Songliao Basin

DOU LIRONG and LI JINCHAO

Research Institete of Petroleum Exploration and Development, *CNPC*, *Beijing*, *CHINA*, 100083

Abstract

The Cretaceous was the peak development time of the Songliao basin. It underwent syn-rift rifting development stage (K_1), post-rift large-lake development stage (K_2 from the Denlouku to Nenjina Formations), and inversion stage (the K_2 Sifangtai and Mingshui Formations), correspondingly forming three types of petroleum systems. Each type of petroleum system has its own tectonic style, its own sedimentary style, its own source rock, and even different petroleum potential. The syn-rift petroleum systems are composed of 32 rifted depressions, each rifted depression is an independent petroleum system. The coal-bearing strata filled in them are served as their source rock. In the central basin the source rock is deeply buried and mainly generates dry gas, whereas on the margin it is shallow buried and mainly generates oil. Drape-anticlines and fault-blocks are the main structural traps for oil and gas. In the Large-lake petroleum system wide dark organic-rich shales and oil shales of sapropelic type are well developed, and morderately buried, so they mainly generate oil. This type of petroleum systems is the major oil system in the Songliao basin and one of the main petroleum systems in East China. The sheet sand beds of large front deltas are the main reservoirs, and inversion structures are the main trap type. The inversion-stage petroleum system is characterized by the immature source rock, biogenetic gas and structural-stratigraphic traps.

Keywords: Structure style, Petroleum system, Songliao basin

INTRODUCTION

The Songliao basin is a large, Late-Mesozoic lacustrine basin in NE China (119°40' to 128°24' E longitude, 42°25' to 49°23'N latitude). It trends NNE, and is 750 km in length and 330-370 km in width, covering an area of 260,000 km². The composite basement of the Songliao basin consists mainly of the Paleozoic and pre-Paleozoic metamorphic rocks and various igneous rocks. The total thickness of the sedimentary cover is more than 11,000 m. It deposited from the Jurassic to Cretaceous time, but mainly during the Cretaceous. The Cretaceous strata are more than 7,000 m and is a main petroliferous measures in the basin[5] .
39 oil fields and 18 gas fields have been found in the Songliao basin. The Daqing oil fields complex is one of the few giant oil fields and the largest nonmarine oil field in the world. The

annual production of oil is more than 55,000,000 tons and becomes an important base of petroleum industry in China.

TECTONIC-STRATIGRAPHIC CHARACTERISTICS

The Cretaceous time is the peak period of the Songliao basin development that consists of three stages of tectonic development (Fig. 1).

During the Late Jurassic to the Early Cretaceous, the Izanagi Plate was subducting towards the Eurasian Plate. The subducting rate of the plate was 30 cm per year during than 500m. Member 1 of the Mingshui Formation contains two upward-fining cycles of thick grey mudstones, sandy conglomerates and dark grey mudstones. Member 2 of the Mingshui Formation is composed of brown and greyish green mudstones interbedded with sandstones.

CLASSIFICATION AND CHARACTERISTICS OF PETROLEUM SYSTEMS

During the evolution of the basin, there are differences in the topography of the basement, structural features of sedimentary cover, history of the tectonic development, and feature of stress field in different parts of the basin. According to the factors mentioned above, the basin can be classified into six first-order structural units: the central depression, southeastern uplift, south-western uplift, western slope and northeastern uplift. Each of first-order unit can be subdivided into many second-order structural units.

12 sags of the Suihua, Binxian-Wangfu (or Yinshan), Qijia-Gulong, Changliang,Sanzhou, Yushu-Dehui, Lujiapu, Zhezhong, Yian, Lishu and Heiyupao are located in different parts and different synthems of the basin. These sags can be classified into four types as follows on the basis of structural and sediments-infill history:

1. Inherited sag type   These sags are mainly distributed in the western depression zone, such as the Qijia-Gulong, Changliang and Heiyupao sags, which belong to the stacking area of the Early Cretaceous faulted depression and the Late Cretaceous large lake basin.

2. The initial development stage sag type   These sags are mainly the Early Cretaceous faulted depressions, for example, the Suihua, Yingshan, Xinshan, and Lujiapu sags, which become slope during the uplifting of the Qingshankou large lake development.

3. The mid-development stage sag type   These sags are mainly formed during the deposition of the Nenjiang Formation (e. g. Sanzhao sag). Part of the present Sanzhao sag was the location of the central paleouplift before the deposition of the Nenjiang Formation.

4. The late development sag type   These sags develop during the deposition of the Sifangai and Mingshui Formations, such as the Wuyur and Yian sags that belong to previous tilting-eastward slope.

The long-term inherited sags are composite development area of the deep-interval gas system and the shallow-middle-interval oil system. The initial-development-stage sags are the devel-

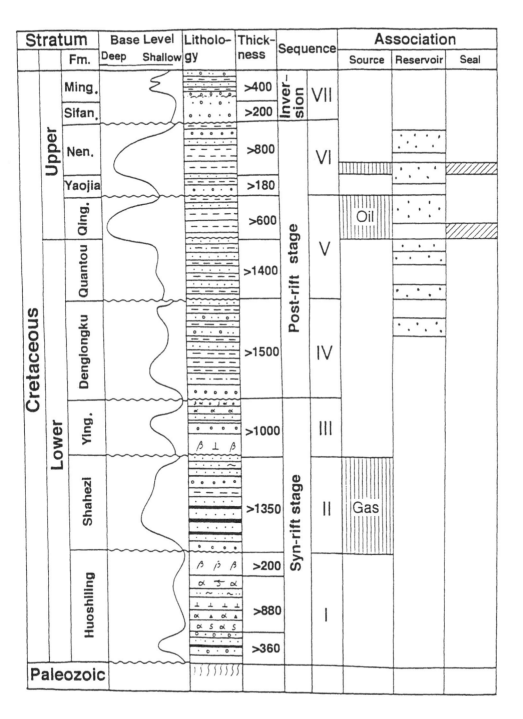

Fig. 1 Tectonic-stratigraphic characteristics of the Songliao basin.

oped area of the deep-interval petroleum system. The mid-development-stage sags are the developed area of the shallow-middle-interval oil system (Fig. 2).

SYN-RIFT GAS SYSTEM

In the deep interval of the central Songliao basin the humic-kerogen source rocks in the syn-rift faulted-depressions reach high mature and over-mature stages because they are deeply buried. Thus, they mainly produce gas as the main gas source of the deep intervals. Now, the industrial gas flows have been found from the Quantou, Denglouku Formations and weathering residue of the base rocks of the structures of the Wangjiatun, Songzhan, Yangcao, Choyanggou, Changcunling, Changde, Maoshan, and Sijiazi. The natural gas is prediminantly composed of methane (more than 90%). Heavy hydrocabon is low. $C_1/(C_2+C_3)$ ranges between 20-1,000. The carbon isotope of methane is heavy, generally ranging between $-25‰$ and $-33‰$, obviously different from that of dissolved gas in shallow-middle-interval oil ($-40‰ \sim -52‰$). The natural gas of sevveral wells has inversion of isotope, for example, the carbon isotope of ethane is heavier than that of propane in natural gas of the Member 3 of the Quantou Formation of Well-501. The He and Hg contents of natural gas are high, e. g., the He content of 140-127 Ma. As a result, the whole NE China entered regional rifting stage. After 127 Ma, the subducting rate of the Izanagi Plate obviously decreased, and the compressional force also decreased towards the Eurasian Plate. The disturbed isotherm of lithosphere in NE China began re-balance. The lithosphere of the stress-concentrated area in the early stage underwent cooling, contracting, and thermal subsidence. Finally, the post-rift large lake basin was formed. Then, the Pacific Plate was subducting westward at the rate of 20. 2 cm per year in 85 Ma, and 10. 4 cm per year in 74 Ma[4]. Changes of plate-subducting directions resulted in the formation of inversion structures in the Songliao basin, depositing inversion supersequence of the Sifangtai-Mingshui Formations. The depocenter shifted westward.

SYN-RIFT STAGE SUPERSEQUENCE

The faulting activities in the Early Cretaceous was very active. The distribution of half-grabens is controlled by the Nenjiang-Shuangwu, and Yilan-Yitong NNE crustal rifts within the basin and basement fault systems in the two sides of the paleouplift in the central basin. Development of the Early Cretaceous rifted depressions underwent three stages of faulted in-fill, expansion, and shrinking, showing a clear, complete sedimentary cycle and succession of volcanic eruption. The se-dimentation rate is high, generally more than 200 m/Ma. The sedimentary features depict coal-bearing strata associated with volcanic eruption rocks. 32 rifted depressions were developed in this stage, the total area of which approximated 50,000 km^2. The thickness of sedimentary rocks varies from 1,000 to 4,000 m.

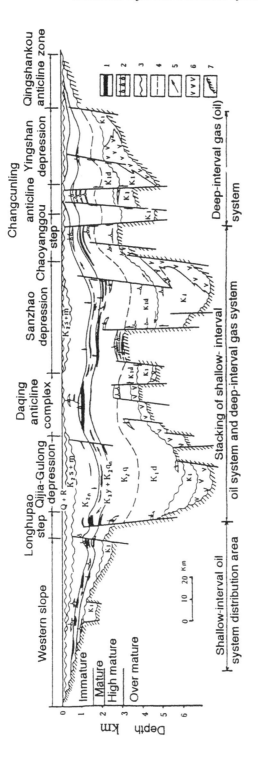

Fig. 2 The cross-section showing the petroleum systems of the Songliao basin

1. Oil reservoir; 2. Gas reservoir; 3. Unconformity; 4. disconformity; 5. Hydrocarbon migration direction; 6. Volcanic and

Volcanic-clastic rocks; 7. Basement

POST-RIFT STAGE SUPERSEQUENCE

The Quantou strata consist of fluvial and discontinuous shallow lacustrine deposits, belonging to infilling stage of red clastic rocks. The dark, deep-lake mudstones are only distributed in the central part of the basin. The vertical succession from bottom to top generally shows from fluvial sandy conglomerate through littoral and shallow lacustrine vari-colored and grey massive fine-medium sandstones interbedded with mudstones, finally to vari-colored mudstones with light grey silts and fine sandstones and thin dark grey mudstones. The fluvial sandstones are predominated. The delta deposits are locally developed.

The climax in the development of the Songliao basin was during the deposition of the Qingshankou and Nenjiang Formations. The sedimentary area was generally increasing, reaching the maximum extent of more than 200,000 km^2 during the deposition of the Nenjiang Formation. The deep lacustrine area in the central depression is about 60,000 km^2 and is divided into the Sanzhou (in the east) and Gulong (in the west) sags which are main depocenters. The important source rocks in the thickness of 500 to 1,000 m were formed during the Qingshankou and Nenjiang Formations. Three main delta sandstones in the southern, northern, and western parts of the basin were surrounded by the mudstones of the Qingshankou and Nenjiang Formations and were composed of a good source-reservoir-seal combination. The Qingshankou and Nenjiang strata are the main petroliferous interval in the basin—the middle petroliferous association that contains most petroleum reserves of the basin.

INVERSION STAGE SUPERSEQUENCE

The depocenter shifted to the western part of the Qijia-Gulong and Changling sags during the Late Cretaceous Sifangtai and Mingshui Formations. The sedimentation rate is 26 m/Ma, showing the feature of the compensated deposition. Deposition of the Sifangtai Formation covers an area of 73,000 km^2 and consists of a set of fluvial and shallow lacustrine sandy conglomerates and massive mudstones, the middle part of which is composed of the greyish green and dark grey sandstones and mudstones formed in shallow lacustrine environment. The interval is thin (300-1,000 m thick). The Mingshui Formation displays conformity-disconformity relationship with the underlying Sifangtai Formation and consists of two intervals on the thickness of more the Member 3 of the Denglouke Formation of Well Shen-1 is up to 384,000 ppm and the Hg content of Member Quan-3 is up to 18,470 ng/cm^3. Distribution of deep-interval gas system is controlled by the regional seal of the Member 1 of the Qingshankou Formation. Gas reservoirs are mainly distributed in faulted blocks and drape-anticline structures of the initial stage of the interior and margin areas of the faulted depressions. The major fault systems developed in the initial and middle stage is the access for gas upward migration (Fig. 3).

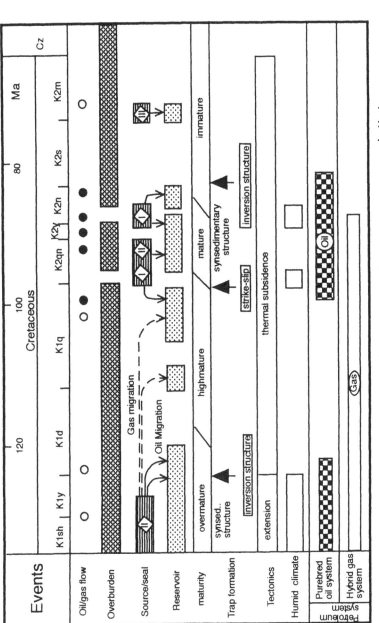

The petroleum system elements and processes are plottet against time on the horozontal axis

Fig. 3   The event chart of the petroleum system in the central part of the Songlioa basin.

SYN-RIFT OIL SYSTEM

In the uplift areas around the Songliao basin, the Early Cretaceous faulted depressions are not continuously buried during the Late Cretaceous. Although Member 1 of the Qingshankou Formation is deposited, it belongs to the immature stage. Thus, the Lower Cretaceous supersequence consists of an independent petroleum system of itself, for example, the Luijiapu sag of the Kailu depression. The source rocks and reservoir rocks are all the sediments of the Juifutang Formation (Fig. 4). The Kerqin and other fields have been found now.

POST-RIFT SYSTEM

The Qingshankou Formation and Member 1 of the Nenjiang Formation deposited during the post-rift large lake stage are the main source rocks of the Songliao basin. The kerogen type in the Members Qing-1 and Nen-1 is sapropel, whereas that in Members Qing-2+3 is sapropel-humic. The hydrocarbon evolution in the Songliao basin is characterized by good potential of source rock, shallow threshold depth of oil geothermal gradient, high heat flow, and the accumulation of the wide drape mudstones of the Nenjiang Formation[3,5]. The area of mature source rock controls the hydrocarbon distribution. The delta front sandstones of the axial large delta complex are interbedded with source rocks or finger-like contacts, which are favorable for hydrocarbon migration. The good co-ordination of the inversion stage and peak period of hydrocarbon generation is the main cause of the huge hydrocarbon accumulation in the structure of the Daqing anticline (Fig. 3).

According to structural features and reservoir-formation history, the post-rift petroleum system may be classified as the western slope, Longhupao-Daan step., Qijia-Gulong-Changling sag, Daqing anticline, Sanzhao sag, and Chaoyanggou-Fuyu anticline plays (Fig. 2). The structural features, source rocks and reservoir styles, trap types, and reservoir-formation history are different in different plays.

INVERSION STAGE BIOGENETIC GAS SYSTEM

The transgressive systems tract of Member 1 of the Mingshui Formatoin during the inversion stage is shallow ($<1,000$ m), being an immature stage ($R° < 0.47\%$). Thus, it is apt to form the biogenetic gas and accumulate reservoirs under the local seal. This type of gas reservoirs is almost all mixed gas from the underlying mature source rocks migrated along faults, for example, the Honggang reservoir[2].

CONCLUSIONS

The Songliao basin underwent three evolved stages during the Cretaceous and formed three types of petroleum system. The syn-rift gas system or oil system in the Songliao basin is a new targeted interval for exploration and is also a new field for increasing hydrocarbon re-

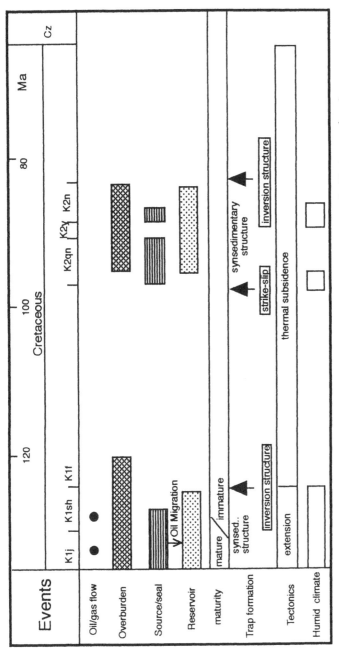

The petroleum system elements and processes are plottet against time on the horozontal axis

Fig. 4   The event chart of the petroleum system in the Lujiapu area of the Songliao basin

serves. The distribution of the petroleum system formed during large lake stage is controlled by mature source center. The western slope is a favorable area for exploration of heavy oil and biodegradation gas. There is a new targeted interval of the reservoirs of the Quan-3+4 migrated downward along faults in the Qijia-Gulong and Sanzhao sags besides Sha-Pu-Gao reservoirs, thus, there will be a large increase of hydrocarbon reserves. Moreover, there is a big challenge for the exploration of hydrocarbon generated by immature source rocks in shallow interval.

REFERENCES

1. Dou Lirong. Geochemical characteristics of coal-type gas in the deep interval of the Songliao basin (in Chinese with English abstract). Jianghan Petroleum Institute Journal, 12(3),9-11,1990.
2. Dou Lirong. Integrated identification of gas origins by using geological and geochemical characteristics (in Chinese with English abstract). Xinjiang Petroleum Geology, 13(4),351-335,1992.
3. Huang Difan and Li Jinchao. Nonmarine Petroleum Generation (in Chinese). Beijing: Petroleum Industry Press,1982.
4. Maruyama, S. and Seno, T. Orogeny and relative plate motions: Example of the Japanese Islands. Tectonophysics, 127,305-329,1986.
5. Yang Wanli (ed.). Petroleum Geology of the Songliao Basin (in Chinese). Beijing: Petroleum Industry Press, 1985.

PART II

NEW PROSPECTS AND TARGETS OF HYDROCARBON IN THE 1990s

Proc. 30th Int'l. Geol. Congr. , Vol. 18, pp. 45~56
Sun Z. C. et al. (Eds)
© VSP 1997

New Oil and Gas Exploration Frontiers and Resource Potential of the Old Cratons in NE Asia

HU JIANYI   ZHOU XINGXI   XU SHUBAO   LI QIMENG
(*Research Institute of petroleum Exploration and Development*, *CNPC*, *P. O. Box* 910, *Beijing*, 100083 *China*)

Abstract

For a long time, the demand for oil and gas resources in the NE Asian countries including China is increasing quickly, meanwhile, this broad region is one of the richest petroleum resource areas in the world. The basins developed on these old cratons (0.8—1.8 billion years ago) are the hot regions for petroleum exploration in 1990s and next century. These areas include the Tarim, Huabei (North China) and East Siberian Basins. The old strata including the Proterozoic and Paleozoic have excellent source rocks which have experienced petroleum generation, migration and accumulation. Multiple tectonic movements and multiple unconformity surfaces made hydrocarbon accumulation, escape and reaccumulation for several times, and formed many unique large unconventional oil/gas traps and distribution models. The inherited subsidence and deposition during the Mesozoic and Cenozoic were favorable for hydrocarbon preservation and also have great influences on the reformation of old oil/gas pools to form some special petroleum systems on these old cratons. The huge sedimentary volume, abundant hydrocarbons of the old strata and epigenetic good preservation conditions show that these basins have great gas and oil potential.

Keywords: *regional seal   petroleum system   exploration frontier   old craton   NE Asia*

FRAMEWORK OF OLD CRATONS IN NE ASIA

Formation and Evolution of Cratons

To the east of Ural Mts. and north of Kunlun—Qinling—Dabie Mountain System, there are three large old cratons in the NE Asia, that is, Huabei (North China), Siberian and Tarim old cratons (Fig. 1). They became such stable blocks as the important part of Asian continent after they accreted and consolidated around the Archaeozoic continental nuclei 1.8, 1.6, and 0.8 billion years ago, respectively.

During the aulacogen development period of these early craton formation stage , abyssal—subabyssal basins around them were filled by several thousand meters of marine carbonate rocks and clastic rocks of the Middle — Late Proterozoic, which are the first set of sedimentary blanket above the Archaeozoic—Proterozoic metamorphic basement (Fig. 2). The Tarim craton was formed later, and its aulacogens were continued to the Ordovician time.

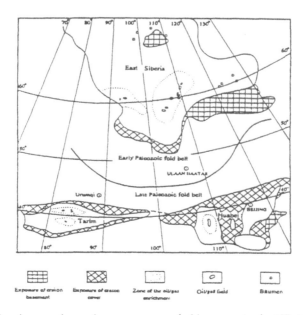

Fig . 1 Sketch map of petroleum resources of old cratons in the NE Asia Uplifts

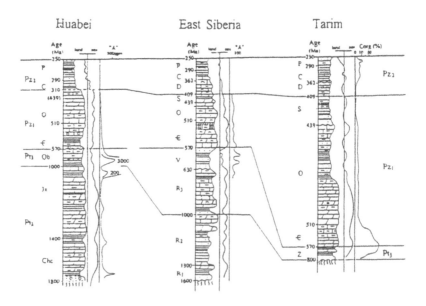

Fig. 2 Comparison of the Proterozoic—Paleozoic stratigraphic columns of
the old cratons in the NE Asia

At the beginning of the Phanerozoic, these three old cratons separated from the

disparated Pal—Asia, and they located in the low latitude area for about 150 Ma
from Cambrian to Ordovician. Several thousand meters of marine carbonate
rocks were in the intracratonic depressions and pericratonic depressions (aulacogens) which contained marine clastic rocks and intercalated gypsum/salt and
phosphorite. The thickness of gypsum/salt strata reachs up to 1000—2000m in
the Cambrian of the Siberian craton.

The Caledonian movement made these three cratons extensively drift northward
and rotate clockwise at a large angle (Fig. 3). The Huabei and Tarim were
bulged and marine regression occurred as the result of compression in the Devonian and Silurian. The Huabei became as a peneplain denotation area, and the
Tarim sedimentary area decreased. The sediments were coastal—neritic facies
with intercalated continental deposits. Inside the cratons old uplifts appeared
due to differential movements.

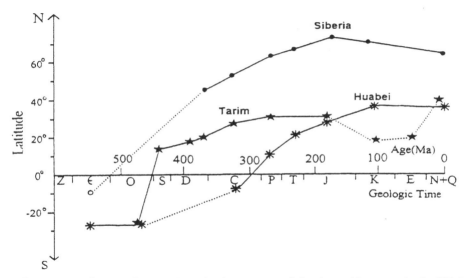

Fig. 3. Map showing the Paleolatitude change trace of the three old cratons in the NE Asia

The Early Hercynian movement at the end of Devonian was violent and caused
the crust rise and widely be eroded. Afterwards marine transgression occurred
again and its highest level in the middle Early Carboniferous, then marine regression happened from Late Carboniferous to Permian. The intracratonic depressions were filled by neritic clastic rocks and carbonate rocks in the lower
part, parallic clastic rocks with coastal swap sediments in the middle part, and
red clastic rocks in the upper part. At this time, old uplifts were finalized basically.

These cratons continued to drift northward and rotate clockwise at a big angle
once again during the Carboniferous and Permian.

The Late Hercynian movement from the end of Permian to the beginning of Triassic is an important geological event. It ended the cratonic development stage,

and caused obvious tectonism and regional elevation and truncation. At the same time, orogenic folding between old cratons occurred successively, sea water receded, and the embryo of the NE Asian continent was formed, as a result, relative locations of these old cratons were fixed basically. Also the Late Hercynian movement caused a fair—sized hypobyssal basic magma intrusion and volcanic activities in the Tarim and eastern Siberia cratons. Correspondingly an important thermal event occurred at that time in these areas.

Although no large—scale drift and rotation occurred in these three craton in the Mesozoic and Cenozoic, tectonic reformation and differentiation were very obvious(Fig. 4.). During the Mesozoic the convergence of cratons in the Asian continents caused the eastern western Huabei and Siberia be uplifted mainly, sedimentary records were only distributed locally in some faulted or foreland depressions, but in the Huabei and Tarim big foreland basins and inland depressions were formed. During the Cenozoic because of the northward extension and collision of the Indian Plate and westward subduction of the Pacific Ocean Plate, the eastern Siberia and western Huabei were dominated by rising and denotation, the Tarim and eastern Huabei, however, subsided strongly. Foreland basins were well developed in the Tarim and a series of rift/downwarped basins occurred in the eastern Huabei.

To sum up, the old cratons in the NE Asia experienced four stages such as the Archaeozoic continental nuclei, basement consolidation, cratonic evolution and tectonic differentiation. The cratonic evolution stage can be divided into two periods: aulacogen development period and stable craton development period. Many sets of good source rocks, cap rocks and reservoir rocks were deposited in these two periods. During the long evolution history many geological events provided energy for hydrocarbon generation, migration and accumulation, thus establishing a good foundation for the formation of petroleum resources.

Geological Development Characteristics of Inherited Cratonic Basins

Huabei old cratonic basin

The Luliang movement 1. 8 billion years ago made the Huabei metamorphic basement be stable, and then produced aulacogens on the northern, western and southern edges of the old craton, which were filled by several to about ten thousand meters of quartize, dolomite and clastic rocks of the Changcheng, Jixian and Qingbaikou Systems during the Middle and Late Proterozoic, containing many alga and spore fossils. These three systems become thinner or even pinch out toward the central craton. The Jinning movement ended the aulacogen development period and the stable development period began. The Sinian is absent or only seen in local areas, and it is less than one hundred meters of arenaceous rocks. The Cambrian and Ordovician are composed of epeiric marine carbonate and/or clastic rocks, with an thickness of 1000—2000m. The regional uplifting in the middle Caledonian Movement made the Silurian, the Devonian and Lower

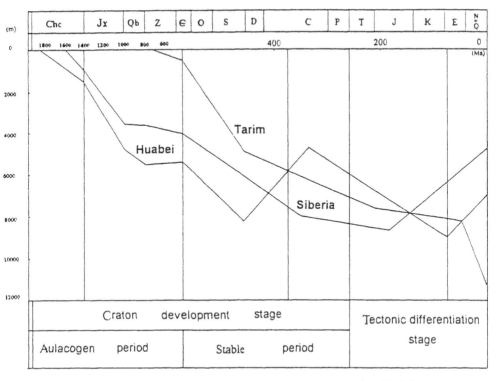

Fig. 4. Subsidence curves of three old cratons in the NE Asia

Carboniferous strata absent in the Huabei. On the weathering crust, there developed laterite residua widely. After then, there deposited bauxitic shales followed by coal—bearing strata of the Upper Carboniferous and Lower Permian were deposited, and gradually changed into varicolored and red terrestrial sandstones and mudstones. The Indosinian movement caused tectonic differentiation of the old craton. The western part continued to subside and received large—scale lacustrine deposits, with an thickness of up to 1000m. However, the eastern part (Bohai Bay Basin) was faulted and elevated/subsided, with limited Mesozoic deposits. During the Himalayan period, the eastern part entered fault —depression evolution period, with well—developed Cenozoic deposits. Further to say, the Paleogene strata, 3000 — 4000m of lacustrine and fluvial clastic rocks, were distributed in rifted basins, and the Neogene were widely distributed in an united Bohai Bay Basin, with 1000—3000m of fluvial—dominated deposits. However, the Cenozoic strata are fragmentarily scattered in the western part.

East Siberian old cratonic basin
In this basin the sedimentary record was began 1. 6 billion years ago. The Riphean is composed of three parts, the lower (R1) is granulite and volcanics, the

middle(R2) is motley clastic and carbonate rocks, and the upper(R3) is carbonate rocks interbedded by sandstones and mudstones. The Riphean sediment thickness changes greatly, from 3000— 5000m in the marginal aulacogens, to very thin in the central craton, and even absent on the Nip—Badubin, and Anabar old uplifts. These old uplifts had strongly inherited feature, and were located near sea level afterwards.

The Vendian deposited on the Riphean unconformably is about 1000m of dark mudstones interbedded by carbonate and clastic rocks. It was covered by 1000—4000m of the Cambrian shallow—sea carbonate rocks, shales, and salt—gypsum layers conformably. The salt—gypsum layers are 1000 to 2000m thickness. The Ordovican, Silurian and Devonian are composed of various limestones interbedded by dolostones. The Carboniferous and Permian are mainly clastic rocks, partly with coal seams.

The Hecynian movement ended the craton evolution stage, and caused the basic magmatic activities to form many hypabyssal basic dikes and veins. The volcanic activities began in the Early Triassic, forming a set of volcanic rocks covering unconformably on the Paleozoic strata. During the Meso—Cenozoic, the craton was mainly uplifted, some foreland basins were only developed on the craton margins, and were filled by very thick clastic rocks.

Tarim old cratonic basin

The Tarim craton was formed relative later, about 0.8 billion years ago. The Sinian strata is its first sedimentary cover, with the thickness of 300 to 500m. The bottom is coarse clastic rocks, sometimes with moraines, and the upper is composed of mudstones and dolostones. The Cambrian and Ordovician are aulacogen sediments in the eastern Tarim Basin, and consists mainly of abyssal, subabyssal, and neritic clastic and carbonate rocks, and the intracratonic depression is covered mainly by carbonate tableland rocks. The Middle—Upper Ordovician contains more pelites and marls. The total thickness is from 1000 to 3000m. The Caledonian movement caused the formation of the Tabei and Tazhong old uplifts form, and the sedimentary area decreases. The Silurian and Devonian are gray—green to brown—red clastic rocks of neritic and coastal facies, unconformably on the Ordovician. The total thickness changes from 1000 to 3000m.

The Early Hecynian movement made the old uplifts further rise and be eroded. The Carboniferous lain unconformably on the different formations of the pre—Carboniferous. It is composed mainly of neritic, parallic and continental clastic rocks, and the thickness is about 500 to 1000m. The upper Lower—Carboniferous is 200—300m of mudstones, limestones and salt—gypsum layers, which are the maximum transgressive sediments. The Permian is mainly continental clastic rocks, and its total thickness is 500 to 900m. There is 100—500m of volcanic rocks in the Lower Permian strata.

The Hecynian movement at the end of the Permian caused the old uplift rise, maded the paleozoic strata be eroded, folded and faulted, thus ended the craton development.

The basin entered the composite foreland evolution time during the Meso — Cenozoic period. During the Triassic fluvolacustrine strata of 500 to 1000m were deposited in the northern part. The sedimentary area increased successively during the Jurassic and Cretaceous. The foreland and pull—apart basins occurred in the southwestern and southeastern margins. They were filled by coal measures and then motley and red clastic sediments of 1000 — 1500m. The continental clastic rocks were widely distributed in the whole basin during the Cenozoic. The transgression only occurred in the southwestern part during the Late Cretaceous and Early Tertiary. The Cenozoic strata are very thick in the foreland basins, about 5000 to 7000m, and become thinner towards the central cratons, generally 1000 to 3000m. The Tertiary is mainly mudstones and siltstones interbedded by sandstones and conglomerates. In the foreland area the Paleogene and Miocene have several ten to several hundred meters of salt—gypsum layers.

FORMATION OF PETROLEUM SYSTEMS IN OLD STRATA

well—developed Source Rocks in Old Aulacogens and Stable Tablelands

The dark — grey shales of the Upper Jixian and Lower Qingbaikou Systems of the Middle—Upper Proterozoic are the source rocks in the Huabei Basin. In the Tarim Basin , the Cambrian and Ordovician are the main development stage of the Manjiar aulacogen, which is filled by thicker mudstones and carbonate rocks acting as source rocks. The Riphean and Vendian mudstones and carbonate rocks are identified as source rocks in the East Siberia. The common features of these source rocks are as follows:

(1) They are thick, generally more than several hundred to thousand meters, and widely distributed, generally several hundred thousand square kilometers. Thus they form huge source volumes.

(2) The source material include procaryotes such as blue—green algae, bacterial, and lower organism such as poly—cell agatic plant, they make up sapropelic kerogen.

(3) The organic richness is higher. For example, the organic carbon richness of the Cambrian and Ordovician is 0. 25%, and the highest is 5. 22% in the Tarim Basin. That of the Proterozoic source rocks is 0. 1 — 0. 7%, and the highest is 6% in the East Siberia. Specially that of shales is higher than that of carbonate source rocks in the Huabei Basin. For example, that of mudstone source rocks is 1. 2%, that of carbonate rocks is 0. 23%. They belong to medium—good hydrocarbon source rocks.

(4) The old source rocks have reached condensate gas and dry gas generation stage. The formed hydrocarbons are mainly dry gas, condensate oil and light oil. Liquid hydrocarbons are characterized by low density, low sulfur content, no bitumen content and high fraction content. The alkane is the main component in the group composition, and is mainly composed of C15 and C17 n—alkane.

Pr/Ph is less than 1. Iso—sterane is rich in the polycyclic cycloparaffinic hydro-carbons.

Good Reservoir Rocks of Old Strata after Long Evolution and Deep Burial

The reservoir rock in the old strata are mostly coastal deposits under high—energy environment, such as coastal sand bar, beach sand, deltaic debouch bar. They are mainly quartzose sandstone or quartzose arkose, with good sorting, high mineral maturity, thick single—layer, and stable distribution. Because of low heat flow and geothermal gradient, the compaction, cementation and various chemical actions in sandstone reservoirs, are restricted despite large depth. Thus they have good physical property. For example, the average porosity of the Silurian and Carboniferous buried at 6000m is 16%—18%, the permeability is generally $(n \times 10 - n \times 10^2) \times 10^{-3} \mu m^2$. The Vendian coastal sand bars with maximum depth of 5000m are buried at more than 3000m at present, but their average porosity is about 14%, permeability is $(n \times 10 - n \times 10^2) \times 10^{-3} \mu m^2$.

The Proterozoic and Paleozoic carbonate reservoir rocks, mainly dolostones and biolithites, are well—developed as the result of paleocarstification and weathering—leaching. The biolithites and oolitic limestones of open tableland marginal beach faces have good petrophysical properties. Under the unconformable surfaces carbonate reservoir rocks are well—developed due to paleocarstification, for example, the effective permeability of the Proterozoic Jixian dolostones is $(100 \text{ to } 300) \times 10^{-3} \mu m^2$ in the Huabei Basin; that of the Riphean—Vendian and Cambrian carbonate rocks is 5%—10%, sometimes reachs up to 10%—20%, and the permeability is $(n - n \times 10) \times 10^{-3} \mu m^2$.

Geological Base for the Formation of Petroleum Systems in Old Strata

Regionally stable and reliable seals

The evolution and preservation of old source rocks experienced several (sometimes strong) tectonic movements need a regionally stable seal. This is the base and premise for formation and preservation of oil/gas pools, specially for the condensate oil/gas and methane gas pools. In the Siberian Basin, more than 1000m of salt—gypsum layers occur in the Middle—Upper Cambrian. Its plasticity increases below 2000m, so it will be favorable for the gas accumulation. The 400—600m of Upper Carboniferous coal measures are stably distributed, specially several ten meters of bauxitic shale at the bottom, is the firm seal for the lower Paleozoic gas accumulation and preservation in the Ordos Basin. For the Tarim Basin, the strength and erosion of various tectonic activities are larger, so there are several unconformable surfaces, and regional faults were open several times, so several regional seals are needed for preservation. The Lower—Middle Carboniferous, Mesozoic and Tertiary mudstones or salt—gypsum barriers distributed in different regions, with the thickness of 150—300m, with strong plasticity, and large entry pressure, are regional excellent regional seals.

Condensate oil and methane gas — —the main hydrocarbon phase behavior
After the long hydrocarbon generation and evolution, liquid hydrocarbons transform into mostly condensate oil/gas or methane dry gas, and partly into light oil. They are the activist and most volatilizable components.

Stratigraphic — lithological traps — —the main petroleum reservior type
Folds are less — developed during the main tectonic activities on old cratons with large rigidity. The drape — anticlines and other anticlines of various origins developed on the base uplifts experienced several — time tectonic faulting and volcanic reformation, so they became uncompleted, and their sealing condition was decreased. Multiple migration and reaccumulation processes of the early accumulated oil and gas took place. Thus it is difficult to form large oil/gas fields. Under the reliable seals, however, the stratigraphic — lithological traps on the flanks of old uplifts are the main petroleum reservior type in the old strata. The reasons are: (1) tectonic movements had less influences on the stratigraphic — lithologic traps; (2) the oil/gas pools are primary, and (3) they have huge dimension. For example, the gas — bearing area of the Jingbian stratigraphic — lithological gas field in the Ordos Basin is 3000 — 5000km2 (Fig. 5) that of the Kevikejin stratigraphic — lithological gas field is more than 3000km2 (Fig. 6) in the East Siberia.

Two models of petroleum systems in old cratons
Simple Model
Under the regional reliable and stable seal, high — mature hydrocarbons (condensate oil/gas, methane dry gas) are the main phase behavior type . The stratigraphic — lithological oil/gas pools are the main oil/gas accumulations on the uplift plunge parts. Large petroleum systems of such type with abundant oil and gas have been found, for example, the Vending — Cambrian petroleum system in the East Siberian Basin and the Ordovician — Carboniferous system in the Ordos Basin (Fig. 7).

Composite Model
Because of the strong tectonic and faulting activities, each seal may be effective in some regions, and open in other regions. The repeated petroleum accumulation — escape — reaccumulation processes took place. The migration scale was very large in horizontal and vertical. Therefore, several sets of seals should occur, generally 2 to 3 regional seals occur in a petroleum system. For example, in the Tarim Basin, the Lower Carboniferous, Mesozoic and Tertiary seal occur successively from the central part to the uplift. The accumulated oil and gas in the old strata remigrated from below older seal and reaccumulated below younger seal, to form secondary oil and gas fields with small — medium size. In this model of petroleum systems, it is speculated that principal oil/gas fields are distributed in the stratigraphic — lithologic traps of old uplift plunge parts under the lowest regional seal.

The petroleum systems in the old strata of old cratons have complex and unique

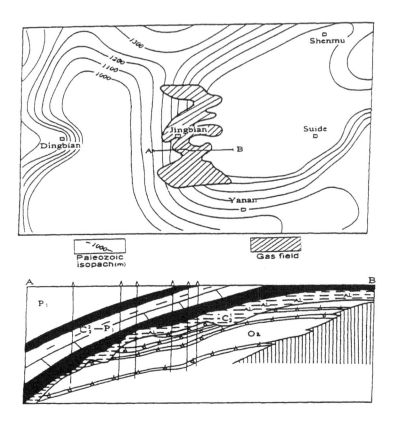

Fig. 5 Synthetical map of the Jingbian gas field, Ordos Basin.

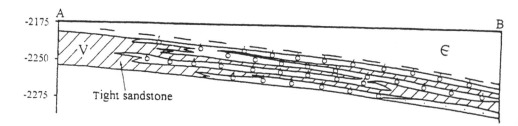

Fig. 6 Cross—section of the Kevikejin condensate gas field, East Siberia.

formation conditions and distribution models, and they have larger petroleum (specially gas) resource potential.

RESOURCE PROSPECT

It is obviously that the inherited cratonic basins of the NE Asia are very wide

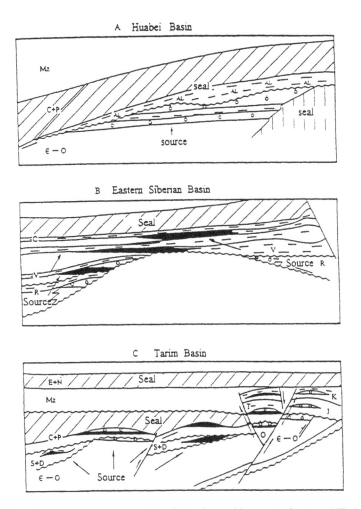

Fig. 7 Petroleum system models of the three old cratonic basins, NE Asia

petroleum—bearing frontiers. The areas of Siberian, Huabei and Tarim cratons are 3. 5, 1. 8 and 0. 72 million km², respectively. They have huge volumes of old strata and source rocks. Although organic matters are lower— organisms such as algae, bacterials, and gill fungi ,the organic content in rocks can be fairly high. The primary hydrocarbon — generation potential of the Riphean source rocks is 0. 3 to 1. 0 million tons per square kilometer, and that of the Cambrian —Ordovician source rocks of the Tarim Basin is 2 to 3 million tons per square kilometer.

During the geological history large—sized petroleum migration and accumulation took place, and the large bitumen and oil sand deposits distributed on the uplifts of basinal margins are the destruction result of large old oil pools due to later tectonic movements. The example is the Vendian—Cambrian bitumen—bearing strata, Aolieniaok large—bitumen deposit, on the northern flank of the Anabar anticline of the East Siberian Basin, with several hundred kilometers in length on the surface. Bitumen becomes maltha below 50m depth. The oil—bearing sequence of the Shuangdong old oil pool in the northern Huabei Basinis another example, its horizon is the Protorozoic Jixian system. This old anticlinal oil pool is exposed on the surface, with liquid oil in some druses and fractures. Also there is an Ordovician — Silurian oil pool on the surface at the Kepin, on the northwestern margin in the Tarim Basin. The bitumen—oil sand zone is 250 km in length. All these indicate that large—sized petroleum migration and accumulation have taken place once.

The initial exploration result shows that the old strata of these three old cratons significant effects have been obtained. More then 40 gas and oil/gas fields have been found in the East Siberian Basin, including the large Kevikejin gas field. Tens of oil/gas fields including the large Jingbian gas field in the Huabei Basin and the Tazhong oil/gas field in the Tarim Basin have been discovered. The exploration dimension is becoming wider.

Petroleum resource assessment result indicates that these three cratonic basins have large petroleum resources. It has been predicted that the resource potential of the East Siberian Basin in Russia is more than 40 billion tons of oil equivalent, mainly gas resource. The prospective resource of the old cratonic basins in China is 30 billion tons of oil equivalent, and also mainly gas resource. The proved and un—proved gas resources in old strata will be half of the total gas resources in China.

The NE Asia is a region rich in oil and gas resources, and also is low—exploited region, specially the old cratonic basins are lower—exploited. With the rapid development of the NE Asia economy, the exploration of old cratonic basins will be sped up. A new increase time of petroleum exploration and development will come in this region.

REFERENCES

1. Antselof, A. S. et al. Petroleum Geology of the Siberia (in Russian). Earth Moscow: (1981).

2. Drobot, D. E. et al. Geochemical evoluation standard of the pre—Cambrian and Lower Cambrian strata in the East Siberia (in Russian). Earth Moscow: (1974).

3. Hu Jianyi and Xu Shubao. Formation of oil/gas pools underwent tectonic regimes (in Chinese). In: Collected works about the research of Zuxia's academic thoughts (ed. by Xun Zhaocai et al.). Petroleum Industry Press. Beijing: 97—91(1993).

4. Hu Jianyi, Xu Shubao, and Dou Lirong. Petroliferous provinces and petroleum resource potential of the NE Asia. In: Symposium on Sino—Russia NE Asia Petroleum Geology and Hydrocarbon Prospect Seminar (ed. by Hu Jianyi and Fancongwu). Petroleum Industry Press. Beijing: 1—11(1994).

5. Liu Shuxuan et al. Reservoirs and phase behavior predict of deep oil/gas pools (in Chinese). Petroleum Industry Press Beijing: (1992).

6. Zalotov, A. I. Structures and Petroleum of old strata (in Russian). Earth Moscow: (1984).

7. Zhai Guangming. ed. 1993. Petroleum Geology of China (in Chinese). Petroleum Industry Press. Beijing: V. 12. 104—109(1993).

Proc. 30th Int'l. Geol. Congr., Vol. 18, pp. 57~70
Sun Z. C. *et al.* (Eds)
© VSP 1997

Optimization of Oil Deposits Prospecting and Exploration at the Stage of a High Degree of Exploration of the Territory
(By the Example of Tatarstan).

R. KH. MUSLIMOV (Tatneft), I. A. LAROCHKINA, R. N. DIYASHEV, E.
R. KIRILLOV, SH. M. BOGATEYEV (TatNIPIneft)
Bugulma, RUSSIA

Abstract

The Volga-Ural oil and gas bearing province, the central territory of which is occupied by Tatarstan territory, is characterized by a high degree of exploration and utilization of reserves. The fact that the area is well investigated (for 15 sq. km there is one prospecting-exploratory well), made it a unique testing ground for geological monitoring and development of the strategy and tactics of oil deposits prospecting in under-explored regions.

The main source of oil reserves replenishment is small, limited in sizes and reserves, formations in Devonian and Carbonic deposits, controlled by different genesis traps. Their search using traditional methods of seismic prospecting is not effective in some cases.

In the conditions of a high degree of the territory exploration detailed revision of the accumulated material with consideration of new data and possibility of new methods becomes a general direction. On the basis of a comprehensive geological analysis new concepts of the regularities of oil deposits location have been developed and perspectives of oil and gas presence in the territory have been specified. Search for new oil fields is carried out using both traditional mehtods, i. e. seismic prospecting, and non-traditional ones.

Application of optimum technique of oil deposits prospecting and exploration for the territories with a high degree of exploration provides for execution of a highly promising geological prospecting in Tatarstan.

Keywords: Oil deposits prospecting, Comprehensive analysis, Optimization, Tatarstan

INTRODUCTION

Tatarstan territory is a consituent of the Volga-Ural oil and gas bearing province, located in the eastern margin of the Eastern-European platform. As to tectonics, it is a central link of the Volga-Ural anteclise.

Deep drilling has been carried in the Republic since 1939, more than 4 thous. exploratory wells have been drilled with total volume about 8 min. m, on the average, for 15 sq. km there is one exploratory well, the density of seismic profiles within the territories under investigation in the most promising eastern part amounts, on the average, to 2.1 run. km/sq. km.

At the modern stage of prospecting the main source of oil reserves replenishment is small, limited in sizes and reserves, formations in Devonian and Carbonic deposits.

Exploration of oil in Devomian deposits presents the greatest problems.

For today the efficiency of drilling in local objects in terrigenous Devonian deposits, prepared by seismic prospecting does not exceed 20%. Low efficiency of the objects' readiness for deep drilling in the reflecting horizons of Devonian deposits can be accounted for by geomorphological multilayering of the being mapped surface, absence of confident stratigraphical referencing of reflecting horizons, ambiguity of relecting waves correlation, etc.

High degree of exploration of the Tatarstan territory made it a unique testing ground for geological observation and development of the strategy and tactics of oil deposits prospecting in insufficiently investigated regions of Volga-Ural province, as well as other regions, for example, Western and Eastern Siberia.

In the conditions of a high degree of the territory explotation, which is as great as 92%, the main and most important source of oil reserves replenishment is small formations in sedimentary mass of Devonian and Carbonic deposits (R. Kh. Muslimove, 1976).

Development of effective technique of oil exploration and prospecting in Tatarstan territory became possible thanks to availability of a great body of accumulated geological and geophysical information, development of new theoretical concepts of formation and location of hydrocarbons, redetermination of perspectives of oil presence in the territory.

Application of optimum technique for oil fields prospecting and exploration provides for high efficiency of geological prospecting in the modern difficult economic conditions in Russia.

MAIN FEATURES OF TECTONIC STRUCTURE OF THE TERRITORY

Main features of tectonic structure of the territory determine several regional structural configurations: Tatarian arch, comprising South-and North-Tatarian arches, Tokmovian arch. Melekesskian trough, Kazanian-Kirovskian avlakogene, and Kamskian-Kinelskian system of troughs (Fig. 1). They can be followed in different ways in the structures of foundation and sedimentary cover stages. (V. E. Khain, 1991).

Modern tectonic pattern of the territory was finally formed in the Alpine stage of tectogenesis, though tectonic movements of Baikal and Caledonian stages set the ground for future structures. All tectonic configurations inherited routes of ancient zones of fractures, locatedin pre-Devonian period. Formation of sedimentary cover dislocations took place under the action of tectonic movements having vibration character. This manifested in repeated change of upheaval and sinking cycles of the territory. Formation of large ancient Tatarian arch with Archean crystalline foundation occurred in Riphean time with the formation of Riphean avlakogenes. During Paleozoic time it was subjected to complicated movements with inver-

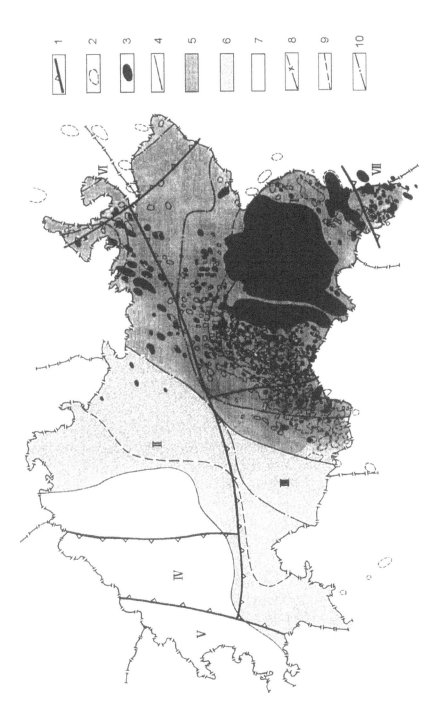

Fig. 1 Tatarstan. Map of oil and gas content perspectives.
1. Boundaries of tectonic elements: I . South-Tatarsky arch; II . North-Tatarsky arch; III . Melekessky depression; IV . Kazansko-Kiroisky avlakogene; V . Tokmovsky arch; VI . Kamsko-Belsky avlakogene; VII . Sergievsko-Abdulinsky avlakogene; Oil Poll. Outlines: 2. in Carbonic deposits; 3. in Devon deposits; 4. Boundaries of earths; 5. highly perspective; 6. perspective; 7. low perspective; Edge zone boundaries of Kamsko-Kinelsky system of troughs; 8. outer; 9. inner; 10. axial.

sion of sign with prevalence of negative movements and accumulation of terrigenous-carbonate mass of sediments.

By the end of Permian time Tatarian arch was formed in its present boundaries. Thus, South-Tatarian arch is a neogenetic structure, originated at Alpine stage of tectogenesis, when Sarajlinskian trough was formed which divided Tatarian block into two parts — South- and North-Tatarian arches.

During the whole geological history North-Tatarian arch occupied a higher position if compared to South-Tatarian one and, actually, it can be referred to the type of arches with inherited development.

Melekesskian trough was formed at Alpine stage of tectogenesis, when against the background of neighboring arched uplifts its downwarping along Prikamskian fault strengthened. During Paleozoic time it was developed as a part of Tatarian arch.

Location of Kazanian-Kirovskian avlakogene took place in Riphean time, during Devonian time it continued to develop rapidly. At Alpine stage of tectogenesis it experienced considerable inversion processes and Vyatskian trough was formed in its place.

In Paleozoic time Tokmovian arch was developed inheritedly as an arched structure and only in Mesozoic time young Ulyanovskian trough began to form. (M. F. Mirchink, 1965).

OIL AND GAS BEARING COMPLEXES

Commercial deposits of oil and oil indications in Tatarstan territory have been determined practically in the section of the whole sedimentary mass — in Devonian, Carbonic and Permian deposits. But both by section and by area distribution of oil presence is irregular. Depending on the character of oil presence, structure of reservoirs and type of cap rocks in the section we recognize six oil and gas bearing complexes in Devonian-Carbonic deposits and three-in Permian deposits, characterized by similar conditions of oil, bitumen and gas accumulation.

In sedimentary mass of Devonian and Carbonic deposits the following complexes are recognized: first-terrigenous Vorobyevskian-Ardatovskian, second-terrigenous Pashiyskian-Kynovskian, third-carbonaceous Semilukskian-Kizelovskian, forth-terrigenous Bobric-Tulskian, fifth-carbonaceous Aleksinskian-Vereiskian, sixth-carbonaceous Kashirskian-Gzelian (Fig. 2).

In deposits of Permian system several independent bitumen bearing complexes are disinguished: low-Permian carbonaceous, Ufimian terrigenous, Kazanian terrigenous-carbonaceous. They are separated by sulfate-carbonaceous cap rocks. These Permian deposits contain heavy oil and bitumen. Their formation is associated with inflow of oil hydrocarbons from the bottom upwards from underlying deposits, Devonian and Carbonic. (I. A. Larochkina, 1995).

Oil and bitumen indications are determined in Melekesskian trough and South-Tatarian arch. Eastern flange of Melekesskian trough and western slope of South-Tatarian arch and adjacent

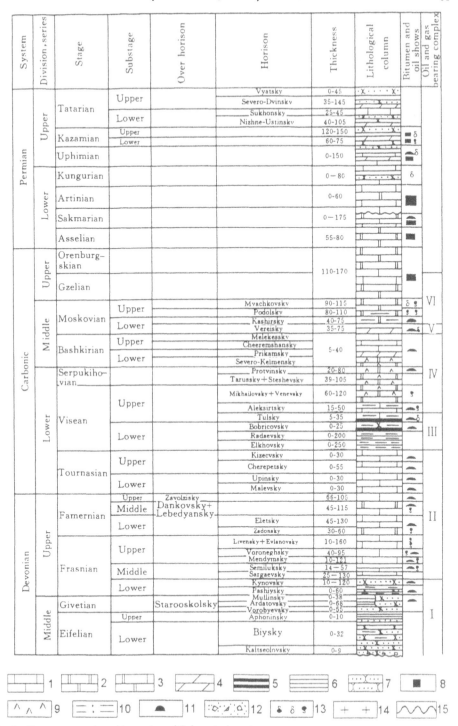

Fig. 2   Summary stratigraphic section of Paleozoic deposits.
1. Limestone; 2. Dolomite; 3. Dolomitic limestone; 4. Marl; 5. Coal and coaly shale; 6. Argillite; 7. Sandstone; 8. Bitumens;
9. Anhydride; 10. Clay; 11. oil pool; 12. Crust of weathering; 13. Oil and gas shows; 14. Rocks of cristalline basement; 15.
Boundary of unconformity.

territories are characterized by the greatest intensity. Tables 1,2 give characteristics of reservoirs and cap rocks of oil and bitumen bearing complexes.

Table 1   Characteristics of Reservoirs of Oil and Bitumen Bearing Complexes in Tatarstan

Oil and bitumen bearing complexes	Lithological type	Type of reservoir	Thickness of reservoirs,m	Porosity factor,%	Permeability, μm^2
Vorobyevskian -Ardatovskian	sandy-aleurolite	porous	0-35	11-26	0. 2-0. 3
Pashiyskian-Kynovskian	sandy-aleurolite	porous	0-52	11-30	0. 2-2. 0
Semilukskian-Kizelovskian	carbonaceous	porous-fractured-cavernous	0-18	7-27	0. 003-0. 1
Bobric-Tulskian	sandy-aleurolite	porous	0-22	10-30	0. 1-0. 4
Aleksinskian-Vereiskian	carbonaceous	porous-fractured	0-20	7-21	0. 003-1. 0
Kashirskian-Gzelian	carbonaceous	porous-fractured-cavernous	0-30	7-20	0. 003-0. 4

Table 2   Characteristics of Caps of Oil and Bitumen Complexes in Tatarstan

Oil and bitumen bearing complexes	Age of cap	Lithological type of cap	Distribution within region	Thickness of cap,m
Vorobyevskian-Ardatovskian	Ardatoskian	carbonaceous-clayey	zonal	0-10
Pashiyskian-Kynovskian	Kynovskian-Sargaevskian	carbonaceous-clayey	regional	5-135
Semilukskian-Kizelovskian	Semilukskian-Elkhovskian	carbonaceous and clayey	zonal and local	0-6
Bobric-Tulskian	Tulskian	clayey	zonal	0-45
Aleksinskian-Vereiskian	Aleksinskian Vereiskian	clayey-carbonaceous	zonal	0-35
Kashirskian-Gzelian	Kashirskian-Gzelian	clayey-carbonaceous	local	0-7
Ufimian	Baituganskian	clayey	zonal	0-30

Vorobyevskian-Ardatovskian complex is represented by sandy-aleurolite beds, 0-35 m thick, their regional decrease occurs from the south-eastern slope of South-Tatarian arch to north-north-west.

Commercial accumulations of oil are determined in the south-eastern slope and in the top of South-Tatarian arch. These deposits are principally small, ranging in size from 3 to 10 sq. km, controlled by tectonic traps, in the basis of which foundation projections are located. As a rule, they are small-amplitude, for this reason oil presence levels do not exceed 15-20 m and are characterized by buried mode of occurrence. The majority of deposits can be referred to the type of structural-lithological. Density of oil in beds regularly increases from the south to the north . In the same direction content of sulfur increases and yield of light fractions decreases.

Pashiyskian-Kynovskian complex is composed of terrigenous reservoirs. This complex contains the largest and most productive deposits. The widest aerial development is characteristic of the D1 bed in the South-Tatarian arch, Melekesskian trough and Kazanian-Kirovskian avlakogene, and to some extent in the North-Tatarian arch.

D1 bed is the main producing horizon in the Tatarstan territory. Here immense reserves of Romashkino, Novo-Elkhovka, Bavly oil fields are grouped, which are being developed for more than 50 years. As far as sizes and reserves are concerned, deposits of Pashiyskian horizon are not of equal worth. Small and tiny oil fields , satellites of gigantic ones, are of widespread occurrence throughout the territory. They are of different sizes, small-amplitude, as a rule, primarily of structural and structural-lithological types. Oil quality deteriorates from the south-east to the west and north-west.

Rock mass of Semilukskian-Kizelovskian oil and gas complex is composed of carbonaceous rocks. Oil deposits are controlled by local upheavals of different amplitudes. Oil presence levels may reach 80-90 m. They are defined geomorphologically most prominently in the flange zones of Kamskian-Kinelskian system of troughs. The deposits are small-sized, from 1 to 3-5 sq. km and are established in the South- and North-Tatarian arches, Melekesskian trough.

Oil deposits in Bobric-Tulskian complex are controlled by sedimentary upper Devonian-low Tournaisian bioherm upheavals, primarily of structural and structural-lithological type. They are recognized in the south- and North-Tatarian arches. Melekesskian trough and are characterized by small sizes, on the average, their area is 3-4 sq. km, oil presence levels may reach 60-90 m (Fig. 3).

Oil reservoirs in Vereiskaian-Bashkirian deposits are controlled by mantled structures of late Devonian-early Tournaisian bioherm edifices. Oil presence levels do not exceed 30-35 m. Oil reservoirs are grouped mainly in two zones—in the western slope of South-Tatarian arch and eastern flange of Melekesskian trough.

The sixth, the upper oil and gas bearing complex includes limestone and dolomites of Kashirskian-Gzelian deposits. This mass does not have a solid cap rock, its place is taken by thin clayey interlayers and benches of carbonates with thickness not exceeding 5-7 m. The deposits are small in size, areal of oil presence is controlled by limits of oil saturated rocks of Aleksinskian-Vereiskian oil and gas bearing complex.

Comparison of average data on physical properties and chemical composition of oils in the main producing horizons points to presence of a definite directionality in their change by the section. From Devonian to Carbonic and Permian deposits increase of oil density takes place as well increase of sulfur, resins, asphaltenes content along with decrease of paraffins, light fractions and gas saturation.

Table 3 shows physical and chemical properties of oils of oil and bitumen bearing complexes.

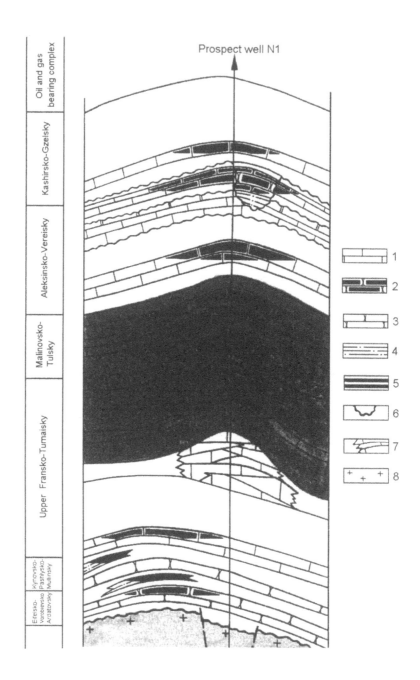

Fig. 3   Geologic section of Tatarstan type field.
1. Carbonate rocks; 2. Oil-saturated reservoir; 2. Sandstone; 4. Alevrolite; 5. Coals and coaly shales; 6. Erosive-karst cut-tings; 7. Reefy limestones; 8. Rocks of crystallinte foundatoin.

Table 3 Physical Properties and Chemical Composition of Oils in Oil and Bitumen Bearing Complexes in Tatarstan

Oil and bitumen bearing complexes	Density g/cm^3	Content, % wt.				Yield of light fractions up to 300 ℃ %	Range of gas factor change, m^3/t	
		sulphur	paraf-fin	resins	asphal-tenes		to	from
Vorobyevskian-Ardatovskian	0.853	1.27	3.52	12.0	3.7	48.5	30.0	71.2
Pashiyskian-Kynovskian	0.874	1.79	4.0	10.6	4.2	42.7	15.2	62.4
Semilukskian-Kizelovskian	0.914	2.95	3.0	13.2	6.8	37.9	8.0	29.0
Bobric-Tulskian	0.924	3.25	2.8	12.5	7.5	36.6	0.33	31.3
Aleksinskian-Vereiskian	0.924	3.33	2.8	20.5	8.3	29.4	0.3	15
Kashirskian-Gzelian	0.929	3.4	3.1	—	6.6	33.7	0.3	7.9
Ufimian	0.945-0.97	3.4-1.3	—	—	4.6-9.3	1.0-4.1	—	—

With reference to spatial distribution, light oils in all producing horizons are confined to the south-eastern slope of South-Tatarian arch. Total tendency is noted that oil becomes heavier in the north-western and western directions to the eastern flange of Melekesskian trough and south-eastern slope of North-Tatarian arch. This phenomenon is accompanied by decrease of gasoline fraction yield. Viscosity of oil increases along the section from the bottom upwards, and by the area-from the south-east to the north-west for each regionally oil bearing horizon. Similar changes are noted in the characteristics of gas saturation: the value of gas factor, 712 m^3/t in deposits of terrigenous Devonian mass in the south-eastern slope of South-Tatarian arch decreases to 15 m^3/t in the middle Carboniferous deposits. Territorial changes of oil properties are in conformity with the change of cap rocks' thickness. Similar tendencies are noted in the character of filling of traps in all producing horizons: from 100% in the south-eastern slope of South-Tatarian arch to 40%～50% in the south-eastern slope of North-Tatarian arch, which is also associated with qualitative deterioration of cap rocks on the whole.

CONDITIONS OF OIL DEPOSITS FORMATION

The basic geological condition, facilitating formation and preservation of oil deposits in sedimentary mass is availability of reservoirs and cap rocks. Thickness of cap rocks exhibited critical influence on distribution of deposits in the section. Regional Kynovskian-Sargayevskian cap rock is characterized by the most high screening properties, which facilitated accumulation and preservation of considerable oil reserves in Pashiyskian-Kynovskian oil and gas complex, and formation, in particular, of such gigantic oil fields as Romashkino, Novo-

Eikhovka, etc. Tulskian cap rock ranks below in qualitative signs, Vereiskiam cap rock exhibits the least isolation properties. In this connection, by hydrocarbon accumulation concentration, further follow Bobric-Tulskian and Semilukskian-Kizelovskian complexes. Main regional regularities of oil deposits location by the section consist in the following. The prevailing part of deposits is controlled by large and small upheavals, though lithological heterogeneity of reservoirs played a considerable role.

Oil deposits of Devonian terrigenous reservoirs from south-east to west and north-west decrease in sizes and reserves due to decrease of total reservoir storage capacity which is a result of sequential omission of Vorobyevskian, Ardatovskian, Mullinskian, Pashiyskian, and Kynovskian beds. Location of deposits in Semilukskian-Kizelovskian and Bobric-Tulskian complexes depends on oil presence in underlying terrigenous structures. Though, maximum concentrations of oil in them are controlled by flange zones of troughs of Kamskian-Kinelskian system, which are characterized by the most favorable combination of a structural factor-local upheavals, Tulskian cap rock and reservoir mass. On the whole, maximum areal of oil presence in low Carbonic deposits in relation to Devonian ones is shified in the western and north-western directions. The field of the most considerable degree of oil content in middle Carbonic deposits is shifted still further to the west and norht-west in relation to oil-bearing areal of low Carbonic deposits. Areals of oil presence of different oil and gas bearing complexes are sequentially connected between each other by gradual transitions. The revealed zoning is caused by asccending migration of hydrocarbons, tectonic and paleotectonic conditions of development, peculiarities of reservoirs and cap rocks structure (R. Kh. Muslimov, R. N. Diyashev, 1996).

The estabished regularities of widespread deposits testify to their formation owing to subvertical migration. Carbonaceous Devonian deposits in the troughs of Kamskian-Kinelskian system, which is located within Tatarstan territory and beyond it, served as oil parental deposits. Leading role belonged to the troughs, located beyond the Republic close to its borders. Favorable conditions for hydrocarbons formation were created not earlier than in Permian-Mesozoic time, when sediments lowered to a required critical depth. Thick sandy beds of terrigenous Devonian and Carbonic mass served as routes of close lateral migration, hydrocarbons shifted along monoclinal slopes from platform troughs and depressions towards surrounding arches. Vertical migration of oil through fractured zones played a leading role in deposits formation. Bitumen and oil shows, developed both along the territory area and by sedimentary mass section testify to repeated reformation and partial deposits collapse caused by active block movements along fractured zones (M. F. Mirchink, 1965).

Geochemical characteristics of Tatarstan oils made on the basis of investigation of qualitative and quantitative composition of individual alkane and polycyclilc biotracers, carried out in Institute of Geology and Prospecting of Fuel Minerals (G. A. Aksenova, 1988) allowed to draw conclusions, first, about organic genesis of oils and bitumens in Tatarstan and, second, about genetic homogeneity of Devonian, Carbonic and Permian oils.

TECHNIQUE OF EXPLORATION AND PROSPECTING OF OIL DEPOSITS AT THE PRESENT STAGE

Carried out geological investigation allowed to zone Tatarstan territory by the degree of promise (Fig. 1). Eastern part of Tatarstan can be referred to high-promising area, eastern slope of Tokmovian arch, Kazanian-Kirovskian avlakogene and a part of North-Tatarian arch top are low-promising. In the context of the above mentioned oil migration, only limited portions of hydrocarbons delivered here because of lithological barriers in terrigenous Devonian mass.

The main promising direction is South-Tatarian arch-upheaval of inversion development, besides, among main directions are also North-Tatarian arch and Melekesskian trough.

The main objects of exploration are oil deposits in Pashiyskian-Kynovskian and Bobric-Tulskian oil and gas complexes with which the highest and most reliable probability of reserves increment increase is associated. In all other complexes deposits shall be revealed by wells, drilled for the key Devonian and Carbonic horizons.

Previously exploration of oil deposits in these complexes was carried out using the first prospecting well, located in optimum structural conditions and prepared for Carbonic deposits. For this reason the well gave, as a rule, negative results in terrigenous Devonian deposits, in some cases it tapped the periphery of oil pool.

In conformity with the developed technique oil prospecting in Devonian deposits is carried out by drilling of special wells on the basis of a complex of paleotectonic and lithological-paleographical reconstructions both of deep drilling data and seismic survey results.

New technique of oil deposits prospecting in terrigenous Devonian complexes is based on the revealed and studied regularities of location of structural forms which control these complexes.

At the first stage regional consedimentary Devonian troughs are followed, since it has been determined that oil deposits in Vorobyevskian-Ardatovskian and Pashiyskian-Kynovskian complexes are controlled by these structural forms and conjugated bar zones. Study of genesis, structure and spread of Devonian troughs allowed to reveal regularities of traps location in terrigenous Devonian complexes. The main morphological indication of Devonian trough-linear plunge in foundation made by sediments. In the flange zones of troughs, complicating them, small buried projections of crystalline foundation are located, which serve as a foundation for traps in terrigenous Devonian complexes.

Resolution potentialities of seismic survey for mapping of fractured zones of foundation and regionally defined Devonian troughs are high enough. For this reason for prospecting of small-amplitude buried traps in Kynovskian-Pashiyskian oil and gas complex at the first stage troughs and conjugated bar zones and flexures are localized. At the second stage of exploration deposits identification is carried out by drilling of single prospecting wells, grouped, mainly, in the flange sections of consedimentary troughs and in structural terraces.

The choice of location of a prospecting well is carried out by a complex of symptoms: structural, lithological and others, the most important of the latter are visual oil shows in the offset previously drilled wells. Differentiation of buried structural forms is carried out using the known technique: local upheavals are associated with closed minimum values of thickness, troughs-with increased values. Difference of depths, fixing possible structural forms in Devonian deposits, are small, as a rule, and do not exceed initial dozens of meters. Using data of drilled deep wells lithological maps are constructed with regard for structure and oil presence in each of the beds. In the promising sections, characterized by favorable combination of the above symptoms drilling of a single prospecting well is performed.

For today the efficiency of drilling in local objetcs in terrigenous Devonian deposits, prepared by seismic prospecting does not exceed 20%. Low efficiency of the objects' readiness for deep drilling in the reflecting horizons of Devonian deposits can be accounted for by geomorphological multilayering of the being mapped surface, absence of confident stratigraphical referencing of reflecting horizons, ambiguity of reflecting waves correlation, etc.

High degree of exploration of the promising eastern part of Tatarstan by seismic survey facilitated the development of a complex technique of prospecting of small-amplitude traps in terrigenous Devonian mass. This technique is based on reorientation of the available stock of geological and geophysical information together with the method of thickness analysis, which allowed to reveal a number of local objects, omitted before while geological prospecting. Thickness analysis is performed also using the above technique with the difference that thickness between two reflecting horizons, foundation and roof of terrigenous Devonian mass, is analyzed.

In a number of cases prospecting of deposits in the lowest, Vorobyevskian-Ardatovskian complex shall be executed independent of overlying Pashiyskian-Kynovskian complex. At present stage it was carried out in the largest oil fields in the Republic, Romashkino, Novo-Elkhovka, and Bavly fields, where underlying producing horizon is poorly explored. Prospecting of these deposits does not call for drilling of separate wells, it is executed by deepening of operating well stock which is drilled for the main being developed horizon, Pashiyskian-Kynovskian. Calculations show that within Romashkino oil field in Vorobyevskian-Givetian deposits possible geological reserves amount to 280×10^6t.

The technique of objects preparation for prospecting in Carbonic deposits is performed with reasonable efficiency by seismic survey using the method of total depth point (TDP).

In Semilukskian-Kizelovskian, Bobric-Tulskian and overlapping oil and gas bearing complexes the overwhelming bulk of oil deposits are controlled by mantled traps of bioherm bodies. These traps are characterized by planned inheritance with gradual flattening upward along the section. For this reason the first prospecting well located in the arch of prepared upheaval for reflecting horizon in the roof of Tulskian deposits provide for, as a rule, through prospecting of all the three upper oil and gas complexes, including Kashirskian-Gzelian deposits. The efficiency of drilling of such prospecting wells reaches 80%-90%. The most efficient is exploratory drilling when the problem of deposits delineation in Carbonic deposits coincides with the problem of prospecting in Devonian deposits and vice versa, which is practiced widely in Tatarstan.

CONCLUSTION

Despite a high degree of exploration and investigation of Tatarstan territory, which reaches 92%, every year considerable increment of resources volume is reached as a result of exploratory-prospecting drilling. Therewith efficiency of exploratory-prospecting wells drilling makes 70%. Promising directions where prospecting and exploration of oil deposits is carried out are South- and North-Tatarian arches, Melekesskian trough or territory of Eastern Tatarstan, priority place holds South-Tatarian arch. Two complexes, Pashiyskian-Kynovskian and Boric-Tulskian possess the largest oil potential, and the technique of deposits exploration and prospecting is oriented towards them.

Search for new oil fields in the old oil producing region is carried out using both traditional methods, i. e. seismic prospecting, and non-traditional ones. Methodological basis of the latter is revision of geological -geophysical information in the context of up-to-data concepts about regularities of oil deposits location, peculiarities of their formation, favorable conditions for generation and conservation of hydrocarbons deposits at different stages of ancient sedimentary basin development. Discovery of new fields is executed at the cost of new or omitted horizons, insufficiently studied sections, located between known fields.

ACKNOWLEDGEMENTS

The authors wish to thank the management of JSC Tatneft who sponsored scientific investigations and generalizations and permitted to submit this paper to the World Geological Congress.

REFERENCES

1. G. A. Aksenova. "Palynological indications of oil migration in Paleozoic deposits of Permian region". *Synopsis of the thesis of cand. of geol-min. sciences.* Moscow, 1988, 18p.

2. A. A. Bakirov, E. A. Bakirov and V. S. Melik-Pashayev. "Theoretical basis and technique for exploration and prospecting of oil and gas accumulations". Moscow, 1987, pp. 205-318.

3. W. C. Krumbein. "Lithofacies maps and regional sedimentary stratigraphic analysis". *Petroleum Geologists. Bull.* 32, 1948, No 10, pp. 1090-1923.

4. I. A. Larochkina. "Principles of optimization of oil deposits exploration and prospecting at the stage of a high degree of exploration of the territory". Thesis for the degree of Doctor of Sciences (*Geol. and Min.*), Moscow, 1995. 310p.

5. R. Martin. "Paleomorphology and its application to exploration for oil and gas (With examples from

Western Canada)". *A. A. P. G. Bull.* 50,1966, No. 10,pp. 2277-2311.

6. E. I. Suleimanov, R. Kh. Muslimov and E. Yu. Mochalov. "Technique of supplementary exploration of small oil fields in Tataria". *Neftegazovaya geologiya i geofizika* , 1976,No 10.

7. V. E. Khain and B. A. Sokolov. "Rifting and oil and gas presence, main problems". *Geological Journal*, 1991,No. 5,pp. 3-16.

8. M. F. Mirchink, R. O. Khachatryan and V. I. Gromoka. "Tectonics and zones of oil and gas accumulation of Kamskian-Kinelskian system of troughs". Moscow, Nauka,1965,214p.

9. R. Kh. Muslimov, R. N. Diyashev, R. P. Gottikh and I. A. Larochkina. "Integrated methods for geochemical prospecting of oil and gas fields" GEO'96 Conference abstracts. *2nd Middle East Geoscience Conference and Exhibition.* 15-17 April 1996,Bahrain.

Proc. 30th Int' l. Geol. Congr. , Vol. 18, pp. 71~78
Sun Z. C. *et al.* (Eds)
© VSP 1997

Geological Characteristics of Central Gas Field in Shan-Gan-Ning Basin, China

YANG JUNJIE

Changqing Petroleum Exploration Burean, China National Petroleum Corporation Qingyang County, Gansu Province, CHINA

Abstract

The Shan-Gan-Ning Basin is a craton marginal basin in China. The giant gas field in central Shan-Gan-Ning Basin discovered by Changqing Petroleum Exploration Burean of China National Petroleum Corporation in 1989 ranks 83th among the world' s largest gas fields.

This paper focuses on the discussions of regional structural baskground, geological characteristics, geological theoretical model and reservoir formation of the Central Gas Field in Shan-Gan-Ning Basin and concludes that the dual and mixed source gas, hinge zone of the paleo-structure evolution, central paleo-uplift with north Shanxi paleo-depression opposite direction development, the coordinated relations of up-dip rock salt-gypsum with down-dip dolomite flat facies in regional slope, the existence of Ordovician top paleo-buried platform subtle trap at Caledonian stage and the development of weathering crust dissolved pores communication body are the main geological factors of Central Gas Field formation. The concepts has important values for the study of craton marginal petroliferous basins and giant paleotopography oil and gas fields exploration.

Keywords: Basin, Grant gas field, Regional geological background, Geological characteristics, Reservoir formation model

The Shan-Gan-Ning Basin that is located in mid-part of the onshore-sedimentary basin in China is also called Ordos Basin. Its north border, Wulanger baserock uplift conjuncts with Hetao graben, to its south, connects with Weihe basin by Weibei flexure, to its east. Jinxi flexure and Luliang Paleo-land and its west , locates with Liupan Mountain and Yinchuan Basin through the over-thrust belt. It is about 600 km from north to south and 400 km from east to west with the total area of 250,000 km². It is a multi-cycle craton marginal basin characterized by stable subsidence, depressional-migration and apparent torsion.

The Shan-Gan-Ning Basin now is viewed as the important area for natural gas industry development in China because of its vast area, high maturity ($R^o > 1.5\%$), excessive coal sediments, stable structural characteristics, dual-layer structural sediments evolving from sea to land in Palaeozoic and in a period of time, it was once in the merging zone of ancient Huabei

Sea and Qilian Sea.

In 1989, Changqing Petroleum Exploration Bureau (CPEB) of China National Petroleum Corporation (CNPN) discovered the giant gas field in central Shan-Gan-Ning Basin and has drawn the close attention and interest of Sino-foreign exploration specialists. The discovered gas field covers the area of Wushenqi, Hengshan, Jingbian and Ansai counties with the proven gas-bearing area of 4148 km², recoverable reserve of 137 bilion m³ and the giant gas field ranks 83th among the world's largest gas fields (Figure 1).

I. REGIONAL GEOLOGICAL BACKGROUND

The basement of Shan-Gan-Ning Basin is hypometamorphic rock in Early Proterozoic and Archaeozoic and the average thickness of sedimentary caprock is 6,000 meters , total volume of sedimentary rock is 1. 5 milion km³. The basin has good development of three hydrocarbon generation systems of Lower Palaeozoic marine facies carbonates, rock-salt and gypsum; Upper Palaeozoic

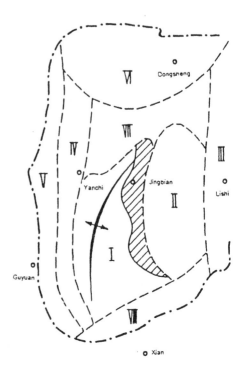

Figure 1   Location Map of Central Gas Field

I . Central Palaeo-uplift; II . North Shanxi depression; III . West Shanxi flexure fold; IV . Tianhuan depression; V . West-margin thrust belt; VI . Yimong uplift; VII . Yimong slope; VIII . Weibei flexure fold; ▨ Central Gas Field

transitional facies clastic rocks and coal measure and Mesozoic fluvial-lacustrine facies clastic rocks and so on. Gas bearing in Ordovician, Carboniferous, Permian systems and oil bearing in Triassic and Jurassic systems are verified.

The evolution of Shan-Gan-Ning Basin was hotter in the early time than in later time. Average thermal flow value is 62. 4 mW/m², gradient is 2. 56 ℃/100m, Moho temperature is 541℃-720℃. Formation of the basin has connections with the over-laid plastic-lithosphere becoming heated and collapsed due to the deep layer asthenosphere uplifting. So, its lithosphere becoming thinner and crust becoming thicker and the two of them has imaging relations.

The Shan-Gan-Ning Basin experienced five evolution stages in the history, they are: Middle-late Proterozoic aulacogen stage; Early Palaeozoic epicontinental-sea platform; Late Palaeozoic littoral plain; Mesozoic inland basin and Cenozoic peripheral subsidence-faulting. In which, the former three stages mainly generate gas and the natural gas resource volume is about 4,000 billion m³, the later two stages mainly generate oil and the oil resource is about

2 billion tons. So, it is a sedimentary basin with great potential of oil and gas coordinative development.

I . GEOLOGICAL CHARACTERISTICS OF CENTRAL GAS FIELD

1. Dual and mixed sources gas

Central Gas Field owns two sets of gas source rocks, Upper and Lower Palaeozoic. In which, Lower Palaeozoic is sapropel-type cracking gas and Upper Palaeozoic is humic-type coal derived gases.

The hydrocarbon generation center of Ordovician is in the area of Yulin and Jingbian. Hydrocarbon source rock is dolomite, limestone and gypsun-bearing dolomite. Its thickness is about 400-500 meters, biological composition is algae and acritarchs, kerogen is mainly lipidic group and belongs to sapropel type. TOC is 0. 08%-0. 64%, average 0. 22%, hydrocarbon content is 40. 48 ppm, R° is 2. 07%-2. 86% and organic matters is in the over-mature dry gas heat evolution stage. $\delta^{13}C_1 - 33. 09\%$, $\delta^{13}C_2 - 32. 4\%$, $\delta^{13}C_3 - 28. 01\%$. Hydrocarbon genetation intensity is $(25-35) \times 10^8 m^3/km^2)$, hydrocarbon displacement intensity is $(7-19) \times 10^8 m^3/km^2$.

The hydrocarbon generation center in Carboniferous and Permian is in the area of Yulin, Jianbian and Fuxian, the source rocks are black mudstone, dark-grey limestone and coal bearing layers with the thickness of around 100-120 meters. TOC is $>0. 61\%$, hydrocarbon content is $>28. 44$ppm, R° is 1. 8%-2. 0% and the organic maters is at the wet-dry gas transit stage of high maturity. $\delta^{13}C_1 - 32. 93\%$, $\delta^{13}C_2 - 25. 87\%$, $\delta^{13}C_3 - 23. 64\%$. Hydrocarbon generation intensity is $(24-32) \times 10^8 m^3/km^2$, hydrocarbon displacement intensity is $(12-16) \times 10^8 m^3/km^2$.

The large area $(75,000 km^2)$ imbedding of Lower Palaeozoic and Upper Palaeozoic hydrocarbon generation center has laid a rich physical foundation for natural gas accumulation in the Central Gas Field (Figure 2). Mr. Chen Anding, Xu

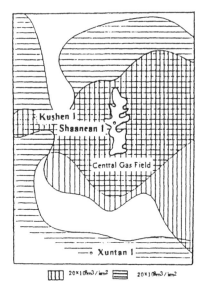

Figure 2 Map of Central Gas Field and the hydrocarbon generation center

Zhengqiu and Wang Keren adopted four methods of ethane carbon isolope and ethane content, methane carbon isotope, ethane composition content and multivariate model map to calculate quantitatively the mixed source rock ratio of Central Gas Field. As result coal derived gas from $M5_1$ gas reservoir takes up 35%-48%; average 41%; the coal-derived gas from $M5_4$ gas reservoir takes up 10%-20%; average 14%. So, it is correct to set the mixed source rock ratio be 3 : 7 for Upper Palaeozoic coal-type gas and Lower Palaeozoic cracking gas.

2. North Shanxi depression—the secondary depression in the Huabei Sea area

The central palaeo-uplift and north Shaanxi palaeo depression are twin brothers in the evolution history of San-Gan-Ning Basin. Mr. Sun Guofan, Liu Jingqing(1987) and Zhang Kang (1989) once stated the existence of rift shoulder uplift from Qingyang to Hongde. Recently, Mr. Zhao Chongyuan and others put forward systematic statements on the causes of the rift shoulder in the central uplift and the balance adjustment mechanism of North Shaanxi palaeo-depression. We think that the rift shoulder developed the central palaeo-uplift and the north Shaanxi palaeo-depression was derived from the central palaeo-uplift. The present structure of north Shaanxi slope, as a matter of fact , is the result of opposite direction development of central uplift and north Shaanxi palaeo-depression at the stage of Yanshanian movement and Himalayan movement. So, we should trace back to factual palaeo-structures when evaluating the gas province geological structures in Shan-Gan-Ning Basin.

North Shaanxi palaeo-depression was covered by Huabei Sea and the basin west margin palaeo-depression by Qilian Sea and the two merged from the area of well Chenchuan 1 to Dingbian and the transit zone width is about 40 km. Huabei Sea was relatively regressive and Qilian Sea transgressive at the Ordovician sedimentary stage and which are the main causes of sediments of very thick salt rock and gypsum formation in north Shaanxi depression (Figure 3).

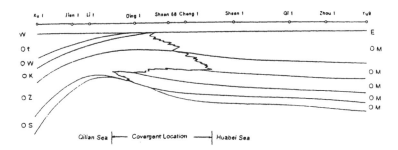

Figure 3   Map showing convergence between Qilian Sea and Huabei Sea

The undoubted facts are: Central Gas Field, in the means of geological structure, locates at the east wing of central palaeo-uplift and the west side of north Shaanxi palaeo-depression. At the period of Cretaceous, when north Shaanxi palaeo-depression rose up and became structural slope. , the Ordovician salt rock, gypsum sedimentary facies formed structural up-dip regional oil and gas barrier. So, the coordinative development of palaeo-uplift with palaeo-depression, palaeo-depression with present slope, migration area with barrier zone formed the basic geological framework of Central Gas Field(Figure 4).

3. Hinge zone the palaeo-structure evolution

The palaeo-structure features of Central Gas Field, uplift in the west and depression in the east, high in the west and low in the east, continued to the period of Jurassic. Till Early Cretaceous, along with the disappearance of central palaeo-uplift and north Shaanxi palaeo-depression, the structural slope, high in the east and low in the west came up for replacement. The balance movement hinge zone formed by the opposite evolution of the above two palaeo-sturctural units has important theoretical and factual meanings for Central Gas Field rich accumulation and development.

The palaeo-structural framework of uplift in the west and depression in the east in the area including two evolution stages:

A. in the period of Palaeozoic, the rift shoulder uplift in the west of the basin is transforming to central palaeo-uplift and in the east, the accompanying marginal basin, formed by crust balance adjustment, also transforming to north Shaanxi depression. This structural framework continued till the end of Permian and then disappeared.

B. In the period of Early Mesozoic, although the palaeo-structural landscape of uplift in the west and depression in the east remained in some way, but the scale and causes of formation are not comparable with the former. Since the forma-

Figure 4   Map of Central Gas Field and the Ordovician salt rock pinch-out line

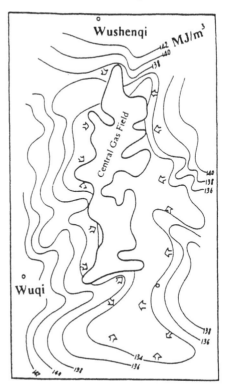

Figure 5   Present gas phase potential map of Ordovician top in central Gas Field

tion uplift and depression in Palaeozoic were caused by deep-layer crust movement, it is structural cause of formation, the formation of uplift and depression in Mesozoic has connections with sedimentary compaction and shallow-layer structural activity.

Take the palaeo-structural evolution of the Ordovician top erosion surface as example: before the period of Permian, palaeo-uplift was in the area of Chenchuan and Wuqi, palaeo-depression was in Lishi and Yichuan area and the accumulated up-rising from Shilou to Chenchuan was 100 meters, before Triassic, the palaeo-uplift axial moved westward to Anbian and Huanxian area and the accumulated up-rising from Yanchuan to Chengchuan was 200 meters, before Jurassic, palaeo-uplift experienced desintegration, its axial was composed of Chenchuan, Dashuikeng, Zhidan and Huangling and the related palaeo-depression retreated apparently at the area of Lishi and Yanchang. At the end of Jurassic, the trace of palaeo-structures disappeared totally and transformed to westward slope.

Therefore, the west high and east low palaeo-structural framework, after two stages development of different causes in Palaeozoic and Mesozoic periods, finally transformed to east high and west low structural slope landscape in Cretaceous and the structural hinge of this balance movement happened to be in the area of Jinghbian and Zhidan. Hydrocarbon migration field potential change controlled by the former is: from east to west at early stage and from west to east at later stage, that is, hydrocarbon was trapped in structural hinge area in the geological history. This embedding function of hydrocarbon migration, as a matter of fact, enhanced the hydrocarbon concentration and rich accumulation in the area and created favourable migration and accumulation environment for Central Gas Field reservoirs.

Upper Palaeozoic were developed, i. e, lower sandstone section in Carboniferous, upper limestone section, Shanxi formation coal bed, Shihezi formation sandstone and mudstone sections in Permian; two sets of reservoirs and caprocks of dolomite bodies under salt layer and weathering crust above salt layer in Lower Palaeozoic were developed.

The 30,000 km^2 dolomite flat facies zone in central palaeo-uplift is the favourable place for the development of karst reservoir body of Ordovician top. The average thickness of main weathering crust body of Central Gas Field is 46 meters, the formation is at the upper section of M_5, its diagenesis including dolomitization, atmospheric fresh water leaching karst and recrystallization etc. , sedimentary environment is mainly dolomite flat, gypsum-contaning dolomite flat and algae dolomite flat, len layers of dolomite reservoirs exist and contains thin bed embedded muddolomite. The set of reservoir can be divided into three categories and described by the following table.

Property Characteristics of Reservoirs

Reservoir type	Porosity (%)	Permeability (μm^2)	Resistivity (Ωm)	Sonic travel time ($\mu s/m$)	Density (g/cm^3)	Saturation of Connate water(%)	Layers
I	6	1×10^{-3}	100-200	160	2.7	<15	m_1, m_4, m_9
I	5-6	$(1-0.02) \times 10^{-3}$	200-400	155-160	2.7	25	m_1, m_2, m_3, m_9
II	2-5	0.01×10^{-3}	300-1000	155-160	2.7-2.8	25-50	m_5, m_6, m_9

The dolomite from dolomite flat facies can be divided into three types[2]

A. Dolomicrite intergrowed with evaporites: It is caused by the evaporative pumping-seepage reflux dolomitization. $\delta^{13}C$ is -13.67‰PDB, $\delta^{18}O$ is -5.223‰PDB, reservoir physical property is poor and they are mainly distributed in Ma_1, Ma_3 and Ma_5^{5-10}.

B. Foggy-center silt-dolomite, meso-crystal residual-grain dolomite and leopard-spot dolomite: It is caused by the dorag dolomitization. $\delta^{13}C$ is -1.067‰PDB, $\delta^{18}O$ is -8.59‰ PDB. The former two have good physical properties and mainly distributed at Ma_5^{1-4}.

C. Coarse-crystal ankerite-dolomite rock distributed along fractures: It is caused by the deep buried compaction flow dolomitization and structural heat fluid and very oftenly intergrowed with gas bearing zone. $\delta^{13}C$ is -1.097‰PDB, $\delta^{18}O$ is -11.2‰PDB.

People often pay more attention to reef facies, beach facies and natural fractures other than dolomite flat facies when looking for oil and gas in carbonates. The discovery of Central Gas Field in Shan-Gan-Ning Basin once again demonstrated the important value of dolomite-flat facies. The geological concept model of such reservoir body is pseudo-porosity bedding network structure, in which, marine carbonates as basis, tidal flat facies dolomite as framework, dissolved pores as communication body and micro-fractures as ending[1].

4. Palaeo-buried platform subtle trap

The Caledonian movement marked the end of the ancient marine environment in Shan-Gan-Ning region and resulted in overall uplifting of the basin and thus the top of Ordovician was exposed to weathering erosion for 130 million years. A peneplain carbonate palaeo-karst morphology, widely-buried platform platform, low and flat buried hill and also near isodistance distributed buried trench was formed (Figure 6)[3].

In conclusion: palaeotopography, tectonics, stratigraphy and lithology are the four major geological factors for the formation of oil and gas traps. The Central Gas Field of Shan-Gan-Ning Basin is located at the large-scale palaeo-buried platform of Ordovician top and belongs to palaeo-lopographic traps. Ordovician top weathering crust reservoir at the east side of the palaeo-buried platform is covered by the Carboniferous bottom allite rock and formed regional monocline uplifting traps along the up-dip direction. It is, of course, the determinative geological conditions for reservoir formation in Central Gas Field.

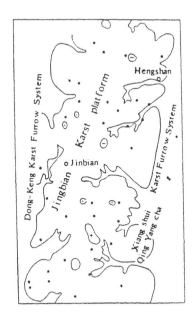

Figure 6   The palaeotopography of pre-Carboniferous in the Central Gas Field of Shan-Gan-Ning Basin (Song Guochu, 1995)

Ⅱ. GEOLOGICAL THEORETICAL MODEL OF CENTRAL GAS FIELD RESERVOIR FORMATION

To summarize the Central Gas Field characteristics on hydrocarbon generation, reservoirs, caprocks, trap, migration and accumulation, its geological theoretical model on reservoir formation can be stated as below.

Depression and widely-distributed multiple hydrocarbon generation centers with significant subsidence in the west, uplift in the middle part of the basin and hydrocarbon migrates and accumulates from all directions to the palaeo-buried platform, salt sags and buried trench were developed in east basin and formed regional multiple barriers and palaeotopographic traps.

In a word, the reservoir types and geological characteristics of Central Gas Field can be compared with those of Hugoton and Mocane-Laverne Gas Fields of Anadark basin in the USA. Mource and Carthage Gas Fields in the Gulf of Mexico and the Croosfield Gas Field of Alberta basin in Canada.

REFERNCES

1. Yang Junjie. Situation and peospect for petroleum exploration of Central Gas Field in Shan-Gan-Ning Basin. China Oil and Gas, Vol. 2,1995.
2. Lan Jianxiong and Zeng Yongfu. Causes of formation of Lower Ordovician dolomite and reservoir characteristics in east Ordos Basin. Oil and Gas Geology,1994.
3. Song Guochu. Analysis of trap conditions of the Central Gas Field in Shan-Gan-Ning Basin. 2nd volume, China Oil and Gas,1995.

Proc. 30th Int' l. Geol. Congr. , Vol. 18, pp. 79~86
Sun Z. C. *et al.* (Eds)
© VSP 1997

The Gas Resource in Tight Sandstone in Sichuan Basin

GUO ZHENGWU

Southwest Bureau of Petroleum Geology, Ministry of Geology and Mineral Resources, Chengdu, Sichuan 610081, *P. R. China*

Abstract

There is a great deal of gas reserve in conventional and unconventional traps in Upper Triassic and Jurassic clastic sediments with a tremendous thickness in the west part of Sichuan basin, forming a special tight sandstone gas-bearing territory. Through a primary exploration, it is shown that a group of gas reservoirs horizontally extending and vertically overlapping reliably exists and an abnormally high pressure zone occurs in a wide scope. As this terriory is located in Chengdu Plain where the agriculture and industry is well developed, intensifying exploration and exploiration of natural gas will have practical economic meaning.

Keywords: basin architecture, source rock, reservoir, tight sandstone, gas pool

INTRODUCTION

Sichuan basin, with the area of $180 \times 10^3 km^2$, is an important gas base in China at present, which has half gas production per year in China (Fig. 1).

There seems to be no doubt that large quantities of gas have been generated in Sichuan basin. At present, most of the discovered gas reserve is in carbonate. The question remains, however, as to how much of it has been traped and how much can be recovered from tight sandstone reservoirs. In recent years, we strengthened the study and exploration on clastic rocks, and there is a good exploration prospect in thick clastic sediments of Upper Triassic and Jurassic in west Sichuan basin.

In the western part of the basin, the tremendously thick clastic rock has generated considerable natural gas and is accumulated in a dynamic-balance fasion. Except that a part of it has been lost through percolation and diffusion in geological history, there are yet sizable reserves stored in conventional and unconvertional traps, constituting a distinctive tight sandstone gas-bearing territory.

BASIN ARCHITECTURE

Sichuan basin is both a tectonic and a sedimentary basin, and still more is a ge-

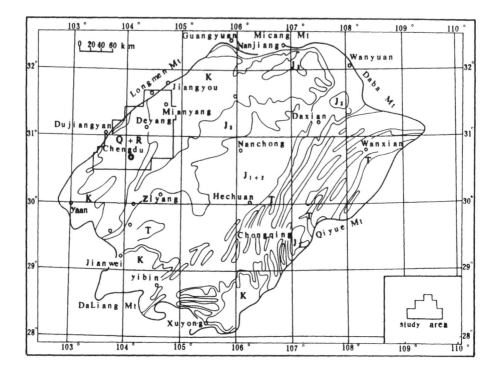

Fig. 1   Oil and gas exploration and development block of Southwest Bureau of Petroleum Geology in Sichuan Basin

omorphological basin during neotectonic stage. In the history of basin evolution, because of the turning from a carbonate-evaporite platform of Upper-Yangtze through major geological events during Indo-China stage to something that made the west margin of the platform transformed into a foreland basin, it built up a basin architecture of multiple type ultimately. Within the basin except for the lacuna of Devonian and partial Carboniferous, the sequence of Lower Paleozoic, Permian, Mesozoic and Tertiary is basically complete, with total thickness up to 12,000 meters, which can be divided into four structure layers macroscopically with major hiatus planes in between (Fig. 2).

It should be noted that there is also marine-facies middle architectural layer composed of Carboniferous in east Sichuan, although its thickness and scale is small, yet it is very important of gas reservoirs. The architectural interior of Sichuan basin obviously includes marine carbonates gas-bearing realm and tight sandstone gas-bearing realm, the latter mainly involving No. 1 and No. 2 architecture layers, in which Upper Triassic is the main part. The Upper Triassic is mainly depressed in west part of the basin. The deposit system of Upper Triassic is clearly shown by the seismic stratigraphic study, in which the lower part is composed of the platform marginal deposit system to the west, and the upper part and Jurassic are of closed lacustrine-fluvial deposit system in the continental basin. The organic matter in the

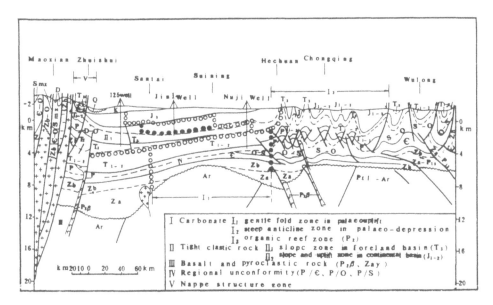

Fig. 2　Sketch map of oil and gas exploration territory in Sichuan Basin

rocks varied from sapropel to humic type. The coal measure is considered as the source rock for generating gas.

The basin evolution history has the characteristics of the foreland basin's development style (Fig. 3).

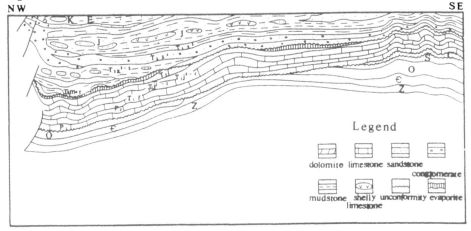

Fig. 3　Sketch map of sedimentation and structure in Sichuan Basin

SOURCE ROCK AND RESERVOIR

The thickness change of Upper Triassic series in the basin is rather obvious (Fig. 3). In western Sichuan it may reaches 3,000 meters and more, thinning dramatically eastward, and it is only of several hundred meters in central Sichuan, then less than 100 meters in eastern

Sichuan and west Hubei. There existis thickness-gradient zone beneficial to petroleum migration and accumulation in the east slope district of west Sichuan depression.

It is determined that Jurassic in western Sichuan depression is red rock without generating gas capability, and the gas producing from Jurassic is similar to the Upper Triassic gas. The natural gas quality is very good with no H_2S content. The Source rock of Upper Triassic is mostly black-grey mudstone, shale and coal-measure, mainly gas-prone. The total thickness of source rock is in the order of 500-1,500m, in which organic carbon content of black-grey shale is roughly 1.2%, the number of thin coal seams is approximately 60, with cumulative thickness of 5-8 meters; organic carbon content of carbonaceous shale amouts to 5%-50%, thickness approximately 100m. The gas source rock has reached or surpassed the mature stage, vitrinite reflectance of which is 0.8%-3.15%. According to the calculation, the total amount of generated gas including coal seam methane and coal-forming gas is 164 trillion cubic meters.

The reservoirs of Upper Triassic and Jurassic are mostly sandstone, the reservoirs are very poor in physical property. The porosity, except in a few reservoirs with average porosity of 5%-13%, is usually less than 5%, and the permeability of matrix is mostly no more than 1 $\times 10^{-3}\mu m^2$. Consequently, the productivity depends on fracture developed in the well, and then porous fractured formation and fractured porous formation become fundamental types of reservoir.

FORMATIVE MECHANISM OF GAS POOL

The generation, migration and accumulation of gas is a continuous and integrate dynamic system. The gas resource amount in tight sandstone reaches one trillion cubic meters in western Sichuan basin. The geological condition and gas pool forming mechanism are included as following:

1. Rich source rock with high pressure anomaly(Fig. 4).

As the gaining pressure phase is caused by gas, the main source rock in Upper Triassic is the base to form high pressure. Because the overpressure leads to relative high energy, so the plentiful gas is favourable for the pressure to discharge outside and fill into the early traps.

2. Timely regional paleouplift is the best place for gas accumulation(Fig. 5).

During late Jurassic, the source rock in Upper Triassic was matured to generate a lot of gas, so just on time, the paleostructures and the lithological traps fromed between Jurassic and Cretaceous structural movement, might form gas accumulation, especially when the rocks has relative high porosity and permeability, a great deal of gas might be accumulated. By using the thickness method and comprechensive geological analysis, the gas accumulation zone could be determined.

3. Favourable reservoir and seal association(Fig. 6).

The gas accumulation is dynamic state all along. The gas accumulated amout surpassing the

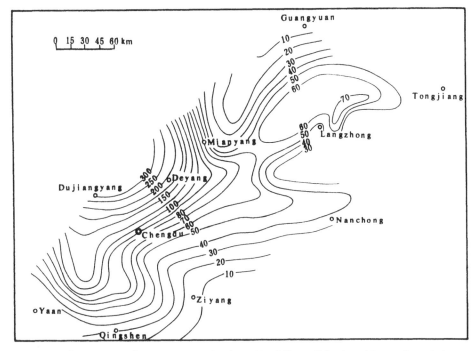

Fig. 4 Map shwing accumulative gas-generating intensify of Upper Triassic and Lower Jurassic in western Sichuan depression and its eastern edge

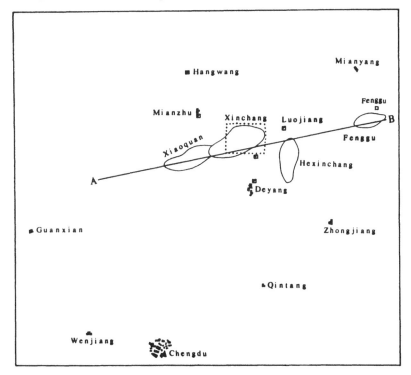

Fig. 5 Gas field distribution in western Sichuan depression

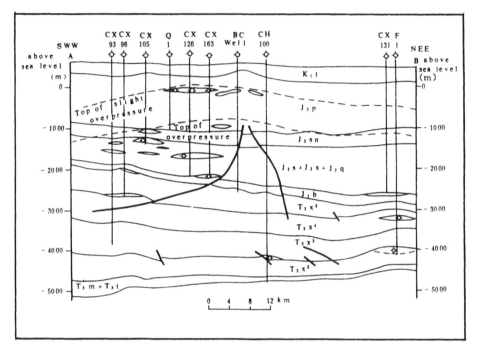

Fig. 6   Gas pool section of Xiaoqan-Fenggu structural zone in western Sichuan depression

dispersal amount may supply the basic conditions to form gas pool. There are five regional seal beds which restrict most gas accumulation.

4. Gas transportation condition

Fault system and erosion surface might be served as the passway for gas transportation, especially the fault and fracture system are important. When the whole rocks were tightened, the suitable activity is necessary to form gas pool.

5. Composite traps

The large scale paleouplift and structural trap composite with stratigraphical trap, which was stacked with fracture network, can form the commercial gas pool.

6. Good protection condition

The suitable deformation and hydrodynamic condition, beside seal condition, are required to form gas pool (Fig. 7).

EXPLORATORY DIFFICULTIES AND COMPLETE-SETTING TECHNIQUES

The sectors of normal pressure, over-pressure and equilibrium pressure in tight sandstone gas-bearing realm of west Sichuan depression reflect three kinds of exploratory target. The exploratory technique is most complicated in over-pressure sector; in normal pressure sector, because of its shallower depth, exploration will be economically effetive, but the gas pool scale is relatively small; in equilibrium pressure sector, although it is relatively stable and

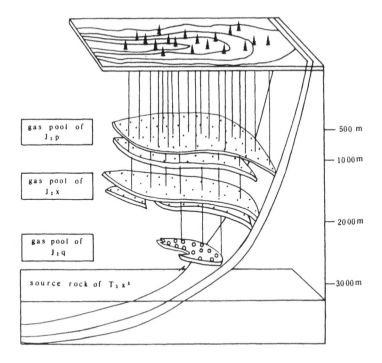

Fig. 7　Geological model of Xinchang gas field

the gas pool scale is larger, yet the burial depth is greater.

The main difficulties in exploration are:

1. Distinguishing fracture zone and effective fracture plays.

2. Comprehensive interpretation of lithology and hydrocarbon using seismic data.

3. Drilling and cementing techniques for high-pressured gas well.

4. Effective stimulation of gas well.

It is necessary to continually deepen the synthetic research of static-state combined with dynamic-state, from the overall to further choose targeting district; in exploratory disposition, a stereo-exploration should be practised, the relation among shallow, middle and deep traps should be correctly handled; in tactical measures, the technical methods and necessary equipments in all techniques links are to be conscientiously resolved. Technical series that form a complete set, suitable for tight sandstone gas-bearing realm should include high precision geophysical and geochemical methods, and correct well programme that can be performed in multi-pressure system, and drilling techniques and testing-logging technology that can protect reservoir of low porosity and permeability in this multi-pressure system. At present, we have achieved some good results on 3D seismic, reservoir description and hydro-sand fracturing.

CONCLUTIONS

Exploratory practice and scientific research that in overall construction of large scale compression-tortional basin, the tight sandstone gas-bearing realm is included which can serves as great target for independent exploration. This realm of Sichuan basin has both unconventional gas pool and approximately conventional gas pools. Through the entire research of the whole realm, the relative concentrative gas-bearing zone is possible to be pointed out. As this territory is located in Chengdu Plain where the agriculture and industry is well developed. The gas market is very good, intensifying exploration and exploitation of natural gas will have a practical economic meaning.

REFERENCES

1. Huang Jiqing, Chen Guoming and Chen Bingwei. Preliminary analysis of the Tethys-Himalayan tectonic domian. Acta Geological Sinica. Vol. 58, No. 1(1984).

2. Guo Zhengwu. Geological framework and hydrocarbon in Sichuan baisn, in Zhu, X. , ed. , The Tectonic and Evolution of Chinese Basins, Science Press(1983).

3. Wang Jinqi, Bao Ci, Lou Zhili and Guo Zhengwu. Formation and development of the Sichuan basin in Zhu, X. , ed. , Chinese sedimentary basins, Elsevier, Amsterdam, 147-163(1989).

4. Ulmishek, Gregory. Geology and hydrocarbon resources of onshore basins in eastern China, U. S. Geological Survey Open File Report, 93-4(1993).

5. Robert T. Ryder, Dudley D. Rice, Sun Zhaocai, Zhang Yigang, Qiu Yunyu and Guo Zhengwu. Petroleum Geology of the Sichuan Basin, China Report on U. S. Geological Survey and Chinese Ministry of Geology and Mineral Resources field investigations and meetings, U. S. Geological Survey Open File Report, 94-426(1994).

Proc. 30th Int'l. Geol. Congr., Vol. 18, pp. 87~101
Sun Z. C. et al. (Eds)
© VSP 1997

Formation and Distribution of Coal Measure-derived Hydrocarbon Accumulation in NW China

ZHAO WENZHI, ZHANG YAN, XU DAFENG and ZHAO CHANGYI

Research Institute of Petroleum Exploration & Development, CNPC, Beijing 100083, CHINA

Abstract

The formation of coal measure-derived oil and gas fields in NW China is the product of the assemblage in spatial and temporal domains of certain tectonic conditions, sedimentology and the accumulation of oil-prone organic materials as well as the effective expulsion of oil and gas from the coal measures. The study made by this paper shows that the following conditions are necessary for the formation of coal measure-derived oil and gas fields in Jurassic sequence of NW China: 1. The Jurassic original basin, which was occupied by lake water or low-postitional swamp during the most Jurassic time, should exist, so that the complete association of source rock, reservoir and caprock was well developed; 2. The original depositional environment was characterized with low and gentle depression and humid climate which are favorable for the development of low-positional marsh land and lake where the oil-prone organic materials were accumulated with large volume; 3. The sequent basin should occur on the Jurassic original basin which is necessary to keep the coal measure source rocks in the condition of progressive maturation; and 4. certain degree of tectonic compression taking place soon after the maturation of source rocks can provide the driving force for the effective expulsion of oil and gas from the coal measures. The coal measure-derived oil and gas fields is horizontally distributed along the inner side of syn-source-rock-deposited lake strandline and vertically occurred above or below the threshold of maturity. Owing to the sharp variation of lithology and facies in coal measures, the lithological and composite types of oil and gas pools play a big role in the petroleum accumulations of coal measure sequence.

Keywords: Jurassic sequence of NW China, Forming conditions of coal measure-derived oil and gas field, Orin_ginal basin, Low-positional swamp, High-positional swamp, Progressive heating history, Tectonic expulsion of coal measure-derived oil, Distribution rule of oil from coal measures.

The North-western China, covering a large area of 2.7 million sq. km, is situated in the north of Kunlun Mountain and bounded in the east by Helan-Liupan mountain and in the west to north by the country's boundary. 46 Jurassic remnant and sequent basins are developed in the region and occupy a total area of 1.27 million sq. km. In the beginning of 1989, commercial oil flow was obtained from the Middle Jurassic interval of Taican 1 well in central Junggar Basin and a large oil field with hundred million tons of reserve, named Shanshan-

Qiuling field was therefore discovered. Since then, a series of oil and gas fields have been discovered from the Jurassic sequence in Tuha, Junggar, Santanghu and Yanji baisns one after another. Moreover, a large quantitative data of oil-to-source correlation and analysis shows that most of the discovered fields comes from the Jurassic coal measures and therefore they are typical of coal measure-derived oil and gas fields. Together with the Jurassic oil fields and oil flows of Qigu, Yiqikelik, Lenghu and Qingtujing which were found in 1950-1960s in Junggar, Tarim, Qaidam and Chaoshui basins, etc, the old and young discoveries demonstrate that the Jurassic sequence of the NW China is the important target for China petroleum industry to find new petroleum reserves in the future. However, in practical speaking, not all the Jurassic coal measures have the capability of forming hydrocarbon accumulation. The formation of coal measure-derived oil and gas fields is the product of the assemblage in spatial and temporal domains of certain tectonic conditions, sedimentology and the accumulation of oil-prone organic materials as well as the effective expulsion of oil and gas from the coal measure. Based on the studies made by the authors in the past ten years, the paper discusses the essential conditions for the formation of coal measure-derived oil and gas fields so as to provide reference for the coming exploration with large scale on the Jurassic sequence of NW China.

CONDITIONS FOR THE FORMATION OF JURASSIC COAL MEASURE-DERIVED HYDROCARBON ACCUMULATION

The necessity of Jurassic original basin existence

Jurassic original basins are referred to be the negative depressions which existed permanently or intermittently during the Jurassic deposition. Lake and marsh land environments developed in these basins during most of the Jurassic time. In view of lithologic assemblege, the sediments of Jurassic original basins is characterized with mudstones interbedding with coals. Sandstones are rare. Original basins have a complete suite of source rock, reservoir and caprock association and are the major area of the accumulation of both coal measure and lacustrine source rocks.

About six subsidence zones of original basins have been verified to exist during the Jurassic along the northern and southern flanks of Tianshan-Qilan mountains of NW China (Fig. 1). The first subsidence zone is mainly situated along Wulungu depression of Jurassic Basin to Santanghu Basin. This zone can also include eastern part of Heshituluogai Basin which is neighbouring to Jurassic. The Jurassic sediments in this subsidence zone are thick and have high percentage of lacustrine and limnetic faices. Toward the southern and northern flanks, the Jurassic sedimentary sequence becomes thinner and the percentage of the limnetic facies gets rarer.

The Second zone is mainly distributed along the southern flank of Jurassic Basin—Turpan depression (Harmy depression is not belonged to the zone). Jurassic sediments in the zone

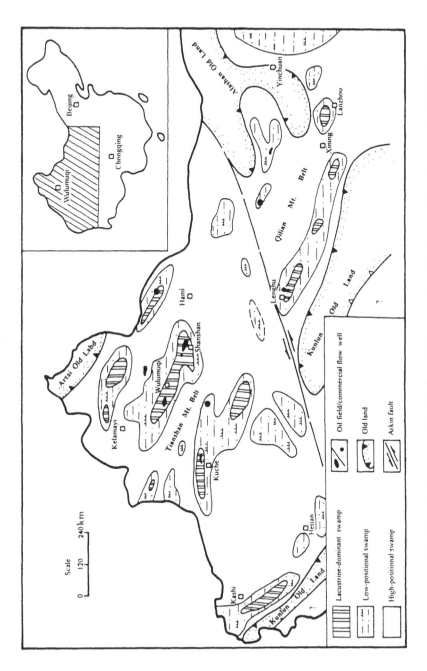

Fig.1 Simplified distribution map of the Early-Middle Jurassic original depositional system in NW China

reach a biggest thickness of 4500m in this zone. During the Jurassic, this zone developed the most unified lake with the biggest water catchment and the longest history of original basin in the northern Xinjiang area. Therefore, the subsidence zone controlled the development of major association of Jurassic source rock, reservoir and caprock in this region. At present, the biggest Jurassic oil fields with the feature of self-generation and self-accumulation have been found in the zone. Fig. 2 gives the outline of the distrubution and variation of coal beds of Middle-Jurassic Xishanyao formation accumulated in the subsidence zone.

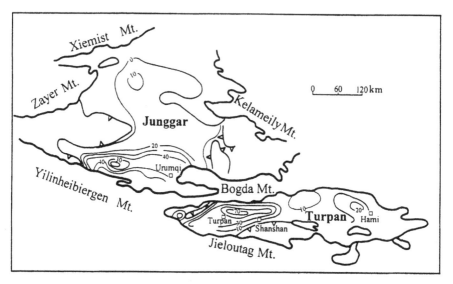

Fig. 2   Coal bed isopach map of Middle Jurassic Xishanyao formation in northern Xinjiang of NW China

The third subsidence zone is mainly developed along Kuche-Yanqi basin, indicating that the marshland environment occupied an important position in the history of Jurassic development in the region. Jurassic sediments accumulated in the zone reach 1500m thick and lacustrine interbedding with coal measures are dominant. Yiqikelike small oil field and the oil flow of Yancna 1 well were obtained in the zone and several gas fields discovered in Tabei uplift along the Yaha-Tiergen faulted fold zone have been verified to come from the Jurassic sequence.

The fourth zone of original basins is mainly consisted of the south-western depression of Tarim Basin, including Keshi, Yecheng sags and the western part of the south-eastern depression of Tarim Basin. The sediments deposited in the zone reach 4000m thick and limnetic facies was well developed. Part of the Jurassic sediments is believed to be overthrusted by the Kunlun mountain. The sediments, exposed at the Hesar coal mine in the western margin of Tarim suggests that the Jurassic in the front of Kunlun mountain is dominated with lacustrine and limnetic facies and no marginal facies has been found, showing that the extent of Jurassic sedimentation might extend further to the south.

The Jurassic source rock is well developed in the south-western depression of Tarim Basin.

Based on the oil/source rock correlation, the crude oil of Kekeya field may come from the Jurassic sequence. It is believed that the Jurassic in the zone has a great potential of hydrocarbon accumulation. However, the exploration targeting directly at Jurassic is difficult because of the burial depth. But the secondary oil pools in proper structures at shallow depth should be paid attention in further exploration. The prospective area to find the secondary oil pools in which Jurassic performed as source rock and Tertiary as reservoir is Kashi depression.

The fifth subsidence zone is developed along the northern edge of Qaidam basin. The thickness of the Jurassic varies from 1800 to 2000m. This zone includes Gonghe basin in the east end and Sugan basin in the west end. Jurassic deposited in the northern edge of the Qaidam basin consists of not only coal measure, but also oil shale. Limnetic facies dominant in the sedimentary sequence. The main part of the Jurassic sediments is developed in Kunteyi-saishiteng sag, Yuka-Hongsha sag, Delingha and Gonghe sub-depressions or subbasins. The general features of the Jurassic sedimentation is that the Lower and Middle Jurassic are thick in the west and thin in the east, whereas the Upper Jurassic-Cretaceous is thick in the east and thin in the west. Therefore, the Kunteyi-Saishiteng sag in the western part of the zone is the main area for further exploration.

The sixth subsidence zone is occurred along the northern edge of the Qilian folded mountain belt, which covers from the Jiuquan basin in the west to the Liupanshan basin in the east and from Minhe basin in the south to the Chaoshui and Yabulai basin in the north. This zone includes several Jurassic original basins which developed along the transitional belt between Qilian folded zone and Alashan massif and is characterized with poor connection in sedimentary system and small scale in basin size. The basin group is getting a higher and higher evaluation in petroleum potential since the low production of oil was obtained from the test in the Jurassic interval of Jiucan 1 well in Jiudong Basin in 1992. Qingtujing small field was discovered in 1950's in the Chaoshui basin which located in the zone, and a lot of oil and gas shows (source rock also including Cretaceous) were also seen in Minghe basin. Jiucan 1 well revealed that the Jurassic sequence has 800m source rock of lacustrine-limnetic facies, which has been proved by source rock analysis and correlation to have provided hydrocarbon in Jiudong Basin. The basin group is worth of exploring in the near futrue, especially in Yabulain, Chaoshui and Yingen basin etc..

The area, in which the assemblage of Jurassic source rock, reservoir and caprock was best developed, should be selected as main targets for oil exploration in the near future.

Low and gentle topography as well as humid climate are favorable for the accumulation and preservation of oil-prone organic materials

The study made by the author in the past ten years shows that low, gentle and negative depressions and humid climate condition are necessary and favorable for the accumulation and preservation with large quantity of oil-prone organic materials. Low and gentle topography is also favorable for the widespread distribution and long term deposition of swamps near the lake and provides the favorable condition for the accuracy of altermative transgression and regression of lake and swamp facies in a large area. Meanwhile, the deltaic sand bodies usually associated with lacustrine-swampy transitional zone are more likely to occur in the environment, which is favorable not only for the accumulation of coal measute source rock with large volume, but also for the formation of the source rock, reservoir and caprock association which might repeat for many times in vertical and lateral directions. The pattern of source rock, reservoir and caprock formed in the lake-related swampy environment is good for the expulsion and accumulation of oil and gas from Jurassic coal measures.

Humid climate is also essential for the development and preservation of ancient vegetation. Based on the analysis of micro-organic materials obtained from the coal meassures, the main sources of oil generation in coal strata are fruit meat cores, leaves of terrestrial plants, suberinite, cutinite, resinite and sporinite and skin of these plant as well as lignocellulose of higher plants which was extensively degraded by bacterium and incorporated the surrounding aliphatic series component to form matrix vitrimite. Vegetation assemblage growing in humid climate is essential for the accumulation of rich hydrocarbon-generating materials. Here the author uses two terms, low-and high-positional swamp, to compare and evaluate the difference of hydrocarbon-generation potential of coal measure formed in different conditions. Low-positional swamp is referred to the paludal facies sediment forming in the low position of original depositional systems. Generally, it is dominated by limnetic facies characterized in lithological assemblage with mudstone and carbargillite interbedding with coal measures (Fig. 3). Most of the sandstones intermediated in the cross-section is the subaqueous part of the delts. The native materials of source rock are mainly composed of matrix vitrimite, suberinite, cutinite, exinite and resinite. The type of native materials mostly I_1-I_2 of kerogen. Hydorgen index (I_H) varies between in 200-500 mgHC/gC. The potential of oil generation is as high as 100-400 mgHC/gC. The reason that low-positional swamp has good condition of hydrocarbon generation is two-fold. One is that its own surroundings favorable for the accumulation and preservation of rich oil-prone materials. Another is that hydrocarbon-rich component distributing in the higher position is easily transported by rain and river to the lower position, which increases the abundance of oil-generating component in low-positional environment. Another reason that low-positional swamp has a better condition of petroleum geology is that besides the association of source rock and reservoir developing in limnetic facies near the lake, the mudstone deposited in frequent transgression of lake can perform as good caprock. Therefore, low-positional swamp is favorable for the occurrence of effective assemblage of source rock, reservoir and caprock.

High-positional swamps is referred to the limnetic sediments occurring in the higher position of original depositional system, which generally consists of fluvial swampy facies. Sandstones interbedding with coal measure are dominant and dark mudstone with lacustrine origin is rare. The sandstone is mostly belonged to the part of delta deposited above water level or

Type of Swamp	Typical Litho Composition	Sed. Cycle	Terrain & Hydrodynamics	Vegetation Community	Hydrocarbon-prone Macerals	Hydrocarbon-generating Potential (S_1+S_2) (mgHC/gC)	Hydrogen Index (mgHC/gC)	Example
High-positional Swamp	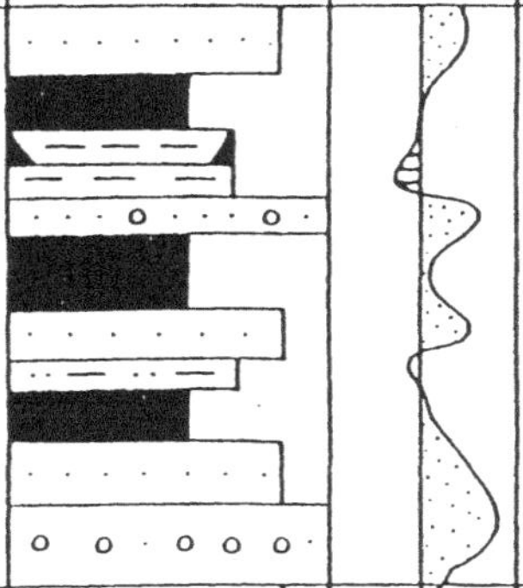		No lake-influenced high slope	Gymnosperm are dominant and pteri, dophyte are rare	1 Vitrinite averagely makes up 65% of the total macerals, among which telinite component accounts for 60% and desmocollinite occupies 10-20%. 2 Inertinite group averagely makes up 30%. 3 Liptinite takes 5% of the total maccrals	50-100	<200	Harmy Depression
Low positional Swamp (Maremma)	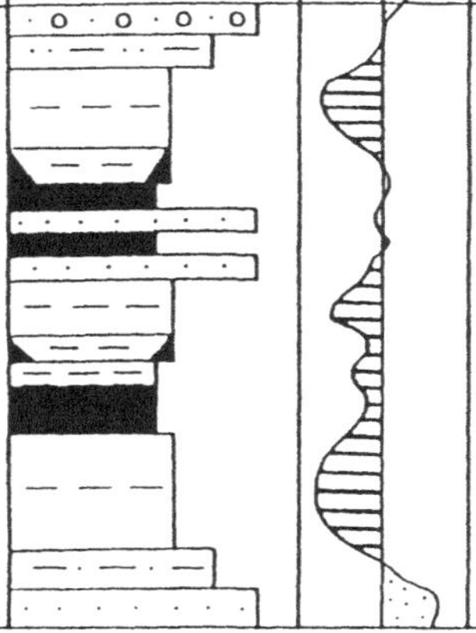		Lake-influenced lade	Petridophyte are dominant and gynnosperm are rare	1 Vitrinite averagely makes up 81% of the total macerals and desmocollinite takes 80-90% of it. 2 Inertinite accounts for 10% in average. 3 Liptinite occupies 10% in average	100-400	200-500	Northern Sag Zone of Turpan Depression

Fig. 3   Difference of Jurassic Coal Measures Formed in Different Depositional Environments of NW China

fluvial facies. The interbedded mudstones are generally not pure in lithology and thus have poor sealing property. Native materials have very high content of fusinite and half fusinite in inertinite. The content of hydrocarbon-rich component is low and the potential of oil generation is poor. Hydrocarbon index of high-positional swamp is less than 200 mgHC/gC and oil generation potential (S_1+S_2) range from 50 to 100 mgHC/gC. The main reason for the poor condition of hydrocarbon generation in high-positional swamp is the extensive oxidation, and part of the oil-prone component was carried by surface radial flow to the relative lower area to accumulate and preserve. Furthermore, mudstone deposited in high position is rare and the quality is poor. Hence it is difficult to form a good assemblage of source rock, reservoir and caprock. So, high-positional swamp is not favorable for hydrocarbon formation and accumulation.

The sequent basin should occur on the Jurassic original basins in order to keep the coal measure source rocks in progressive heating and maturation.

After the deposition of Jurassic coal measure source rock, the superimposition of sequent basins is absolutely necessary for the maturation of native materials and hydrocarbon preservation. In Turpan basin which is located in the Northwest corner of China, the sealing of huge lacustrine mudstone formed in the Late Jurassic and the sequent development of Cretaceous-Tertiary basins are the important factors for the formation of coal measure-derived oil and gas fields so far discovered. The study made by the author also shows that the occurrence of independent depressions developed during the Late Jurassic in Turpan is vital for the maturity of coal measure source rock and the timing of hydrocarbon expulsion. The function of successive basin development on coal measure source rock maturation and oil expulsion can be clearly proven by the difference in hydrocarbon providing condition of Jurassic coal measure between Tokexin and Shengnan sub-depressions in western Turpan basin. The two sub-depressions, i. e. Tokexin and Shennan, developed in the south-western part of the Turpan basin. The Low-Middle Jurassic deposited in the two subdepressions is uniform and the sediments are the same and thicker in Tokexin. But the oil generation condition of coal measure in the two sub-depressions is quite different. The fundamental reason is that the Jurassic in Shengnan subdepression has been already matured and provided oil and gas for Shennan-Shanquan structure which is located to the east of Turpan city, whereas the Jurassic in Tokexin sub-depression is not matured and no oil and gas have been revealed in wildcats which penetrated the Jurassic sequence.

The isopach maps of different intervals from the Upper Jurassic to the top show that Shengnan sub-depression had independently developed in Late Jurassic and sequentially developed during Cretaceous-Tertiary. Whereas the Low-Middle Jurassic coal measure in Tokexin area has not been deeply buried until Tertiary time. Therefore, hydrocarbon generation history of source rock was greatly delayed. In NW China, the superimposition of sediments on Jurassic can be generally summarised into three types based on their burial curves. They have different petroleum geological condition and richness of petroleum resource.

(1)"Thick meat and thin skin" type: that is, the Jurassic source rock is much thick and the overlying sediments are moderate in thickness (Fig. 4a). The overlying sedimentation kept the source rock to a depth sufficient to be mature but not made it buried deeply. Turpan depression and Yanji in NW China are belonged to such kind of examples. After the deposition of the Low-Middle Jurassic coal measure in Taibei sag zone of western Turpan basin, the development of the sequent basin has the following characteristics:

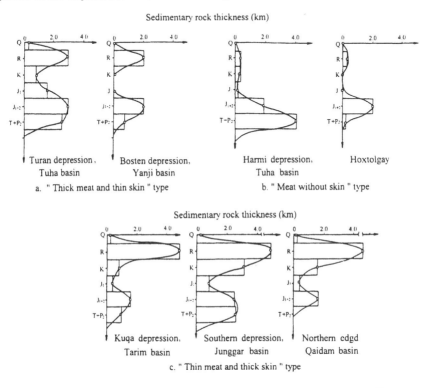

Fig. 4 Development curves of the inherited basin superimposed on the Jurassic in NW China

(a) The late Jurassic is a huge depositional formation of near 2000m thick and mainly consists of lacustrine mudstones. This huge mudstone sediment, like a heat-proof plate, prevented heat from diffusing upward and created the condition to make the coal measure source rock mature to generate oil. (b) the sequent subsidence from Cretaceous to Tertiary kept the Jurassic source rock in a progressive buried history, which maintained the source rock in a progressive heating procedure and therefore kept the source rock storing enough energy to expel hydrocarbons. Meanwhile, moderate load of sediments did not put the target bed deeply buried and thus decreased the difficulty of drilling.

The superimposition of "thick meat and thin skin" type is the best one in oil-bearing potential.

(2)"Thin meat and thick skin" type (Fig. 4c): the Jurassic source rock is relative thin and the sequent sediments are too thick, especially the interval received after the Neogene. This kind of superimposition is quite common in NW China. For Instance, the southern margin of Junggar basin, Kuche depression and the South-western depression of Tarim basin, and the northern Qaidam basin etc. are belonged to the group. The subsidence of the Jurassic original basins mentioned above have a common feature: the thickness of Jurassic source rock is small (ranging from 300m to 500m) and the sediments of Neogene are too thick (over 4,000-5,000m).

The superimposition of "thin meat and thick skin" type of sequent basin is favorable for the maturity of source rock and the expulsion of hydrocarbon. But the target bed was deeply buried and the reservoir property became worse. In many cases, we have to look for the secondary oil and gas fields, in which the source rock is Jurassic. Therefore, this superimposition has petroliferous prospecting but is relatively difficult for exploration.

(3)"Meat without skin" type: the source rock of the Jurassic well developed, but it had not been sufficiently buried after Jurassic deposition. Either Cretaceous and Tertiary of only Tertiary and Quaternary, both with thin thickens, sit on the Jurassic source rock(Fig. 4b). The source rock did not reach maturity because the overlying caprocks were not well developed. In addition, the sealing condition is also poor.

The Jurassic featured with "meat without skin" association experienced a decreasing thermal history. That is, the source rock underwent a higher temperature during Jurassic time due to relative high geothermal gradient and thick sedimentation than later. The basin was generally in and uplift setting after Jurassic deposition as the overlying sediments are not thick enough. This buried hsitory resulted in a very low efficiency of hydrocarbon generation. The petroliferous prospecting of this superimposition is the poorest one because of the poor ptential of hydrocarbon generation in NW China.

Tectonic compression soon after the maturation of source rocks is the major driving force for the effective expulsion of oil and gas from the coal measure

The maturation and expulsion of hydrocarbon from the Jurassic coal measure the source rock in NW China have the following features: (a) The threshold depth of maturity is generally deep. For instance, the threshold in Turpan basin is 2400-3200m in depth and exceeds 3000m in Kuche and South-western depression of Tarim basin. In Yanji and Santanghu basin, the threshold depth reaches to 2400m. In the south marginal depresion of Junggar basin, the threshold is deeply buried to 4200m. The threshold in Jiudong basin is 2800-3200m. Therefore, relatively intense compaction took place before the source rock reached to the threshold, resulting in difficulty of the hydrocarbon migration. (b) The source rock of coal measure has high absorptivity to oil and gas, and the decrease of coal porosity and pore water were mainly occurred in early diagenesis, e. g. in the low maturity stage of $R^{\circ} <$ 0. 6%. Whereas in the mature stage after $R^{\circ} > 0. 6\%$, most of the porosity maintained in coal measure is microporosity with the diameter of pore less than 12 Å and the first migration of hydrocarbon did not have carrier and driving force. (c) The coal measure source rock is not good sealing rock. So, the abnormal compaction zone is difficult to occur in coal measure source rock and the condition for hydrocarbon expulsion of coal measure is relatively poor.

Bacause coal measure source rock has the disadvantages in maturation and oil expulsion, the tectonic compression taking place soon after the maturity of source rock played an important role in the effective expulsion of hydrocarbon.

The study on the hydrocarbon generating history of the three major basins in Xinjiang Uygar Autonomous Region of NW China reveals that the peak period of oil and gas generation occurred correspondingly with the happening of tectonic compression. The study also shows that the formation of oil/gas pools in the three major basins of Xinjiang were mostly occurred after the peak period of hydrocarbon generation, roughly happening during the Silurian-Devonian, Permian-Triassic, Jurassic-Cretaceous and Tertiary, which corresponded respectively to Caledonian, Hercynian, Yanshanian and Himalayan movements, which indicates that the tectonic compression played an important role in hydrocarbon expulsion from coal measures.

The analysis on the formation of Shanshan field in Turpan basin shows that the migration and accumulation of hydrocarbon is closely related with the tectonic compression happened at the end of Yanshanian and Himalayan movements. The study on the hydrocarbon generation history of coal measure source rock of Turpan depression suggests that the peak period of hydrocarbon generation occurred in Qiudong and Taibei sags in the ends of Jurassic and Cretaceous, which provided oil and gas for Shanshan structure. Compressional tectonic movements in the late and end of Yanshanian orogeny took place just after the two peaks of hydrocarbon generation, which provided not only traps for hydrocarbon accumulation, but also driving force for oil and gas migration. The Shanshan structure which formed before the two peaks of hydrocarbon generation increased the structural magnitude and was not destroyed in the movements. Hence, it received oil and gas during the two compressions to form oil pool with high oil-bearing column(Fig. 5).

Determining the role of tectonic compression in the hydrocarbon expulsion and migration of coal measure source rock enables us to give deferent evaluation to the oil-prone property of the Jurassic in different basins of NW China. The exploration of Jurassic coal measure-derived oil and gas fields should be in the area which meets the above conditions. Meanwhile, tectonic compressions with moderate magnitude in these area, not very strong and not very weak, can provide traps for oil accumulation but does not destroy the sealing condition.

DISTRIBUTION OF COAL MEASURE-DERIVED OIL AND GAS FIELDS

The distribution of coal measure-derived oil and gas fields has its own characteristics and is constrained by original depositional environment, sedimentary association and expulsion condition.

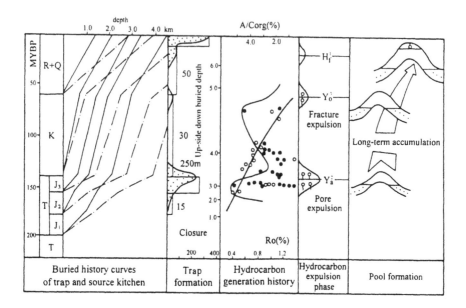

Fig. 5   Analytical diagram showing the formation procedure of Shanshan oil pool, Turpan Basin

The horizontal distribution of oil and gas fields is along the inner side of lake strandline

The Jurassic coal measure in NW China mainly developed in the environment of low-position-al lake-related swamp and high-positional river-related swamp facies. The lake-related swamp facies are the major source rocks and the river-related swamp facies has poor condi-tion for hydrocarbon generation. In the area of lake-related swamp facies, the deltaic front sand bodies and beach sands are either fingeredly interbed with coal measures and lacustrine mudstone in lateral direction or superimposed each other in vertical direction which form a excellent association of source rock, reservoir and caprock and are the major position of hy-drocarbon accumulation. The discoveries so far discovered in NW China suggest that the Jurassic hydrocarbon accumulation are mainly distributed along the inner side of lake strand-line. The reason is that source rocks are closely associated with reservoirs in the swamp envi-ronment near the lake strandline and the fact that the migration distance of oil is short in coal measure is another reason to keep oil in the area near lake strandline (Fig. 6). Therefore, the oil exploration on the Jurassic in NW China should focus on analysing and determining the position of lake strandline of the main target formation, and combining the palextructure analysis to determine trap formation and to map out the distribution of sand bodies. In this way, the new target of drilling can be selected.

The vertical distribution of oil and gas mainly occurs above or below the threshold of maturi-ty

In normal compaction, the reservoir porosity of coal measure decreases to less than 5% in

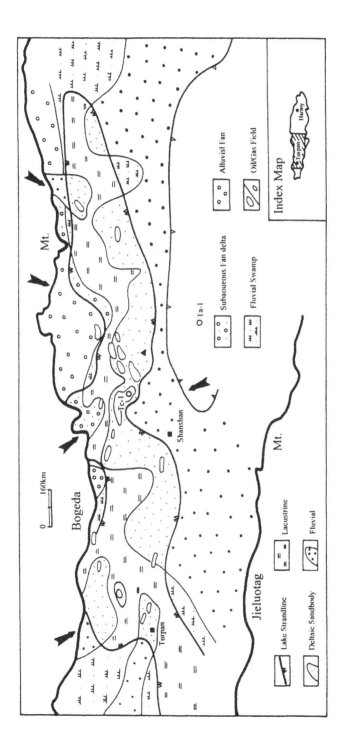

Fig. 6 Map Showing Relationship between Coal Measure-derived Oil/gas Accumulation and Sedimentary Facies

the depth of 2000-2500m and loses reservoir capability. Whereas dissolution occurring in early diagenetic stage can modify the physical property of deep reservoir. The dissolution usually occurred above or below the threshold of hydrocarbon generation because large amount of
organic acid was released in association with the procedure of hydrocarbon generation, which
results in the dissolution of parts of the grains and cements of the reservoir. Thus, the significant oil and gas resource are distributed above or below the threshold of hydrocarbon
(Fig. 7).

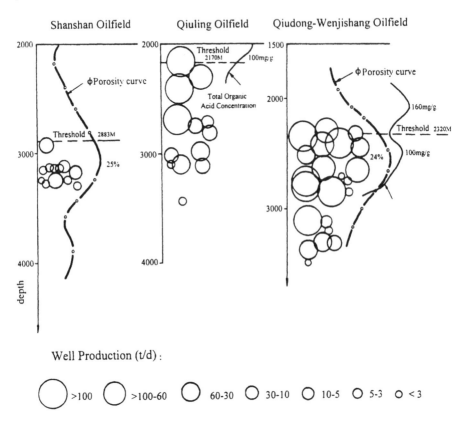

Fig. 7   Diagram showing relationship between production and threshold of maturity, Turpan Basin

Lithological and composite types of oil and gas pools have a significant percentage in hydrocarbon accumulation

In the discovered oil and gas pools which have been counted in the three major basin of Xinjiang, the lithological and composite types of oil and gas pools make up significant percentage
in hydrocarbon accumulation, accounting to 30%-40%. The big portion of lithological and
composite oil/gas pools in coal measures is determined by the sharp changes of source rock
and reservoir in three dimensions. Along with the further exploration, more and more lithological oil pools will be discovered. Therefore, the exploration on Jurassic sequence in NW
China should correctly choose the excellent target area at first. Then it is necessary to ex-

plore the lithological and composite oil/gas pools. The lithology-related targets in the Jurassic sequence which is near source rocks should be selected for active drilling.

CONCLUSIONS

1. Not all the Jurassic coal measures in NW China can form oil and gas accumulations. The following controls are essential for the formation of Jurassic coal measure-derived oil and gas fields in NW China:

(1) The Jurassic original basins should exist and the complete association of source rock, reservoir and caprock is well developed. During the Jurassic, six large original basin zones were developed in NW China. Two zones are situated in both sides of Tianshan mountains and another two zones developed in the southern and northern sides of Qilian mountain.

(2) The original depositional environment in which the low-positional swamps were dominant and humid climate was prevailing is favorable for the accumulation and preservation of oil-prone component with large volume. High-positional swamps are not favourable for oil generation.

(3) The sequent basin should occur on the Jurassic original basin in order to keep the coal measure source rock in progressive heating and maturation. In NW China, the superimposition of the Jurassic with the overlying sediments can be divided into three types: "thick meat and thin skin", "thin meat and thick skin" and "meat without skin". The first two types have a good petroliferous prospecting and the last one is the poorest in oil-bearing potential.

(4) The tectonic compression soon after the maturity of source rock is the major driving force for oil expulsion from coal measures.

2. The oil and gas fields are horizontaly distributed along the inner side of synsource rock-deposited lake strandline. Tectonically, they are situated in the area where compression is moderate, the faults cutting to source rock exist and destruction is relative weak. Coal measure-derived oil is less occurred in stable intracontinental depressions and slope areas. In vertical, the oil/gas fields are mainly occurred above or below the threshold of maturity.

Quite a part of oil/gas accumulation in coal measures are in lithological and composite traps. Therefore, it is necessary to develop the technique to describe subtle traps in advance, so as to find more reserves from Jurassic coal measures.

REFERENCES

1. Zhao Wenzhi et al.. The formation and distribution of oil and gas pools in three major basins of Xinjiang. Oil and Gas Geology,1994,15(1):1-11

2. Zhao Wenzhi et al.. Features of coal measure-derived hydrocarbon formation and accumulation of Turpan basin and the exploratory countermeasures. Natrual Gas Industry, 1995,15(4)

3. Cheng Keming (edit.). The hydrocarbon formation in Tuha basin,. Petroleum Industry Press,1994

4. Beaumont E A. Creativity in petroleum exploration. AAPG Lecture Notes, 1990

5. Price L C. Basin richness and source rock disruption-a fundamental relationship. Jourmal of Petroleum Geology, 1994,17(1)

Proc. 30th Int' l. Geol. Congr. , Vol. 18, pp. 103~122
Sun Z. C. *et al.* (Eds)
© VSP 1997

Oil and Gas Generation Potential Resources and Prognosed Highly Perspective Objects of Prospecting and Exploitation Activity in Precambrian and Cambrian Sedimentory Complexes of Siberian Platform.

D. I. DROBOT and E. V. KRASYUKOV

"RUSIA Petroleum"Company, 53 *Gogolya str. Irkutsk*, 664039. *RUSSIA.*

Abstract

In the context of sedimentory migrational theory of naphtids genesis the conditions for oil and gas generation in ancient sedimentory complexes of Siberian Platform are considered. The leading role of Riphean complex is shown in this process, the share of which makes up 44. 5% of total aggregation of scattered organic matter of Platform mantle and more than 90% of total hydrocarbon aggregation, migrated from PreCambrian and Cambrian sediments. With Riphean-Vendian complex it is connected the generation of the largest accumulation belts of migratory naphtids, such as: Angaro-Anabarsky. Baikisko-Katangsky and Aldano-Maisky. The features of Kovyktinsky gascondensate field are brought, its structural model and field characterisitics of productive formation are considered, the forecast of specific well outputs is proposed, the conclusion is made that Kovyktinskoye gascondensate field should be considered as extremely large lithologically sheilded deposit with perspective area covering 6500 km², prognosed resources of which are evaluated in the range of 3. 5 BCM.

Keywords: Riphean, Vendian, Cambrian, Oil-gas bearing provinces, genesis of naphtids, hydrocarbon gases, bitumens, reservoir rocks, catagenesis, accumulation belts, naphtids.

On a huge territory of ancient Siberian Platform between Lena and Enisei rivers a new Russian oil and gas producing province is under formation now. The lands perspective for oil and gas on this territory make 3400 thousand square kilometers.

Rather complex geological structure in the region, a wide range of oil and gas bearing formations from Riphean to Cretaceous, intensive shows of trappppean metamorphism, presence of thick halogen-carbonaceous complexes allow to consider Siberian Platform as a unique object for many key problems of petroleum geology solving. First of all it concerns the estimation of old sedimentary mass - Upper Proterozoic and Lower Cambrian. On other platforms the oil and gas bearing complexes are much younger.

No doubt it is possible to assert that Siberian Platform is for the time being the unique region in the world where the commercial reserves of oil and gas in Riphean-Vend and Lower Carmbrian sedimentary complexes are discovered. For last 20 years these sediments have been the main target of geological prospecting which resulted in discovery of 35 oil and gas fields. Among them there are objects which according to the proved reserves are considered to be large and the largest and coming up to the giants: Verkhnechonskoye, Kovyktinskoye, Chayandinskoye, Talakanskoye, Yurubcheno-Takhomskoye and other fields (Fig. 1).

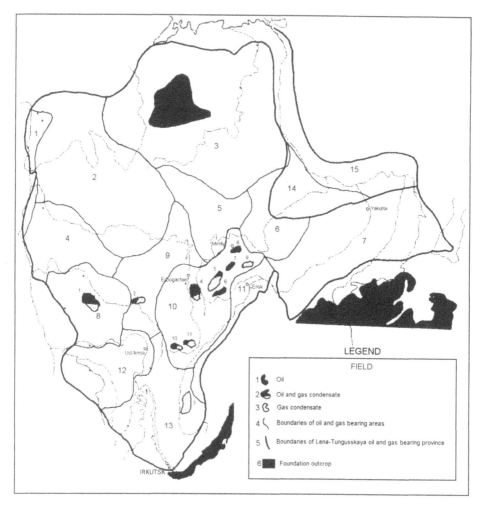

Fig. 1 Scheme of Oil Gas Bearing Areas of Siberian Platform. Big and Biggest Oil and Gas Fields Location
Oil and Gas Bearing: 1. Turukhand – Norilskaya; 2. Severo-Tungusskaya; 3. Anabarskaya; 4. Yuzhno-Tungusskaya; 5. Sugozherskaya; 6. Zapadno-Viluiskaya; 7. Severo-Aldanskaya; 8. Baikitskaya; 9. Katangskaya; 10. Nepsko-Botuobinskaya; 11. Predpatomskaya; 12. Presayano-Eniseiskaya; 13. Angaro-Lenskaya; 14. Viluiskaya; 15. Predverkhoaynskaya. Oil and Gas Fields: 1. Yurubchend-Takhomskoe; 2. Sobinskoye; 3. Kovyktinskoye; 4. Verkhnechonskoye; 5. Chayaandinskoye; 6. Talakanskoye; 7. Srednebotuobinskoye; 8. Irelakhskuye; 9. Verkhne-Viluchanskoye; 10. Yaraktinskoye; 11. Dulisminskoye.

The particularity of oil and gas generation process on Siberian platform is that Riphean sediments have been the source beds of most hydrocarbons (HC). The Riphean deposits mostly consist of grey coloured marine aleurite-clay and marlaceous mass, enriched with OS (organic substance). Riphean sediments thickness varies in a wide range from zero and first dozens of meters within the old anticlise and shields up to 5-12 km on the areas of periclinal subsidences.

The generation potential of Vend terrigenous-carbonaceous formations is much lower. The Vendian sediments are almost distributed everywhere,combining the deposits of yudom complex and fill the typical platform structures such as syneclises and anticlises. The thickness of Vendian complex varies from $150 \sim 2000$ m, reaching its maximum in Lena-Patom depression. Cambrian sediments showed even lower potential of oil and gas generation. On Siberian Platform the Cambrian system is presented by three series that form single sedimentary complex closely connected according to the formation conditions with the underlying beddings. The generation of thick halogen-carbonate formation (up to 3. 5 km) that constitutes the most part of the internal field is associated with Lower and Middle Carmbrian.

The Riphean and overlaying sedimentary complexes within the pericraton zones experienced significant changes of old structures that negatively affected the preservation of hydrocarbons (HC) in them. The structure of slab megacomplex kept in its basis the inherited development during long formation history of sedimentary cover of the region. All that, along with the broad distribution of regional salt masses covering the most productive complexes on the most part of the territory created favourable conditions for generation of zones of oil and gas accumulations and their preservation. The modern structure of sedimentary cover includes Nespko-Botuobinskaya, Baikitskaya, Aldanskaya and Anabarskaya anticlises, Katangskaya and Sugdzherskaya saddles, Bakhtinski megaprojection, central parts of Angaro-Lenskaya step (Fig. 2).

Analysing the problem of oil and gas accumulations in old sedimentary complexes of Siberian Platform on the basis of naftidogenesis sediment migration theory the following points should be marked out:

source beds have the regional character of spreading and include the wide spectrum of grey coloured terrigenous and carbonate sediments;

oil and gas potential of sedimentary complexes is generally determined by the thickness of source beds, by the original (initial) content of DOS (Dispersed Organic Substance) in them and by its diagenetic and catagenetic transformation (metamorphosis);

the reliable indication of source beds (oil generating) is the presence regional traces of bitumoil and appropriately directed (oriented) changes of DOS content and its bitumiol fraction, that is caused by the hydrocarbon components migration.

The investigation allowed to establish that catagenetic transformation of DOS in old sediments was accompanied by the generation of more or less liquid and gaseous hydrocarbons and other volatile products and by the rapid migration of hydrocaarbons components at the spectific phases of organic substance transformation. At the final stages of catagenetic evolution (paleodepth 5. 0-6. 5 km) DOS generates methane in big quantities. The process of DOS transformation is finished by the generation of appreciable volume of volatile products-CO_2 and N_2(paleodepth 7-9 km). Though these processes had the same trend, the terrigenous

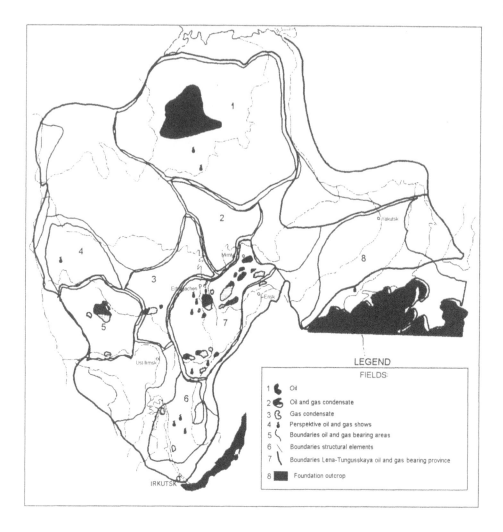

Fig. 2   Scheme of the Biggest Positive Elements Location in Siberian Platform Sedimentary Cover Within which the Regional Zones of Oil and Gas Accumulations Formed Oil and Gas Fields Have Been Diskovered and Perspektive Oil and Gas Shows Defined

Structural elements: 1. Anabarskaya antyclise; 2. Sugdzherskaya saddle; 3. Katangskay saddle; 4. BaiKitinski projection; 5. Baikitskaya antyclise; 6. Central parts of Angaro-Lenskaya step; 7. Nepska-Botuohinskaya; 8. Aldanskaya antclise.

and carbonaceous rocks with different DOS concentration had different intensity of HC fluids generation and emigration, the depth of main oil and gas generation phases manifestation (Fig. 3). Their significant similarity with Precambrian and Phanerozoic formations has been found according to the depth trend and oil and gas generation processes development activity. The Riphean formations possessed exceptionally high ability for oil generation among sedimentary complexes of the Platform according to the rock organic potential and scale of Organic Substance accumulation. They are mostly located along the peripheral zones of the re-

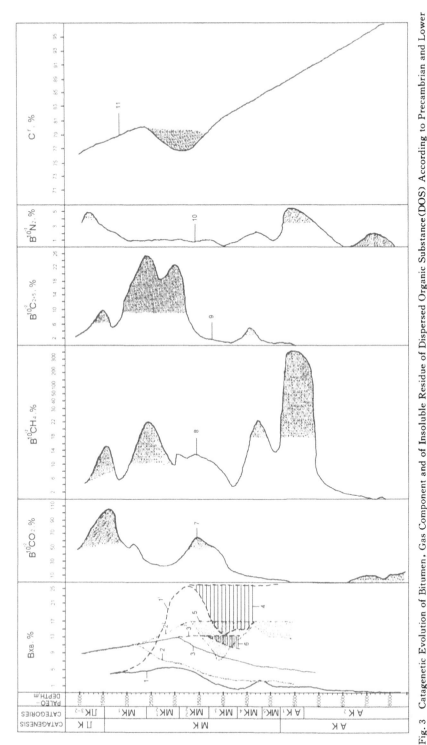

Fig. 3   Catagenetic Evolution of Bitumen, Gas Component and of Insoluble Residue of Dispersed Organic Substance(DOS) According to Precambrian and Lower Cambrian Sedimentary and Lower Cambrian Sedimentary Complexes Depth.

Lines 1,2,3. Average (Actual) Values of Bchloroform/bitumen in the Despersed Organic Substange(DOS)

1. in Precambrian and Lower Cambrian Terrigenous Deposits with the Content of Corg>0. 2%; 2. in Clay-Carbonate Rocks with Corg>0.1%, and 3. in Clay-Carbonate Rocks with Corg <0.1%; 11,2², 3³ .Intial Values of Bitumen Coeffigient (B initial); 4,5,6. Intervals of Pitch-Asphaltic components Fall-out into the Insoluble Fraction of DOS; 7,8,9,10. Lines of Average Gas Components Concentration in Rock DOS; 11. Line of Average Corg Concentration in insoluble Residue of Organic Substance(OS)

gion where in the pericraton depressions they form carbonaceous complexes up to 2. 0-7. 5 km thick with the weighted average content of C org from 0. 5 to 2%-3. 5% at maximum value from 5%-10% and up to 22% (Fig. 4,5). The Riphean rocks contain 44. 5% of all Dis-

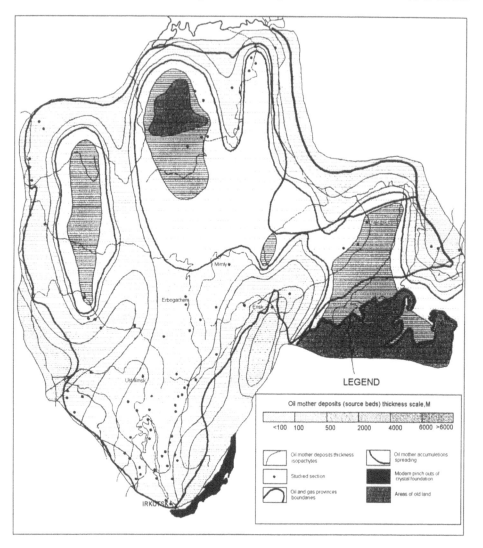

Fig. 4   Thickness Map of Siberian Platform Riphean Oil Mother Deposits (Source Beds) and Folded Struktures Ringing Them

persed Organic Substance of Siberian Platform sedimentary cover and 76. 5% of its whole mass, which is concentrated in Riphean, Vend and Cambrian sediments. Accordingly, Riphean source beds have more than 90% of the total mass of hydrocarbons emigrated from Precambrian and Cambrian deposits (Fig. 6). The belts and centres unique in oil and gas generation and migration scale are related to the Riphean systems of pericraton depressions. These belts are as follow: Eniseisko-Turukhanski in the West and Prebaikalo-Patomski in the

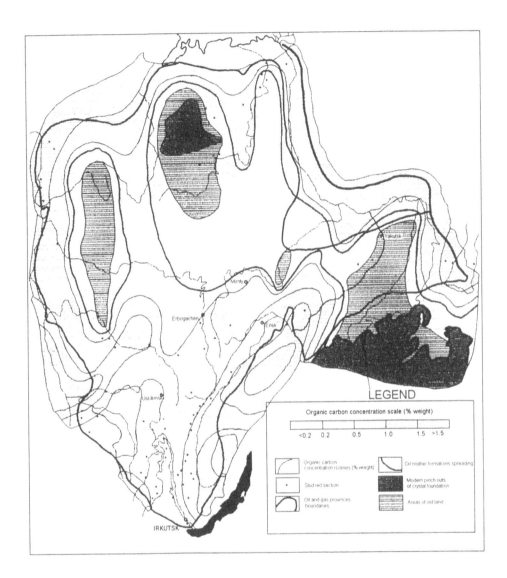

Fig. 5   Map of Organic Carbon Initial Concentration Spreading in Riphean Oil of Siberian Platform and Folded Structures Ringing Them Mother Deposits

south-east of the platform; Anabaro-Olenekski (Arctic) in the North and Predverkhoyanski in the north-east of the region are prognosed. Among big centres Irkineevo-Chadobetski in the West, Uchuro-Maiski in the north-east and Presayanski in the south-east of the Platform are estabished (Fig. 7).

The regional oil and gas potential of basal Ripheon-Vendian sediments is defined by a set of-geochemical indices that coincides well with a very high oil generation possibilities of Riphean megacomplex and it is proved by the stable confining of big zones of bitumen and oil-gas shows and discovered commercial oil and gas reserves to the basal Riphean-Vendian sediments.

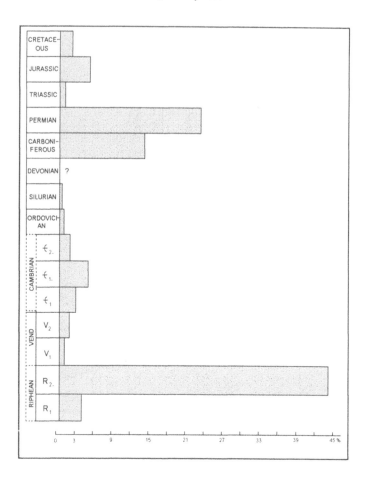

Fig. 6   Distribution of Relative Organic Carbon Mass in Sedimentary Late Precambrian and Phanerozoic Complexes of Siberian Platform

The formation of biggest migrated hydrocarbons accumulation belts is connected with the Riphean-Vend megacomplex of Siberian Platform: Angaro-Anabarski, Baikitsko-Katangski and Aldano-Maiski (Fig. 8). Within their limits the main mass of hydrocarbon fluids emigrated from the ancient zones of oil and gas generation to ancient zones of oil and gas accumulation has been accumulated. The development of two biggest regional zones of bitumen and oil accumulation: Nepsko-Olenekskaya, Aldano-Lenskaya and of some others smaller regional hydrocarbon accumulations is clearly mappped in Upper Vend carbonaceous complex (Fig. 9).

The intial potential geological resources in old sedimentary complex of Siberian Platform are assessed at 68. 2% billion tons including 33. 5 billion tons of oil and 34. 7 trillion cubic meters of gas. The most perspecitive targets for prospecting work at the present stage of region exploration are Nepsko-Botuobinskaya and Baikitskaya anticlise, Angaro-Lenskaya Step and Katangskaya Saddle. More than 90% of oil discovered fields and of oil highly perspective oil

Fig. 7   Riphean Belts and Oil and Gas Accumulation Centers

Oil and Gas Formation Belts Defined: 1. Eniseisko-Yurukhanski; 2. Prebaikalo-Patomski Prognosed; 3. Anabaro-Oleneks-ki; 4. Predverkhoyanski Oil and Gas Accumulation Centers; 5. Uchuro-Maiski; 6. Irkineevo-Chadobetski; 7. Presayanski

and gas shows are located within these structures (Fig. 2).

According to the dynamics of oil generation process development in sedimentary cover of Siberian Plotform the fields formed near big oil generation centres and in which the deep phase of gas generation took place will have gas (gas condensate) saturation. Well expressed tendency of liquid phase increase (expansion) is clearly traced depending on the distance from the zone of ancient active warping (depression) along the regional productive seams pitch in the direction of pre arch sections of anticlises. The example of it is Verkhnechonskoye oil and gas condensate field. And the second example may be Kovyktinskoye gas condensate field. According to the explored and prognosed reserves it ranks among the category of giants. We will envisage the history of its formation more in detail.

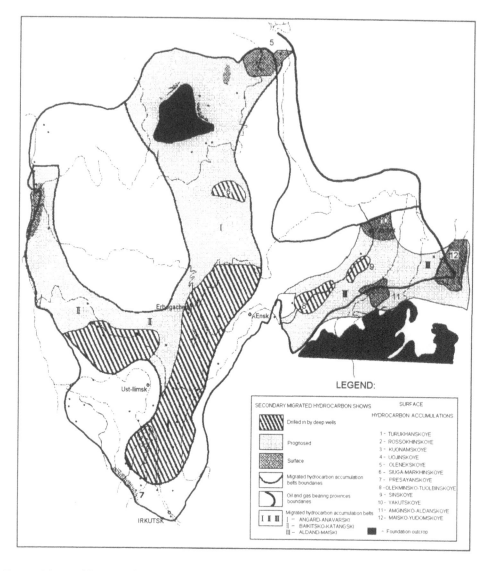

Fig. 8   Scheme of Bitumen of Bitumen Accumulations and Regional Zones of Migrated Hydrocarbon Accumulations in Upper Vend Complex of Siberian Platform

This field in its modern structural aspect along the cover of Nizhnemotskaya subsuite is confined to the gentle monoclinal slope of Kovyktinsko-Zhigalovskaya structural terrace of Angaro-Lenskaya Step with the bed inclination gradient in western, north-western direction 1. 45-2. 12 m/km. On the east and west the structural terrace is limited by the isolines of submeridional course 2100 m and 2300 m relatively (Fig. 10). The structural plan of Parfenovski horizon layer I cover is of a satisfactory conformity with the mapped regional structure of Nizhnemotskaya subsuite cover (Fig. 11). Within the drilled part of the field simple monoclinal subsidence of layer I cover with the isolines orientation to submeridional direc-

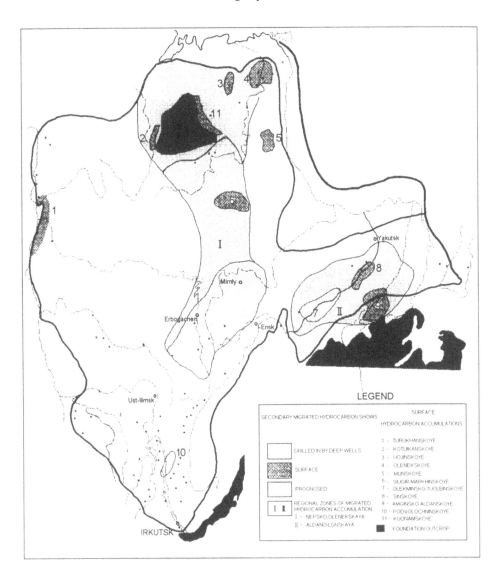

Fig. 9 Scheme of Bitumen Accumulations and Regional Zones of Migrated Hydrocarbon Accumulations in Upper Vend Complex of Siberian Platform

tion is observed. Insignificant structural complications are only marked on the site of wells number 1,22,25 and 28. Gas water contact is fixed of 2283,5 m and according to the structural particularities of the bed its submeridional trend is prognosed (Fig. 11).

The absence of oil fringe in gas deposit is confirmed by the one (indivisible) and significant methane content of natural gases (CH_4 90%-92%) both in the zone of gas water contact and in the remote areas, as well as by the obsence of oil components in condensate obtained from pre contour zone of the deposit (well[1] 11-Gruznovskaya).

At the period of field formation Nizhnemotskaya subsuite cover had another structural plan.

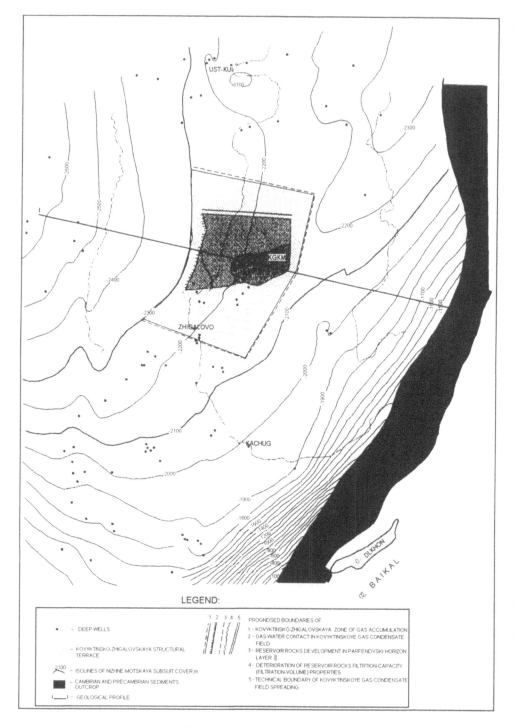

Fig. 10   Structural Map of Nizhne-Motskaya Subsuit Cover of Kovyktinsko-Verkhnelenskaya Zone of Angaro-Lenskaya Step of Siberian Platform

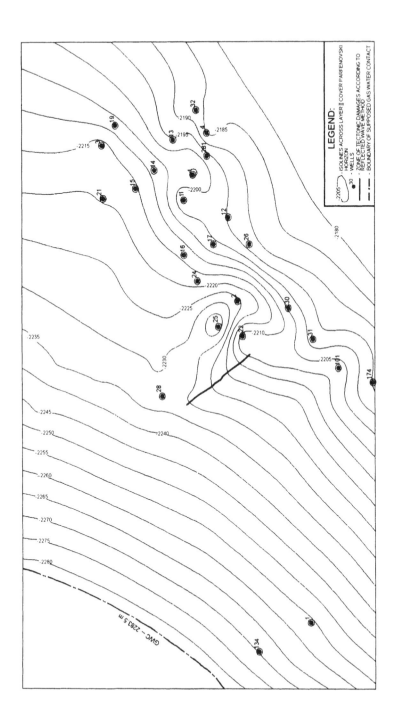

Fig. 11 Structural Map Parfenovski Horizon Layer **I** Cover of Kovyktinskoye Gas Condensate Field

It was inclined in the northern direction till Ordovician with the inclination gradient increasing as approaching the old Prebaikalski paleodownwarping (paleodepression) which was characterised by the active inherited development during Riphean, Vend, Cambrian and Ordovician. So Kovyktinskoye field formed close to the biggest Prebaikalski oil and gas generation centre that contributed a lot to the formation of biggest gas accumulation zone on the south of Kovyktinsko-Zhigalovskaya platform (Fig. 12).

At the first stage of field formation by the end of Atdabanski period (the stage of main oil generation phase completion) a big oil accumulation is formed in I Parfenovski horizon layer. The second layer of the horizon was filled by formation water.

With further inherited development of Prebaikalski paleodownwarping and with the arrival of Riphean-Vend complex in the main zone of gas generation, paleopetroleum deposit by the end of Ordovician period was completely destroyed by gas condensate system actively migrated from Prebaikalski gas generation centre. High concentrations of migrated hydrocarbons (1%-5.5%) such as kerites, anthraxolites, that fill pore space in sandstone of layer I, and practically are absent in layer I. The processes that caused the destruction of petroleum deposit in layer I negatively influenced on filtration-volume properties of reservoir rocks transforming them into unproductive rocks. It was the main reason why gas condensate fluids formed a big gas condensate deposit in layer I displacing water from it.

During Post Ordovician inversion stage of Angaro-Lenskaya Step development Nizhnemotskaya subsuite cover experienced significant structural reconstruction causing the formation of modern sedimentary cover structure (Fig. 12). There is a reason to think that the structural reorganisation didn't result in general destruction and reformation of Kovyktinskoye gas condensate deposit. The main reasons of its conservation appeared to be the consolidation and secondary cementation of sandstones that more actively displayed on the territory of Prebaikalski paleowarping development.

On the essential part of its territory catagenesis gradations approached MK_4-AK. In the region of Kovyktinskoye field location they were much lower (MK_2), reaching MK_3 in the zone contiguous to the Prebaikalski paleowarping (Fig. 13). That's why despite the regional rise of productive layer in the direction of Prebaikaiski paleowarping the rocks transformed from productive into unproductive have been the main reason of Kovyktinskoye gas accumulation conservation, it is also supported by data analysis of area effective thicknesses spreading, porosity, pemeability, specific (capacity) volume of reservoir rocks of Kovyktinskoye gas condensate field layer I. The materials show that moving from Prebaikaiski paleowarping to the north to the area of wells[1] 28,15,3 these indexes are much better (Fig. 14,15). There is a reason to suppose that not the most perspective part of Kovyktinskoye gas occumulation zone has been explored by drilling. Till present time more than 70% of its most perspective territory is still unexplored.

The conclusion is that Kovyktinskoye gas condensate field shouldn't be considered as a limited reservoir but as a rather big lithologically shielded deopsit (gas accumulation zone) with the total area of perspective territories of about 6500 km². According to the prognosed capacity-filtration properties of reservoir rocks and their effective thicknesses the prognostic com-

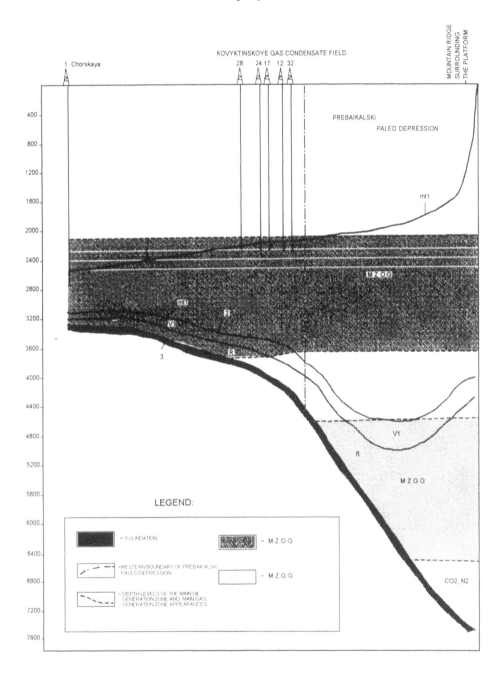

Fig. 12   Modern(1) and Paleostructural Profiles of Nizhnemotskaya Subsuit Cover(2) and of the Foundation(3) at the End of Ordovician Period; Depth Levels of Main Zone of Oil Generation and Main Zone of Gas Generation in Sedimentary Cover of Kovyktinsko-Verkholenskaya Zone of Angaro-Lenskaya Step of Siberian Platform

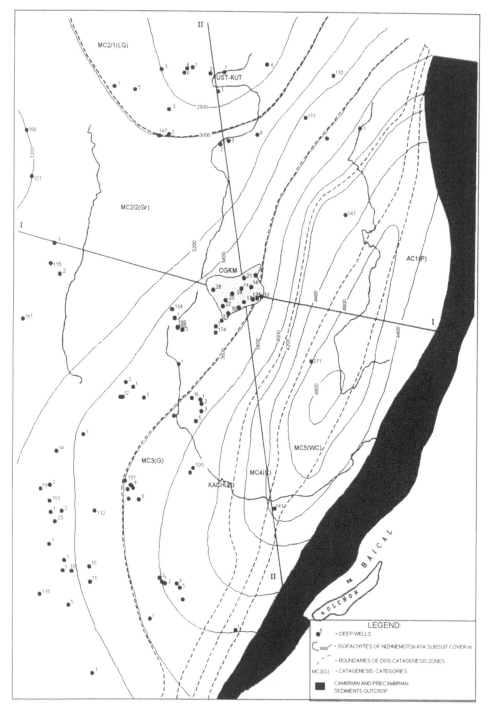

Fig. 13   Paleostructural Map of Nizhnemotskaya Subsuit Cover at the End of Ordovician Period and Scheme of Despersed Organic Substance (DOS) Catagenesis in Nizhnemotskaya Subsuit Cover of Kovyktinsko-Verkholenskaya Step of Siberian Platfrom South

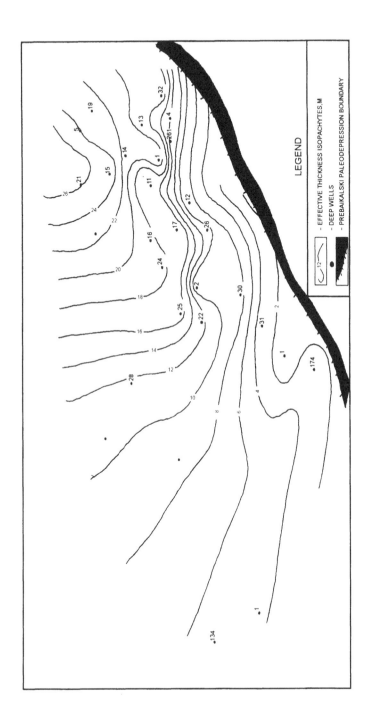

Fig. 14 Map of Effective Thickness of Sandstone Reservoir Rocks in Layer **I** of Parfenovski Horizon of Kovyktinskoye Gas Condensate Field

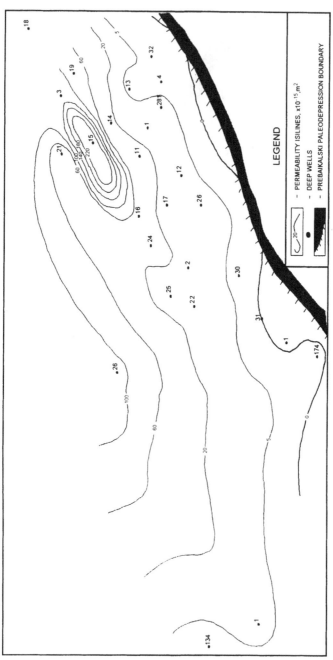

Fig. 15   Prognostic Scheme of Permeable Zones Location in Sandstone Reservoir Rocks of Parfenovski Horizon Layer Ⅰ of Kovyktinskoye Gas Condensate Field

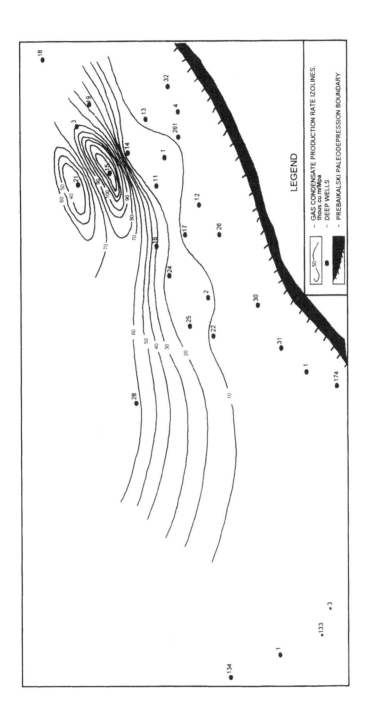

Fig. 16 Prognostic Map of Gas Condensate (Specific) Production Rate from Parfenovski Horizon Layer **I** of Kovyktinskoye Gas Condensate Field with Wells Working at DP 3.8 MPA

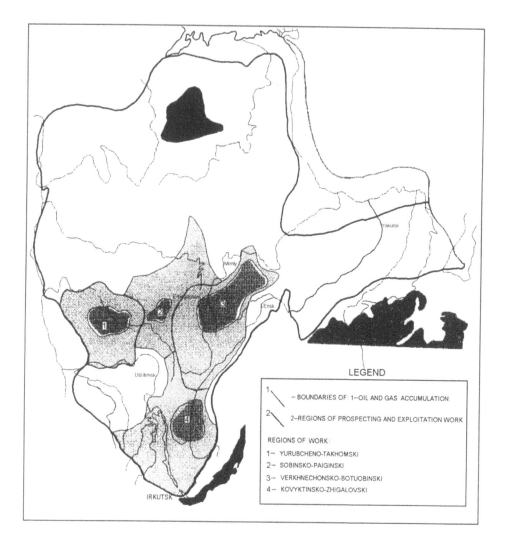

Fig. 17   Main Regions of Prospecting. Exploration and Exploitation Work on Siberian Platform

mercial reserves in this zone are assessed up to 3.5 trillion cubic meter. The additional exploration of unprospected part of the zone will allow both significantly extend the commercial reserves of Kovyktinskoye gas condensate field and to receive on its territory much higher well
rates (Fig. 16).

From above-stated the main commercial oil and gas bearing complexes in sedimentary cover
of Siberian Platform are Riphean terrigenous-carbonate, Lower Vend terrigenous and Vend-
Lower Cambrian carbonate. The main objects for prospecting and exploration are Verkhne-
chonsko-Botuobinski in Nepsko-Botuobinskaya anticlise; Kovyktinsko-Zhigalovski on Angaro-Lenskaya Step; Yurubcheno-Takhomski in Baikitskaya anticlise and Sobinsko-Paiginski
region in Katangskaya Saddle (Fig. 17).

Proc. 30ᵗʰ Int' l. Geol. Congr. , Vol. 18, pp. 123~129
Sun Z. C. *et al.* (Eds)
© VSP 1997

The Pattern of Combination and Industrial Division of Gas-Bearing Basins in China and the Strategic Thinking for Natural Gas Exploration

QIAN KAI, WANG ZHEHEN
Langfang Branch of Research Intitute of Petroleum Exploration and Development, CNPC, Langfang, Hebei, 102800, CHINA
DENG HONGWEN
China University of Geoscience, Beijing, 102200, CHINA

Abstract

The Meso-Cenozoic gas-bearing basins in China can be divided into 5 basin-types, which can be grouped into 4 combination types. The 4 combination types respectively are the combination of continental margin basins, the combination of intracontinental rift basins, the combination of intracratonic depression basins, and the combination of composite and superimposed basins. Consulting the ratio of gas resources to total resources, >60% is gas province, 60%-15% is oil-gas province, <15% is gas-oil province. So, Oil-gas basins in China can be divided into 4 petroleum province. From east to west, they are respectively the offshore oil-gas province, the eastern China gas-oil province, the central China gas province, and the western China oil-gas province.

In each province, the conditions for gas enrichment are different. The explorative prospect in each province is also different. The strategy for natural gas exploration in China should be focusing on the central part, developing the west, paying more attention to the offshore and intensiving explorating the east.

Keywords: Combination, Oil-gas basin, Province, Exploration, Strategy

THE PATTERN OF COMBINATION OF GAS-BEARING BASINS IN CHINA

The gas-bearing basins in China can be divided into: (1) cratonic depression basin; (2) intracratonic rift basin; (3) continental margin basin; (4) intermontane basin; (5) foreland basin. The five types of basins can be grouped into several combination types according to their temporal and spatial distribution pattern (Fig. 1).

Combination of Contionental Margin Basins in Offshore Areas of Southeastern China

The combination of continental margin basins is composed of the South Yellow Sea, the East

Qian et al

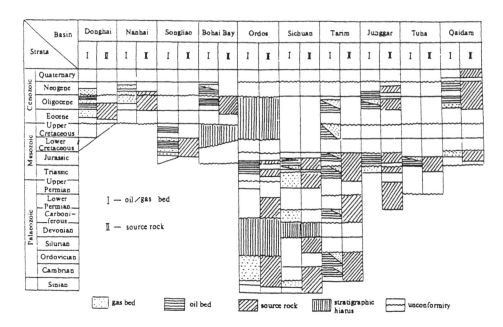

Fig. 1 Tectonic evolution and correlation of source rock of main gas-bearing basins in China

China Sea basins, etc. Their evolution is controlled by the opposing movement between the continental and oceanic crusts during Meso-Cenozoic. The basins are characterized by faulting in the earlier stage, but downwarping in the later stage, filled by continental deposit earlier, but marine deposit later.

Basins in the East China sea are typical examples. Their evolution is closely related to the subduction of the Pacific Plate. Three evolution stages can be recognized: (1) faulted-depression from Upper Cretaceous to Eocene, filled by littoral-shallow marine deposit from 1000 to 5000 m thick; (2) depression stage from Eocene to Miocene, accumulated by fluvial-lacustrine deposits from 2000 to 7000 m thick; (3) the stage from Pliocene to Quaternary. The basins are filled by fluvial-open marine deposit from 1000 to 2000 m in thickness due to regional subsidence of the shelf of the East China Sea[1].

Combination of Intracontinental Rift Basins in the East of China

It mainly includes the Songliao, Bohai Bay, Jianghan basins formed on the Xinan-Mongolia Hercynian fold belt and the North China platform. These basins underwent two rift cycles: Mesozoic rift cycle and Cenozoic rift cycle.

Well-developed Mesozoic rift basins are chiefly distributed in the north section, the Songliao basin can be an example. Faulted subsidence occured during Jurassic, and totally downwarping during Cretaceous, forming depression which was filled by stable lacustrine deposit with large area[2]. Compared with the rifting in Jurassic, the rifting in Cenozoic was not very intense.

Different from the Songliao Basin, the rifting of the Bohai Bay initiated in Mesozoic, and underwent two times of transforming from faulted-depression to depression in early Jurassic to later Jurassic and from early Cretaceous to later Cretaceous respectively[3]. Deposit in the basin is dominated by coarse clastic rock, coal and volcanic rock. Stable lacustrine deposit did not accumulate until Cenozoic. Well-developed rift baisn formed in Cenozoic. Some bsins on the Yangtze Platform and south of Yangtze Platform, such as the Jianghan basin, underwent mainly compressive stress in Mesozoic, but extensional stress in Cenozoic, leading to a complete fault-depression cycle composed of stable lacustrine deposit in Cenozoic.

Combination of Intracratonic Depression Basins in the Central Part of China
It is dominated by the Ordos, Sichuan and Chuxiong basins developed on the old cratonic basement. Similar tectono-sedimentary evolution occurred in these basins, showing ① complete stratigraphic sequences, including paltform sequence, foredeep sequence and inland depression sequence; ② multi-grade of depositional cycles, multi-phase tectonic unconformities and paleouplifts; ③ dominant foredeep sequence in the west basin; ④dominant fluvial-lacustrine deposit in inland depression.

Taking the Sichuan basin located in the west part of the Yangtze Platform as an example. The platform sequence from 6000 to 12000 m thick can be divided into three main cycles: ① the Caledonian cycle from Sinian to Silurian which consists of carbonate rock in shallow marine and black shale. In the stage of Silurian, the elevation of the basin basement led to formation of large scale plaeouplifts, and companied extensive erosion; ② the Hercynian cycle, the greater part of the Sichuan basin went up extensive transgression did not occur untill Permian, resulting in carbonate rock deposition of platform type; ③ early Indosinian cycle composed of enclosed bay deposit. The foredeep sequence comprised upper Triassic strata, the lower part of which includes darkcolored interbeded mudstones and sandstones with minor coal, and black shale, sandy conglomerate and carbonaceous mudstone and coal in the upper part. Intracratonic depression comprise of Jurassic-Cretaceous terrestrial-lacustrine deposit from 2000 to 5000 m in thickness.

Combination of Composite and Superimposed Basins in the West of China
It includes the Junggar, Tuha, Tarim, Qaidam basins and the basins in Tibet. The Tarim basin is a typical example. The evolving history of the basin can be divided into seven stages, i. e. aulacogen stage, foreland stage, intracratonic stage, etc. In the same stage of the basin evolution, several different types of subbasins may occur together. For example, the Tarim basin was composed of cratonic margin depression, intracratonic depression and intracratonic rift during Carboniferous-Permian[4].

INDUSTRIAL DIVISION OF PETROLEUM PROVINCES AND CONDITIONS FOR GAS ENRICHMENT

Oil-gas basins in China are divided into 4 petroleum provinces according to distribution of Meso-Cenozoc combination types of oil-gas bearing basins, and absolute value and relative value of oil or gas resources. Consulting the ratio of gas resources to total resources, $>60\%$ is gas province, 60%-15% is oil-gas province, $<15\%$ is gas-oil province.

From east to west, the petroleum provinces in China are the offshore oil-gas province, the eastern China gas-oil province, the central China gas province, and the western China oil-gas province (Fig. 2). These provinces correspond well to the basin combination type above mentioned.

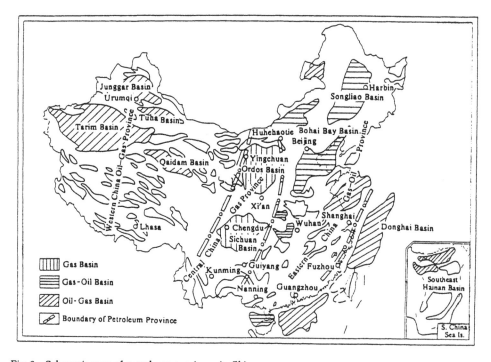

Fig. 2   Schematic map of petroleum provinces in China

The Offshore Oil-gas Province

The province is the part of Tertiary oil-gas bearing zone of the western Pacific Ocean[5], includes the East China Sea, the Zhujiang Mouth, Southeastern Hainan, the Yingge Sea, the Beibu Gulf oil-gas basins.

Explorative practices have shown rich oil-gas resources and good explorative prospect in this province. Considering the ratio of gas resources to total resources, the Southeastern Hainan basin and Yingge Sea basin are higher, 40. 98% and 44. 80% respectively; the East China

Sea basin is second, 31. 56%; the Zhujiang Mouth basin and Beibu Gulf basin are relatively lower, 16. 25% and 8. 13%, respectively.

Up to mow, 5 gas fields and 25 gas-bearing structures have been found in this province (not including Taiwan sea and Nansha sea). Meanwhile, there are 3 zones of gas-enriched have been proved in the Yingge-Southeastern Hainan basin, the East China sea and the north of the Liaodong Bay, in which pay beds are Pliocene, Miocene, Oligocene and Eocene.

The controls on gas enrichment in the province mainly include: (1) gas-forming sags provided abundant gas resources; (2) large gas fields always have been found in large scale draping anticlines, which connected with these sags; (3) Oligocene of Upper Eocene mudstones are good regional seals.

The Eastern China Gas-oil Province

The province mainly includes Songliao, Bohai Bay, Jianghai, Nanxiang, Shiwandashan basins etc. It is an important oil industrial base in China, of which the Daqing oil field is a very large scale oil field.

In the province, the ratios of gas resources to total resources are less than 15%. In basin scale, Songliao basin's, Bohai Bay basin's and Jianghan basin's are 6. 13%, 10. 11% and 8. 41%, repectively. All discovered gas field are mainly medium-small scale, the reserves of gas fields are generally $(5\text{-}50) \times 10^8 m^3$, some gas fields are more than $50 \times 10^8 m^3$.

Geneally, the vertical distribution pattern of oil and gas is "gas-oil-gas" in this province i. e. the deep strata mainly formed cracking gas, owing to high maturity [6]. Structural traps and stratigraphic traps on both sides of the basement faults, which control the formation of faulted depression, are favorable areas for gas enrichment because deep gas migrated verically along these faults. Middle-upper strata mainly produce oil. The nature of the gas in the middle-upper strata is oil-type gas. The nature of the gas in the upper strata belongs to secondary dry gas. The gas in the middle-upper strata formed by lateral migration along faults from the deep centre to the periphery. The type of the gas-pools chiefly is faulted-block.

The Central China Gas Province

This province mainly includes the Sichuan basin and the Ordos basin. A series of large and medium gas fields have been found, such as Wubeiti gas field in the Sichuan basin and the Center gas field in the Ordos basin. Gas resources, up to the end of 1994, account for more than 30 percent of the total gas resources all over the country. But their oil resources are only 4 percent of those of China. Gas resources are more than 60% of oil-gas total resources, the Sichuan basin's are 86. 63% and the Ordos basin's are 68. 65%, showing they are typical gas-rich areas.

The favourable conditions for gas enrichment are as follows:

(1) Abundant gas sources, including marine carbonate rocks and dark shales, fluvial-lacustrine coal-bearing formations with big thickness and extensive distribution, in which high

thermal evoluting degree of organic matters leads to the formation of dominant gas.

(2) Paleouplifts existing for a long term and weathered crusts contribute to gas accumulation.

(3) Regional distributed seals which overlay favourable reservoirs are important to preservation of large gas field's.

The Western China Oil-gas Province

This province mainly includes the Tarim, Junggar, Tuha, and Qaidam basin, etc. Gas resources of the Tarim basin is the biggest in oil and gas-bearing basins. The ratio of gas resources to oil-gas total resources is 15%-60%. The Tarim basin's is 43.81%. Junggars is only 15%.

This province is characterized by ① multi-sets of source rock formed in both marine and no-marine strata[7]; ② having three genetic types of gas, including cracking gas, degrading gas and biological gas; ③ Gas accumulation occured during several stages of basin evolution, which leads to the ultimate formation of gas fields in late stage. Discovered gas fields distributed along piedmont are controlled by the Cenozoic overthrust structure belt. The types of gas pools are mainly anticlines, fault blocks, fault-anticlines and structure-stratigraphic traps. Gas accumulation in the interior of basins is commanded by long-term developing paleouplifts and regional distributed unconfomities.

THE STRATEGY FOR NATURAL GAS EXPLORATION IN CHINA

Gas exploration in China should follow the general strategic thicking, i. e. "focusing on the central part, developing the west, paying more attention to the offshore, intensive exploration in the east".

First-focusing on the central part. Gas resources of the two biggest gas basins in the central part, the Sichuan and Ordos basins, make up more than 30% of the total resources in China, but the discovered reserves only account for 5.43% and 4.55% of the total reserves in China. So it is very important to focus on the natural gas exploration in the center of China.

Second-developing the west. The exploration for oil and gas in the west area is not intensive, but the resources of the basins in Xinjiang and the Qaidam basin in Qinghai province amount to 27% of the resources in China. Therefore, actively developing strategy in natural gas exploration must be adopted so as to make the another west base of natural gas industry.

Third-paying more attention to the offshore. The exploration and the researches show that the offshoree areas (such as the Yingge Sea, Southeast Hainan, the East Sea, the Zhujiang mouth, and the Bohai Bay basin, etc.), have abundant resources and good conditions for the formation of large-medium size gas fields.

The last, further exploration in the east should proceed in the following target strata and regions: the deep strata of Mesozoic and Cenozoic, the Palaeozoic, the offshore areas.

REFERENCES

1. Wang Shanshu. Offshore gas enrichment belt of China and the explorative direction. Natural Gas Industry, Vol. 14, No. 2, 8-13(1994).

2. Ying Jiliang. Study on the formation and distribution of oil-gas reservoirs in the north part of Songliao Basin, Study on Oil & Gas Reservoirs of China, Edited by Petroleum Institute of Chinese Petroleum Society. The Petroleum Industry Press, Beijing, China, 62-75(1985).

3. Tong Xiaoguang. Study on the spacial distribution of oil reservoirs in Bohai Bay Basin, Study on Oil & Gas Reservoirs of China, the Petroleum Industry Press, Beijing, China 97-104(1985).

4. Jia Chengzao, Yao Hujun, Wei Guoqi, Li Liangcheng. Plate tetonic evolution and geologic features of major structural elements in Tarim Basin, Collected Papers of Oil and Gas Exploration in Tarim Basin, edited by Tong Xiaoguang, Liang Digang, Xinjiang Science, Technology and Hygiene Press, Urumqi, China, 207-225(1991).

5. Qian Kai, Deng Hongwen. Lacustrine sedimentation and hydrocarbon accumulation in the Western Pacific oil bearing belt. Oil & Gas Geology, Vol. 3, 273-280(1990).

6. Xu Shubao, Hu Jianyi, Wang Weijin. Types and distribution of complex oil-gas accumulation zones in eastern China. Study on Oil & Gas Reservoirs of China, The Pertroleum Industry Press, Beijing, China, 50-61(1985).

7. Liang Digang, Wang Huixing. Problems about oil sources in Tarim Basin, Collected Papers of Oil and Gas Exploration in Tarim Basin. Xinjiang Science, Technology and Hygiene Press, Urumqi, China, 321-330(1990).

Proc. 30ᵗʰ Int' l. Geol. Congr. , Vol. 18, pp. 131~141
Sun Z. C. *et al.* (Eds)
© VSP 1997

Coalbed Methane in China: Geology and Exploration Prospects

CHEN XIAODONG and ZHANG SHENGLI

North China Bureau of Petroleum Geology, Funiu Road, Zhengzhou, Henan,

450006,*PRC*

Abstract

China has given priority to coalbed methane (CBM) in developing new energy resources for recent years in its effort trying to alleviate energy shortage and air pollution. As an unconventional natural gas, CBM adsorbed on the internal surfaces of micropores and fractures in coal seams is quite different from conventional oil and gas in exploration and development. This paper discusses the CBM geology and assessment and predicts its exploration prospects in China

Three factors of evaluating CBM potential are proposed, coal abundance, gas resource, and producibility, which are controlled by coal depositional environments and later tectonic evolution. China is endowed with vast coal resource, which provides substantial base for methane generation and storage. The CBM abundance in China is largely controlled by regional coal metamorphism because metamorphism directly determines caol gas-generating and -adsorbing capacities. Burial depth is the secondary factor controlling CBM abundance. CBM producibility primarily depends on cleat permeability of coal seams and is a crucial variable in evaluation of CBM potential.

North China Bureau of Petroleum Geology has made a signifcant breakthrough in CBM exploration at the Liulin CBM pilot site, the Shanxi province, China, which demonstrated that the areas with optimistic integration of the above three evaluation factors exist in China. Those areas are geologically comparable to Black Warrior Basin and San Juan Basin in USA, where commercial CBM developments have been successful. The paper reviews the CBM exploration and development activities in China and concludes that CBM industry will fundamentally take shape in China with the advent of the 21st century.

Keywords: China, Coalbed Methane, Exploration and Development, Prospect

INTRODUCTION

As the largest producer of coal in the world, the People's Republic of China is also poised to become one of the leading producers of coalbed mehtane in the world. China possesses a vast coalbed methane resource. With the fast development of economy and, as a result, the increasing need of energy resource, China will accelerate the development and utilization of a new clean energy — coalbed methane. The development and utilization of coalbed methane can improve the economy of the nation, reduce air pollution from methane vented from coal mines and emission from coal-burning plants, and increase mine safety.

The energy producers and a number of government agencies in China have made great efforts and a breakthrough in the exploration and development of coalbed methane in China for the past several years. The practices of coalbed methane exploration in China indicate that geological appraisal of target areas is crucial to successful exploration and development of coalbed methane. The geological appraisal mainly includes studies of coal abundance, methane resource, and producibility. The North China Bureau of Petroleum and Geology under Ministry of Geology and Mineral Resources funded by UNDP has successfully established the first seven-well pilot site of coalbed mehtane exploration and development at Liulin of the Shanxi province. The results of the pilot site demonstrate the exciting prospects in the coalbed methane exploration and development in China.

COAL RESOURCES AND OCCURRENCE OF CHINA

China is very rich in coal resources. The results of the national coal resource investigation in 1990 indicated that the in-place coal resources of China are totally up to 5.33 trillion tons at depths up to 2000m.

The major coal-accumulating periods in the world also existed in China, such as early Carboniferous, late Carboniferous-early Permian , late Permian, Late Triassic, Early Jurassic, Middle Jurassic, Early Creataceous, and Tertiary, etc.. In late Paleozoic, North China and South China were humid and dominated by a broad epicontinental depression in which tremendous peats were developed forming the first important coal-accumulating period in China. In Mesozoic, with favorable coal-accumulating climatic zones migrating toward north from late Triassic, early and middle Jurassic to early Cretaceous, the South China, North China, Northeast China, and Inner Mongolia coal-accumulating zones were formed accordingly. Based on tectonic settings, five coal-accumulating regions were distinguished in China, which are Northeast China, Northwest China, North China, Southwest, and South China regions (Fig. 1). The occurrences of coal resources are different in the five coal-accumulating regions resulting in the difference in exploration prospects of coalbed methane resources.

The most abundant coal resource exists in the North China coal-accumulating region, which accounts for 53.6 percent of the total coal resource of China. The North China region is in urgent need of natural gas to develop economy and improve environment, therefore the region is most prospective in developing coalbed methane in China. In view of gas markets and geological conditions, the Northeast China region is an important area for exploration and development of coalbed methane as well, though the coal resource of the region only accounts for 3.4 percent of the total of China. The Northwest China region has secondary coal resource in China, accounting for 37.5 percent of the total, therefore, it has a high potential for coalbed methane exploration and development as well.

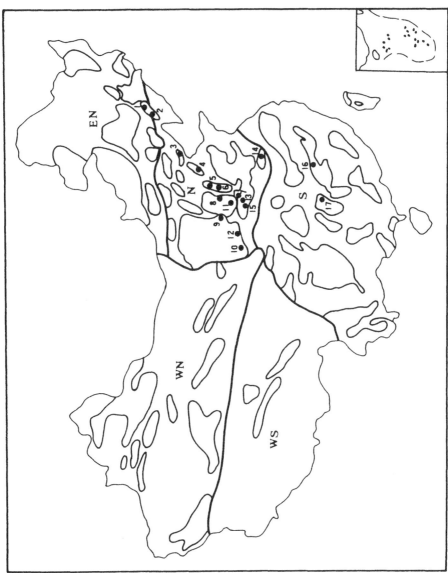

Figure 1 Map of locations of coal-bearing basins and coalbed methane project sites.

1. Shenbei; 2. Hongyang; 3. Tangshan; 4. Decheng; 5. Fengfeng; 6. Anyang; 7. Jiaozuo; 8. Qinshui; 9. Hedong Liulin; 10. Binchang; 11. Jincheng; 12. Weibei; 13. Xinggong; 14. Huainan; 15. Pingdingshan; 16. Fengcheng; 17. Lengshuijiang

In Paleozoic, the North China region was a mega-craton basin where the migrations of coal-rich zones were controlled by the regional sea water elevation and subsidence and delta-clastic shore zones of mega-epicontinental sea. The lithology of the coal measures is simple and the occurrences of the coal seams are continuous due to consistent sedimentation. Giant intracontinental depressions, such as Ordos Basin, Junggar Basin, etc., were developed on the stable platform basement in the early Yanshan movement, where the main coal-accumulating zones were developed in huge shallow lake shores and delta systems. The basins of that kind, in which usually deposited very thick coal seams, similarly have high potential for coalbed methane development, because the thick caol seams can compensate for low permeability and gas content. The late Jurassic-early Cretaceous basins and Tertiary coal-accumulating basins are isolated small or medium-scale fault basins. Though the basins occupy small areas, they possess very thick coal seams of low sulfur content and moderate ash content and, therefore, also have high potential for coalbed methane exploration and development. Some similar basins in USA have been or are going to be important areas of coalbed methane exploration and development.

Most of the late Paleozoic coal-bearing basins have experienced strong tectonic deformations. In North China, the late Paleozoic basins west of the Taihang mountain generally have gentle folds, but those east of the Taihang mountain have been altered into fault blocks or faulted folds by the Yanshan and Himalayan movements. In South China, folds and faults were highly developed in the late Paleozoic coal-bearing basins. The Mesozoic coal-bearing basins generally have some gentle and broad folds. Usually, tectonic movements were accompanied by magma movements in most of the coal-bearing areas, which resulted in high coal rank or destroyed connectivity of coal seams. The tectonic deformations mentioned above usually have significant effects on resources and producibility of coalbed methane, therefore they should be noted in selection of coalbed methane target areas.

Of the total coal resources of China, medium and low volatile bituminous coal most favorable for coalbed methane development accounts for 25.8 percent; high volatile bituminous coal with high potential of coalbed methane development for 54.1 percent; semianthracite and anthracite coals for 13.6 percent. The coalbed methane development potential of semianthracite and anthracite coals needs further study and verification. In short, the coal resources favorable for coalbed methane development are abundant, therefore, there are sufficient areas for looking for high productive coalbed methane fields in China.

COALBED METHANE RESOURCES OF CHINA

China is one of the largest coalbed methane resource countries in the world. It was estimated that the in-place resources of coalbed methane to a depth of 2000m range from 10.6 to 25.36 trillion m^3 (Feng Fukai and others, 1990) or from 30 to 35 trillion m^3 (Zhang Xinmin and others, 1991).

Like the coal resources, the distribution of coalbed methane resources is uneven. Coalbed methane resources are very variable not only between coal-bearing basins, but also between the different parts of a basin. The main geological factors affecting coalbed methane resources include coal resources, coal metamorphic degree, burial depth, cap rock, and tectonic movement.

Coal metamorphic degree (coal rank) is an important parameter for evaluating and predicting coalbed methane resources. Usually, methane content of coal in-

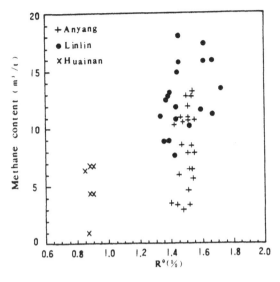

Figure 2 Methane content versus vitrinite reflectance in Anyang, Liulin and Huainan areas.

creases with increasing of metamorphic degree (Fig. 2, Tab. 1). That phenomenon exists in most coal-bearing basins.

Table 1 The gas contents of the coal seams at similar depths in the Hedong coal-bearing area.

locations	$R°$, %	methane contents, m³/t	depths, m
northern segment	0.56 to 0.8	$\dfrac{2.11 \text{ to } 2.73}{2.42^*}$	950 to 980
middle segment	1.2 to 0.8	$\dfrac{6.65 \text{ to } 14.5}{11.9^*}$	948 to 978
southern segment	1.6 to 2.0	$\dfrac{8.5 \text{ to } 26.1}{16.9^*}$	674 to 1020

* average methane content

Methane content of coal seams increases with increasing of metamorphic degree because the higher the coal metamorphic degree is, more methane is generated and the higher the methane sorption capacity of coal.

Burial depth is another factor affecting methane contents of coal seams. Methane contents of coal seams usually increase with depth. For example, in the Jiaozuo coal-bearing area of the Henan province where only anthracite occurs, the methane contents of the coal seams increase with depth except for the areas near faults (Fig. 3). The reason is that with increasing of depth (formation pressure), methane sorption abilities of coal seams increase and coal permeability decreases resulting in better preservation of methane. In numerous cases, methane contents are not linear with depths. Methane content gradients increase, then decrease and

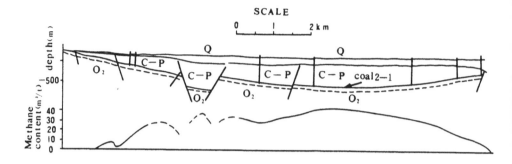

Figure 3   Methane content of the coal seam versus burial depth in the Jiaozuo coalfield.

finally trend to be zero with depths. In another word, methane contents increase little at deeper depths.

However, what is worthy of note is that it does not exist in all areas that methane contents of coal seams increase with depths. On contrast, methane contents of coal seams do decrease with depth in some areas. It was found that methane contents are related to the residual thickness between coal seam and unconformity above and have nothing to do with current coal seam depth (Fig. 4). The reason is that during geological times, coal seams were once elevated resulting in erosjon and mehtane release, then subsided again, but the coal seams did not reach depths at which more methane could be generated. Therefore the current methane contents of coal seams remained those before subsidence and were controlled by the residual thickness of the stratum covering coal seams before resubsidence.

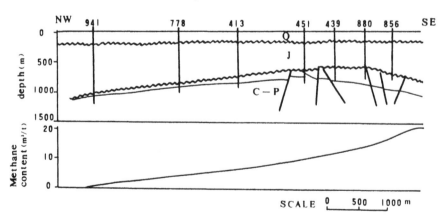

Figure 4   Relationship between methane content of the coal seam and burial depth in the Hongyang coal-field. The methane content is in direct proportion to the thickness between coal seam and unconformibil-ity, but in inverse proportion to burial depth.

Cap rocks are necessary for formation of conventional oil and gas reservoirs. Cap rocks are not necessary for formation of coalbed methane reservoirs but they do affect the methane contents of coal seams. Mehtane contents of coal seams are mainly controlled by formation pressure and lithotype because they are mainly composed of absorbed gas, but the methane diffusion never stops during geological time. The only difference is the diffusion rate. The diffusion rate is indirectly controlled by cap rock quality. As a result, the methane contents of coal seams covered by impermeable roof rocks usually are higher than those covered by permeable roof rocks (Tab. 2).

Table 2 Comparison of methane contents and roof and floor lithologies between the Fugu mine located at the northern segment of the Hedong coal-bearing area and the Longfeng mine at the Fushui coalfield, indicating the influence of cap rock on methane content of coal seams.

mines	ages of coal beds	coal rank	methane content m^3/t	roof rock	floor rock	depth m
Fugu	C−P	long-flame or gas coals	$\dfrac{2.11 \text{ to } 16.6}{2.42^*}$	mudstone or sandstone, <25 m thick	mudstone or sandstone, <19 m thick	950 to 980
Longfeng	E	long-flame coals	$\dfrac{8.5 \text{ to } 16.6}{10.35^*}$	tight oil shale, 50 to 190 m thick	tight basalt, >100 m thick	545 to 609

* average methane content

A lot of actual data indicated that faults had a great effect on methane contents of coal seams. The influence was related to the types and scales of faults. Normal faults often form ways through which gas escapes, resulting in decrease of methane contents of coal seams near the faults (Fig. 2).

Besides those mentioned above, other factors including macerals, moisture, inorganic materials, etc., have influences coalbed on methane contents as well.

To sum up, there are a number of factors affecting methane contents of coal seams but their effects are different. Of them, coal metamorphic degree is primary because it determines not only methane generation but also mehtane storage in coals. Burial depth, cap rock, and faults are secondary, which determine preservation of coalbed methane. They are crucial to methane content of coal seams if sufficient methane has been generated. Materials, moisture, inorganic materials, etc., have less influence on coalbed methane contents and they can not essentially determine the coalbed methane contents.

COALBED METHANE PRODUCIBILITY OF CHINA

The development prospect of coalbed methane depends, to large extent, on its producibility. Producibility mainly depends on permeability and isotherm adsorption of coal seams.

Because coal is pore-fractured reservoir, its permeability mainly depends on fractures and stress. The studies indicated that two types of fractures were developed in coal seams: cleats and tectonic fractures. Cleats formed during coalification were mainly related to lithotype and coal rank. Cleats density is highest in bright coals of $R° \approx 1.3\%$ (Zhang Shengli, 1995) (Fig. 5, Fig. 6). Therefore permeability of medium-rank coals is proposed to be high. Tectonic fractures were formed by tectonic stress, therefore they are controlled by local structures. In areas where cleats and tectonic fractures are overlapped, the permeability of coal seams should be high.

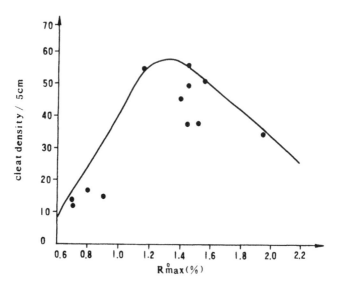

Fig. 5   Cleat density of vitrain versus vitrinite reflectance in the Hedong coal-bearing area.

Stress has a great effect on permeability of coals. Low stress is favorable for forming open fractures and high permeable reservoirs. On the contrary, high stress could close fracture and cause decreasing of coal permeability (Fig. 7). For example, the Huainan coal-bearing area in East China is a synclinorium with two great thrusts at the north and south boundaries, where the permeability of coals is very low, less than $1 \times 10^{-3} \mu m^3$, due to the high compressive stress. On the contrary, in the Liulin and Weibei areas of Ordos Basin, the permeability of the coals is high, maximally $20 \times 10^{-3} \mu m^3$ (Fig. 7), due to existence of extensive stress.

In addition, in the areas where gliging tectonics were developed along coal seams destroying origin textures of coal seams, the coal permeability usually is very low due to the disappear of the fractures.

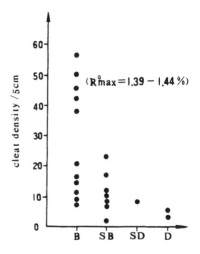

Figure 6   Cleats density of the different lithotypes in the Hedong coal-bearing area.

The case exists in the Xinggong coal-bearing area in the Henan province.

PROSPECTS FOR COALBED METHANE EXPLORATION IN CHINA BY THE RESULTS OF LIULIN COALBED METHANE PILOT SITE

The vast coalbed methane resource in China has attracted the government of China and a lot of native or overseas energy companies to invest in the coalbed methane development. More than 70 wells for coalbed methane have been completed at about 20 coalfields or areas in China (Fig. 1) and some encouraging and exciting results have been achieved from 1990 to 1995. The achievements made by North China Bureau of Petroleum Geology are most remarkable. Having conducted geological evaluation of the eight coal-bearing areas or basins

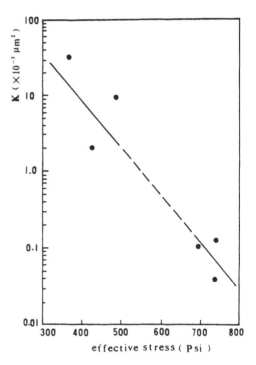

Figure 7   Coalbed permeability versus closurestress in the Liulin coalbed methane pilot site.

in the North China basin, North China Bureau of Petroleum Geology selected four areas to drill for coalbed methane and made a breakthrough at Liulin pilot site. Up to now, seven wells have been drilled for coalbed methane at the Liulin coalbed methane pilot site in the Shanxi province. At the Liulin pilot site, all of the seven wells have achieved commercial gas flows. The average depth of the target coal seams in the seven wells is only about 400m, while the average single well gas production is about 2000m³/d, maximally over 7000m³/d (Fig. 8).

The Liulin coalbed methane pilot site is located at the middle segment of Hedong coal-bearing area (Fig. 9). The Hedong coal-bearing area covering 16500 km² is located at the eastern margin of Ordos Basin. The coal-and coalbed methane--bearing formations occur in Carboniferous and Permain strata. The formations totally contain eight to sixteen coal beds that are 5 to 30 m thick, but only two to five coal beds are laterally continuous. The thickness of each coal bed is over 2 m. The coal ranks increase from northeast to southwest. Tectonic and structures are relatively simple and the coal beds are well preserved in the Hedong coal-bearing area.

The Liulin pilot site is located on the southern limb of the Lishi nose-like anticline at the middle segment of the Hedong coal-bearing area. The coal beds 4#, 5#, and 8# are the main coalbed methane targets, cumulatively 8 m to 10 m and individual 1.7 m to 5.2 m

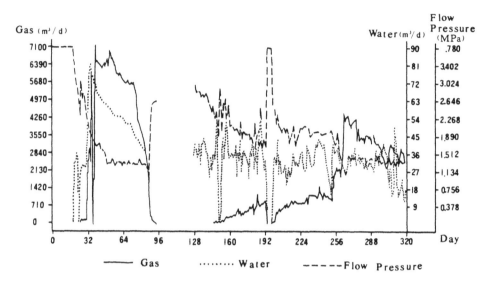

Figure 8   Flowing pressure versus gas and water production rates of a well at the Liulin coalbed methane pilot site.

thick. The coal beds are 343 m to 409 m deep, dipping west at angle of 3 to 8 degrees. The lithology is mainly composed of bright-and semi-bright coals with 10 to 20 percent of ash and 60 to 80 percent of vitrinite. Coke coal dominated and vitrinite reflectance value ranges from 1.4 to 1.7 percent. The methane content of coal seams ranges from $10m^3/t$ to $20m^3/t$ and the abundance of coalbed methane resource in place is up to $1.60 \times 10^8 m^3/km^2$. The gas saturation and sorption time of the coal seams are over 90 percent and 1 to 5 days respectively. The well testing permeability of the coal seams is maximally $12.1 \times 10^{-3} \mu m^2$.

The results of the Liulin coalbed methane pilot site indicated that China has coal-bearing areas of high

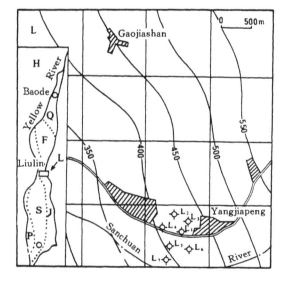

Figure 9   Structure map of the Liulin coalbed methane pilot site.

H: Hedong coal-bearing area; L: Liulin pilot site; Q: gas coal; F: fat coal; J: coke coal; S: lean coal; P: meager coal; 500: elevation (m) of coal #5 bottom; L1~L7: coalbed methane wells.

coalbed methane resources and producibility, which are geologically comparable to the San Juan basin and Black Warrior basin in USA.

It still has a long way to go to develop coalbed menthane in China. To explore for and develop a new kind of energy resource is not easy. It requires to do a lot of hard work. The advances on coalbed methane exploration and development of China represented by the Liulin pilot site are exciting and encouraging and are a good beginning but they are the results of great efforts for many years. The geological conditions of China are complex. As conventional oil and gas, coalbed methane exploration and development have to go through a course of understanding-practicing-reunderstanding-repracticing. It can not be finished in one move.

CONCLUSIONS

The vast coalbed methane resource has set up a solid foundation for coalbed methane development in China. The successes achieved to date in coalbed methane development in China have demonstrated that China has good prospects for coalbed methane development. It is proposed that coalbed methane industry will fundamentally take shape and result in good social and economical effects with the advent of the 21st century.

REFERENCES

1. J. C. Close. Natural fractures in bituminous coal gas reservoirs. GRI Report 91/0337 (1991).
2. G. E. Eddy, C. T. Rightmire, C. Boyer. Relationship of methane content to coal rank and depth. SPE 10800 (1982).
3. Feng Fukai et al. Study and evaluation of natural gas (including coalbed methane). Scientific Report (1990).
4. Hu Tianyu, Li Runling and others. Prospect prediction of coal resources in China. Geological publishing house (1995).
5. Rightmire C. T., Eddy G. E., Kirr J. N.. Coalbed methane resources of the United States. AAPG Studies, in geology 17 (1984).
6. Zhang Shengli and Li Baofang. Mechanism of coalbed cleat and its significance in potential evaluation of coalbed methane exploration and development. Coal geology of China. 8, 72-77 (1996).
7. Zhang Shengli and Li Baofang. Occurrence and affective geological factors of coalbed gas of the Carboniferous-Permian in the eastern margin of Ordos. Experimental Petroleum Geology. 18, 182-189 (1996).
8. Zhang Xinming et al. Coalbed Methane in China. Shaanxi Scientific Press (1991).

Proc. 30ᵗʰ Int' l. Geol. Congr. , Vol. 18, pp. 143～153
Sun Z. C. *et al.* (Eds)

Petroleum Geological Features of Tertiary Terrestrial Lunpola Basin, Tibet

LEI QINGLIANG   FU XIAOYUE   LU YAPING

Central-South Bureau of Petroleum Geology, MGMR, Changsha 410007,

CHINA

INTRODUCTION

Lunpola basin is located in Bange county, Xizang autonomous region with the area of 3600 km² and elevation of more than 4500 m. It is one of the terrestrial basins developed in Tertiary with relatively favorable petroleum geological conditions. The commercial oil flow has been gotten from it primarilly.

Lupola basin is a early Tertiary terrestrial basin developing on the folded basement of Yanshan cycle. Its formation, evolution are relative to the movement pattern of Banggong-Lujiang great fault. Due to the dextral wrench movement in early Tertiary, extensional stress occurred to the NEN-SWS direction in the region and then a strike-slip extensional basin came into being. There developed Eocene Niubao formation and Oligocene Dingqing formation in the basin, respectively representing two sedimentary evolution stages during which the rifts changed into depressions. In Niubao formation, there developed three sub-sequences, primarily reflecting the course of rifts evolution from its early stage through stably developing stage to shrivaling stage. during the period of depression, although Dingqing formation can also be divided into two sub-seqences, the whole difference and cyclicity of the lithology are relatively unobvious. The two sub-sequences are all made of half-deep to deep lake dark mudstone, reflecting the starving sedimentary regime on the background of the subsidence rate being faster than the deposition rate(Fig. 1).

The current structural framework of Lunpola basin is charaterized by the belts which are divided in the south-north direction and patches in the east-west direction (Fig. 2). Their formation is not only related to the sedimentary and structural conditions of the original basin, but to strong deformation. The current exploration targets are mainly concentrated in the central-east region (Jiangriaco-paco sub-depression) with fairly favorable oil-generating geological conditions and oil and gas accumulation belts founded mainly occur in the north of the region. The present data suggest that the geological features and the oil and gas generating laws of Lunpola basin are closely related to the properties of strike-slip-extension possessed by its own.

Stratum System				Thickness	Lithology	Seqence Division		Sedimentary System
System	Series	Formation	Member	(m)		Grade1	Grade2	
Quaternary								
Lower Tertiary	Oligocene	Dingqinghu Formation	2nd~3rd Member	800 | 1000		I	I — 2	Deep lake sedimentary system interlayered with subaquatic fan
			1st Member	200 | 400			I — 1	Half-deep lake sedimentary system interlayered with subaquatic fan
	Eocene	Niubao Formation	3rd Member	700 | 1100		I	I — 3	Beach-lake shallow lake sedimentary system interlayered with riverine sediments
			2nd Member	800 | 1200			I — 2	Beach-shallow-lake ~half-deep lake sedimentary system interlayered with storm rocks
			1st Member	>400			I — 1	Terresterial~lake-bank-lashing-styled sedimentary system

Fig. 1   Stratigraphic sequence and sedimentary features of Lunpola Basin

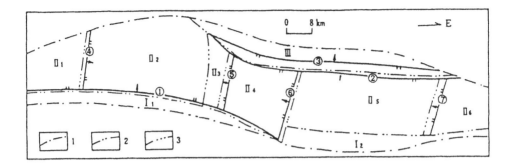

Fig. 2   Division of structural units of Lunpola Basin

1. Division of the grade 1 belt; 2. Divisions of the grade 2 belt; 3. Divisions of the grade 3 belt; I. Southern thrust-uplift belt; I₁. Dingqing thrust belt; I₂. Changshan uplift belt; II. Central depression belt; II₁. Shanbare depression; II₂. Jiangriaco depression; II₃. Tonglulongguo transform belt; II₄. Jiangjiaoco depression; II₅. Paco depression; III. Hongxingliang overthrust-napping structural belt; ① Changshan normal fault; ② Hongxingliang normal fault; ③ Dayushan overthrust fault; ④ Jiangriagongyu transform fault; ⑤ Dingkashen transform fault; ⑥ Papa transform fault; ⑦ Luqinghai transform fault

ANALYSIS ON PETROLEUM GEOLOGICAL FEATURES

"Pod-shaped" basin and the distribution of oil-generating depression

Lunpola basin is a extensional rift basin with the feature of strike-slip (Fig. 2). It takes the shape of pod and belongs to the basic pattern of strike-slip basin whose formation is related to branching of strike-slip fault system[1,2]. According to the analysis of the boundary stress which resulted in the "pod-shaped" basin (Fig. 3), there mainly developed two inclined corresponding rifting sedimentary regions in Lunpola basin: one is Jiangriao sub-depression belonging to the pattern of south-faulting-north overlapping and the other Paco sub-depression to the pattern of south-faulting-north-overlapping and Paco to the pattern of north-faulting-south-overlapping. Between the two sub-depressions, there formed a bi-faulting graben sub-depression of Jiangjiaco(Fig. 4). In addition to that, because of the restraint of "pod shaped" basin, the strata are sticking up toward the east and west and the thickness is gradually decrease along the basin axis.

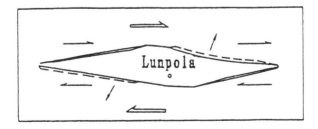

Fig. 3   Schematic diagram of stress analysis of Lunpola Basin

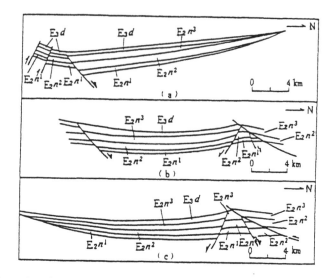

Fig. 4   Schematic section of sedimentary and structural framework of Lunpola Basin

a. Pattern of south-faulting-northlapping(Jiangriaco sub-depression); b. Pattern of bi-faulting graben(Jiangjiaco sub-depression); c. Pattern of north-faulting-south-overlapping(Paco sub-depression)

According to the data of Hongwei No. 1 well in the east basin, Niubao formation are the coast-swamp and riverine sediments with coarse granulity and red colour, belong to basin-edge sediment.

Due to the controlling of "pod-shaped" strike-slip-extension basin, oil and gas generating conditions in the basin have two significant features: One is that the oil-generating depressions occur mainly in three sedimentary sub-depressions of Jiangriaco, Jiangjiaco, Paco, Only in Niubao formation developed the source rocks oil-generating stratum more than 1000m. The other is that due to the differential boundary conditions which control the depressions, the inner structural framework of three sub-depression are fairly various, thus there formed three independent oil-generating units (Fig. 4), among which Jiangriaco is a main oil-generating sub-depression as well as a depression with the deepest rift and the thickest sedimentary strata such as Niubao No. 2 member up to more than 1700m. The inner structure of Jiangriaco sub-depression may be divided furtherly into three structural units of steep belt、depression belt and gentle slope belt. Basing on petroleum results we have obtained currently, the north gentle slope belt of the sub-depression is the main zone to accumulate hydrocarbon. The inner structure of Jiangjiaoco sub-depression is relatively simple. Affecting by the bi-faults, the oil-generating depression is mainly located at the central part of the sub-depression and the thickness of source rocks is thinner than that of Jiangriaco sub-depression. Oil and gas shows mainly concentrated in the zone from the north steep belt to Honggxinglang overthrust nappe belt. The inner geological structure of Paco sub-depression are primarily similar to that of Jiangriaco, but the distribution of oil-generating depression or boundary fault controlling the depression are opposite to it, belonging to the pattern of north-faulting-south-overlapping. Oil-generating depression occurs in the north and part of it has been hidden under the nappe (Fig. 5). This sub-depression is the secondary principle oil-generating depression in the basin. The thickness of source rocks of Niubao formation is about 700-900m, underlying a bed of mudstone, shale of Dingqing formation about 900m. According to the analysis of geological burial history of the basin, the primary sedimentary thickness of Dingqinghu formation is 1500-1700m. Due to the fairly high temperature gradient (up to about 5.5-6 ℃/100m), there still developed a bed of efficient source rocks more than 400m in the formation.

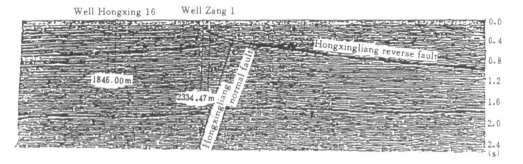

Fig. 5   Seismic profile of L_1-93-16Z_1

The conditions for oil and gas preservation of Paco sub-depression are comparatively superior to the two others. The first reason is that Dingqinghu formation developed well. In addition to the oil and gas generating conditions of its own, a cap formation has been formed regionally, which is very important for Lunpola basin which has undertaken serious uplift as well as intense erosion. The second is that part of oil-generating depression of the sub-depression has been hidden under the overthrust nappe, so, although the late-term structural deformation is great voilent, the basin primary body still possesses fairly good preserving conditions.

"Late coming and early leaving" double-layered structure and two petroleum systems
Lunpola basin formed under the background of strike-slip fault still reflects the features of double-layered geological structure derived from the results of changes of fault into depression, which are possessed by patterns of terrestrial petroliferous basins of the most of rifting-valleys in the east China. Since the rifting of Lunpola basin are mainly formed in Eocene of early Tertiary, in the terms of forming times, it is significantly later than the terrestrial petroliferous basins in the east China; On the contrary, during the course of Lunpola basin transforming from rifts into depression, Dingqinghu formation which representing the sediments of the stage abruptly paused or disappeared only after there developed mainly the half-deep lake to deep lake dark mudstone, shale. The sediments of the shallow water enviroments are obvious absent in addition to that no upper Tertiary in the basin can be found,

thus, it is also suggested that the basin possesses the character of early-ending-deposition and quick-returning.

The short history which Lunpola basin has undertaken during the course of its generation and development, reflecting by "late coming and early leaving" double layered structure, is related, on one side, to the regional geological background of Chinese Tethys sea[3,4], on the other side to the character of strike-slip-extension of Lunpola basin.

The differential geological structure of hydrocarbon basins can result in differential effects on oil and gas controlling, which is a prevading feature of terrestrial hydrocarbon basin in China[5], Lunpola basin is not an exception too. Only because of being affected by "late coming and early leaving" double layered structure, it appears more voilent and obvious and therefore there developed two significant differential petroleum systems, namely lower one represented by Niubao formation—the sediments deposited during the course of rifting and the upper one represented by Dingqing formation—the sediments deposited during the course of depression.

As a whole (Tab1), the lower petroleum system possesses the features of thick efficient source rocks, favorable matches of source rocks, reservoirs, caprocks, high resistivity and favour entrapment, superimposed pools etc, while the upper petroleum system possesses short deposition stage with shallow burrial and absence of matching condition of the overlying stratum in addition to violent erosion in late stage, resulting in active groundwater and relative intense cold morphomism, therefore, the whole hydrocarbon potential is inferior to that in the lower petroleum system.

Three sub-sequences developing in Niubao formation—in the lower petroleum system represent respectively three sedimentary evolution terms from early stage of rifting through stably developing stage to shrivaling stage during the period of rifting, correspondingly reflecting the sedimentary features of coarse-thin-coarse, red-dark-red on the terms of lithology and color. Basing on the views of low water level, lake-transgressing level, high water level of sequence stratigraphy, there are two points we can recognize: (1) The second member of Niubao formation representing the sediments deposited during the stage of lake-transgressing-level stably developing term is the main developed interval of the source rocks in lower petroleum system with the most sedimentary thickness of more than 1600m. It mainly consists of dark-grey mudstone, shale interlayered with thin siltstone, containing residual organic matter generally over 0.5%, belonging to kerogen 1. There are many crude oil shows in the breaks of mudstone, reflecting that the interval possesses relatively great hydrocarbon potential. (2) Sandstone reservoir can be divided into three types. ① Beach delta sandstone with the burial of 900-1600m, located in the lower part of the third part of Niubao formation, belonging to high water level system ② Lake-beach sandstone interlayered in dark-grey mudstone、shale of the second member of Niubao formation, belonging to lake-transgressing system, with the burial of 1200-2300m; ③ Channel sandstone, fan-delta sandstone and so on, located at the bottom of the second member and in the upper part of the first member of

Table 1 Comparion of oil-forming geological conditions between the upper and lower petroleum system

Patroleum system	Effective source rocks				Pattern of accumulation and combination					
	Thickness (cm)	C %	Kerogen	$R°$ %	Types of Reservoir	combination of Source rocks reservolrs, cap rocks	Hydrocarbon-Draining combinations	Capturing patterns	combination of pool-forming	Oil & gas shows
Upper petroleum system (Dingqinghu Formation)	200-300	0.96	I	0.6	Sub-aquatic fan	Lateral changes-fault	Dominated by vertical plus lateral	Low resistivity	Single	Dominated by bitumen
Lower petroleum system (Niubao Formation)	1000-1500	0.46-0.49	I	0.74-1.14	Rivershore, delta, coastal barrier	Interlay-fault	Dominated by lateral plus vertical	High resistivity	Overlapping	Dominated by oil

Niubao formation, with the burial of more than 2300m, belonging to lower water level system. Among the three types of reservoirs, the hydrocarbon potential of the first is limited due to the shallow burial, low mineralized degree of strata-water, low content of Cl^- (only up to 1200×10^{-6}, belonging to oil-bearing water layer by tests, although it possesses thick reservoir of more than 5m with fairly good physical property. The mineralization of the strata-water of the second type obviously increase with the general content of Cl^- of 5000×10^{-6} suggesting that the reservoir possesses relative favorable sealing condition. It belongs to oil-water layer and water-bearing oil layer according to the evaluations on test. The most visible thickness is about 2m, generally less than 1m. From the side of efficient thickness, the reservoir is thin and its hydrocarbon potential is not optimistic. The hydrocarbon potential of the third one is not proved by drills up to now. According to regional data, the reservoir is relative great thick, with coarse lithology. The overlying dark-grey mudstone, shale of the second member of Niubao formation is not only the main oil generating formation but fairly favorable regional cap formation. In addition to that, the burial of reservoir is relative deep without any unfavorable factors due to the massive uplifts of Qingzhang plateau, so its hydrocarbon potential and producibility can not be underestimated. In the future, to deepen the drilled well to reveal the hydrocarbon potential of the first member of Niubao formation should be a direction of exploration for Lunpola basin to get significant oil and gas breakthrough.

Petroleum geological significance of the transform structure in weak deformed domain (central depressed belt)

The transform structure which is distributed in the NEN and NWN direction close to the vertical axis of the basin and developed in the week deformed domain, namely the central depressed belt of Lunpola basin (Fig. 2), is a kind of important structural style of the basin, which takes the role of controlling the division of the sub-structural units and the distribution of the sediments and hydrocarbon in the basin to certain degree, the reason for the formation of the transform structures in Lunpola basin are related to the properties of strike-slip-extension of its own. On one side, N-S extensional boundary (or the normal fault controlling the basin) of the basin distributes in oblique-symmetric direction (en echelon arrangement), its end and overlapping area need adjusting by lateral structures newly produced due to the changes of stress conditions; On the other side, it's related to the variation of intensity of extension along the trend. The established data suggest that the intensity of the north extension decreases gradually from the east to the west while the south one does the same from the west to the east. The heterogeneous regime of the stress field need balancing laterally to form transform fault zone, just like first factor.

The transform structures of Lunpola basin mainly developed in the middle, late Eocene, dominated by transform faults such as Tanglulongguo conjugate-convergent-overlapping fault. They have played great role of controlling the distribution and thickness of the strata

(Fig. 6,7) as well as the third grade structural units in the basin as the boundary fault and structural belt (Fig. 2). The recognizition of transform structures of Lunpola basin are that the faults have been considered as late-term normal faults for a long time. The central depressed belt is the secondary structural unit possessing relatively favorable oil and gas preserving conditions on the views of geological evaluation. Under the background of unfavorable traps and unclear reservoir distribution laws, clear technical thoughts of how to analyze possible favorable belt for oil and gas accumulation and how to choose suitable way to carry out geological evaluation are not still formed. As the development of seimic resolving power, a new conclusion has been reached gradually that the transverse faults close to N-S direction has been formed during the stage of rifting, taking the role of controlling the strata thickness of both flanks of faults while the ratio of sand/mud in the strata along the extending direction of fault is higher than that in other area in the central depressed belt (non-transform structural area) (e. g. Niu No. 4 well), reflecting that the fault plays the role of controlling the source of material, it's suggest that the area are the main area for lateral river development.

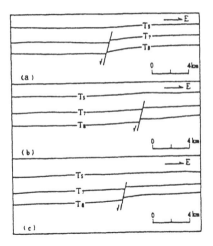

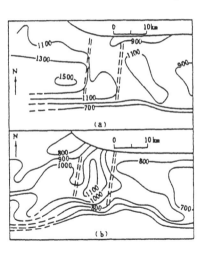

Fig. 6   Cross-section of transform faults of Lunpola Basin

a. Tonglulonggao transform fault; b. Dingkashen transform fault; c. Papa transform fault

Fig. 7   Isopach map of transform structure of Lunpola Basin

a. Isopach map of $T_5 - T_7 (E_2 n^3)$; b. Isopach map of $T_7 - T_8 (E_2 n^2)$

According to the generallizition of the above, the transform faults or belts developing in the central depressed belt in Lunpola basin, playing the role of adujusting laterally, are a king of important contemporaneous structures formed during the course of the development of primary basins and have very important meanings in petroleum geology. The authors consider with reasons that Tanglulongguo、 Dingkasheng and Paco transform faults in the central depression are three favorable structures for controlling oil and gas. They are mainly located in oil-generating depression and between the three secondary oil-generating depression and possess great favorable structural, depositional, oil and gas supplying and reservoir conditions. In the transform belts, there developed three important types of traps: ① "skylight-styled" trap developing in the jointing part of the normal fault controlling the baisn or depression and transform structure; ② narrow range drape trap developing on the transform structure; ③ lithological trap. All those should be payed close attention to in future work.

Deformation of the basin evaluation on and favorable petroleum zones

Since Eocene, deformation with obvious evidences of Lunpola basin have mainly undergone two terms.

At the end of Eocene, namely the late term of rifting the whole area underwent a weak heterogeneous rising and falling strucutural movement (equal to the second term of Himalaya movement). The whole expression forms of the movement are the rising in the north and falling in the south, Narrow range uplift in the north basin (to the north of seimic profile LE-91-47) results in broard fold. The third member of Niubao formation is partially eroded and the relationship denuded contact between the third member of Niubao formation and overlying Dingqinghu formation can be seen in the profile. The south part of the basin subsided relatively. The sedimentary thickness is generally greater than that of the north part for more 200m and the relationship between the upper and lower strata is conformity. The great petroleum geological significance of the structural movement is to accelarate the conversion of hydrocarbon-generating regime into hydrocarbon-draining regime. The regional direction of hydrocarbon migration moves toward the north uplifting belt. The broad anticline developing contemporanously with the uplifting belt also formed timely traps and made a foundation for current main oil and gas accumulating belts and trap patterns.

The third term structural movement of Himalaya after Oligocene is the most violent deformation which Lunpola basin underwent. The structural movement resulted in the structural framework of overthrust belt in the north, thrust and uplift belt in the south and affected greatly the re-allocation and preservation of early oil and gas. It mainly expresses such like these: On one side, the south belt is dominated by the style of folded-basement and the north by detached cap in due to the influence of deformed styles therefore, the conditions for pool-forming in the north belt is better than that in the south as the real data say. On the other side, the differential structural associations of south, north belts result in varions petroleum geological significances. The north belt belongs to the type of overlapping complex structural associations, forming vertically the structure of "three story-styled" (Fig. 8) while the south belt is the structural association of fault nose and fault uplift whose relationship is juxaposition and the opening degree of fault in the south belt is higher than that in the north, so, on the whole, the geological conditions of the north for pool-forming is superior to that of the south belt.

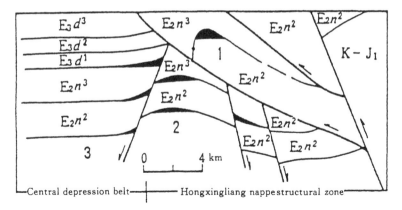

Fig. 8   Schematic diagram of reservoir combinations of complex structure area in the northen Lunpola Basin

1. Anticline pools of the upper side of the overthrusts; 2. Block-styled pools of the lower side of the overthrusts; 3. Nose fault pools of the downthrow of the normal faults

One point which need pointing out is that oil and gas preserving conditions of the central depressed belt, being as a weak deformed area, are better than those of south, north belt because the central depressed belt, being under the background wholly of compressional structure is absence of late-term extensional faults. The key of oil and gas exploration direction is to catch transform structure to solve the problems of absence of the development of reservoirs, structures. But, according to the data obtained from the three belts and the petroleum geological conditions of them, the north belt of Lunpola basin is the initial target for oil and gas exploration and petroleum fruits can be easierly reached.

REFERENCE

1. Qu Jiawei. Some major problems of strike-slip faulting. Earth Science Frontiers, 1995,2(2):132-133

2. Niu Hepu. Classification of Earth Dynamics of Sedimentary Basins and Analysis on Structural Patterns. Earth Science-Journal of China University of Geosciences, 1993,18(6): 716-717

3. Zhu Zhanxiang. Rock Stratum and Event Stratum of Tibetan Tertiary Period. Tibet Geology, 1993,9 (1): 20-26

4. Huang Jiqing, Chen Bingwei. The Evolution of the Tethys in China and Adjacent Regions. Geological Publishing House, 1987, 55-56

5. Yan Kesheng. A Way to Find Oil in Tarim Basin by Structural Analysis of China Sedimentary Basin, 1987,55-56

6. Hu Wangshui. Transform Structure in Rifting Valley Basin and Its Geological Significance. Oil & Gas Prospecting Aboard, 1994, 6(2): 145-154

7. Hu Wangshui, Wang Xiepie. Transform Structures in Northern Part of Songliao Basin and Its Significance in Petroleum Geology. Oil & Gas Geology, 1995, 15(2): 164-172

Proc. 30ᵗʰ Int' l. Geol. Congr., Vol. 18, pp. 155~166
Sun Z. C. *et al.* (Eds)

Assessment of Oil & gas Resources by 3-D Basin Modeling of Tarim Basin, China

WANG YINGMIN, ZHAO GUANJUN, DENG LIN, LIU LING,
DONG WEI, JIANG JIANPING and LUO LIMIN
Chengdu Institute of Technology, Chengdu, 610059, P. R. CHINA

Abstract

Tarim Basin, the largest oil and gas basin in China, are a superposed basin with multi-cycles. Most of its oil and gas resource are from the lower Paleozoic group, which has the nature of residual basins that have undergone strong modification by multiphase tectonic activities. To evaluate the oil and gas resources in Tarim Basin, a software system for three-dimensional modeling of oil and gas basins has developed. The influence of tectonic movements on the generation and migration of hydrocarbon in Tarim Basin has summarized, and the amount of oil and gas lost in tectonic activities has calculated. Thus, the existing resource of oil and gas in Tarim Basin has figured out by deducting the loss of oil and gas in the movements.

Keywords: 3-D basin modeling, Tarim Basin

Tarim Basin, located in the West of China, is the largest Chinese oil and gas basin. It has an area of 560,000 km² and sediments of 15,000 m in thickness that range in age from Cambrian to Quaternary. Much hope has held out for sizable oil resources in Tarim Basin, but achievements in search of oil and gas so far are much less than expectations. Some geologists have lost their confidence in exploration prospects for the basin. It is difficult for geologists to know the extent of oil and gas resources in Tarim Basin and what the exploration prospects are.

The cause of errors in appraisal of the exploration situation is the complexity of geological and hydrocarbon evaluations. It is now recognized that Tarim Basin is a superposed basin with multicycles (Fig. 1). Unconformity in Tarim Basin developed markedly in early Hercynian movement (latest Devonian), late Hercynian (latest Permian), and late Yanshanian (latest Cretaceous). The greatest erosion occurred in early Hercynian, with a maximum removal of 4,500 m. The erosion in late Hercynian removed with a maximum of 2,400 m. In late Yanshanian, the maximum eroded thickness was 2,400 m. The main erosion events developed in different parts in the basin, as one strong, another weak, just like a see-saw. The erosion in early Hercynian was strong in the southwest part of the basin, and weak in the

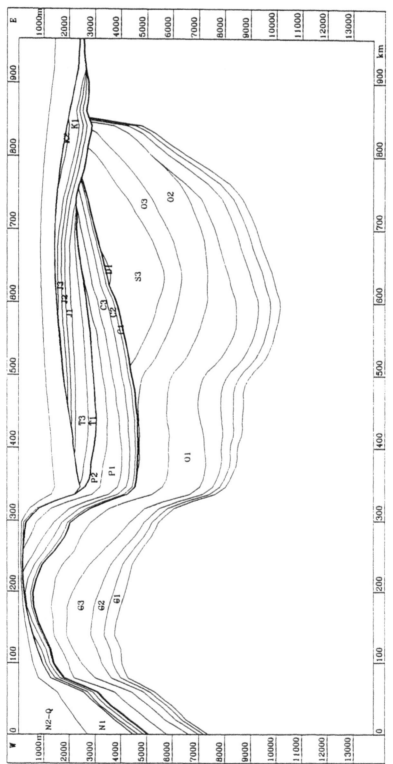

Fig. 1   Structure profile of Tarim Basin, show the superposition of unconformities.

northeast part. In late Hercynian, the erosion was strong in the northeast part, and weak in southwest part. In late Yanshanian, erosion was again strong in the southwest part and weak in the northeast part. At many parts of the basin, formations have been denuded many times, which resulted in the superposition of multiphase unconformity. As shown in figure 1, the early Hercynian unconformity superposed that of late Hercynian in the East of the section, and Jurassic strata came into contact with Ordovician and Cambrian strata forward the East.

The resource rock were mainly in Ordovician and Cambrian formations and them underwent strong multiphase tectonic activities, so the Paleozoic basin has the nature of residual basin. The histories of structure, sedimentation, geotemperature and reservoir development, and of the generation, migration and accumulation of hydrocarbon in the residual basin are much more complex than those of primary basins. Considerable amounts of oil and gas scattered that has generated early in the residual basin. Large amounts of hydrocarbons migrated from destroyed pools to secondary traps producing new pools. Because of geological heterogeneity, some primary deposits of hydrocarbons still exist. In addition, the remaining source rocks generated hydrocarbons again after tectonic movements, so that pools filled with late generated hydrocarbons developed in old formation. Thus, Tarim Basin has the character of multiphase generation, migration, scatter and accumulation of hydrocarbon.

How much did hydrocarbon generate, migrate, and accumulate before each tectonic movement? How much hydrocarbon remained? How much did hydrocarbon generate, migrate, and accumulate again after tectonic movements? These are questions for the resource evaluation of oil and gas in a basin with a complex of process evolution. All of the geological processes requre great attention for the reasons. Geological processes are the result of interactions of a number of geological factors. Characteristics of a geological process can be determined only through systematic study of the interaction of structure, sedimentation, reservoir, geotemperature, and the generation, expulsion, and migration of hydrocarbon. The method of basin modeling is key technology to reach this aim.

OVERVIEW OF THE 3-D BASIN MODELING SYSTEM

We developed a three-dimensions basin modeling software system for the purpose of mentioned above. The influence of tectonic movements on the generation and accumulation of hydrocarbon in Tarim Basin is summarized, and the amount of oil and gas lost in the movements is calculated. Thus, the extent of existing oil and gas resources in Tarim Basin is figured out by deducting the loss of oil and gas in the movements.

Our basin modeling system is mainly composed of 5 parts. These are database of basin modeling, basin description, basin modeling, and assessment of oil and gas resource (Fig. 2). All the models are three dimensional except the simulation of oil and gas migration, which is two dimensional.

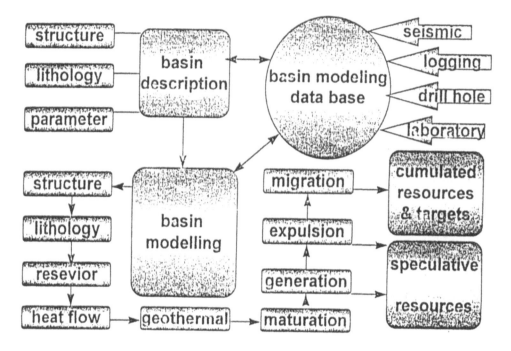

Fig. 2   Flow chart of 3-D basin modeling.

The studied area is 1000 km long in the E-W direction and 520 km wide in the N-S direction and includes almost all the structure units of Tarim Basin. All geological formations from Cambrian to Quaternary are included in the model and the maximum depth studied reaches 14,000 m. There are 450,500 cells in a three-dimensional data unit. The value of each cell is 20 km (E-W)×10 km (N-S)×80 m.

We have developed some new methods to meet the geological character of Tarim Basin in the software system. The method of interactive restoration of structural superposition surfaces can be used to describe the structures, to calculate the denuded thickness by stage, and to simulate the processes of structural evolution, under the condition of superposition of multi-phase unconformity. On the modeling of reservoir evolution we considered the controlling factors of compaction, secondary porosity and the dissolution near surface synthetically. We approved a nonlinear deductive model to deduce paleo-heat flow by $R°$. The methods of paleo-temperature simulation was by an equation of heat conduction and convection in three dimensions. In light of the simultaneous production of oil and gas and strong degradation from oil to gas and from gas to carbon, we proposed a model of the cumulative amount of hydrocarbon generated in a three-dimensional cell. The stage oil generated amounts, stage gas generated amounts, stage oil-gas-degraded amount and stage gas-carbon degraded amount can be calculated by use of the method. A model of multiphase and multi-component hydrocarbon expulsion has proposed to account for the complicated relations among the phases or compositions of kerogen, oil, gas, water and carbonized bitumen in the highly evolved basin. On the simulation of hydrocarbon migration, we proposed that the whole basin was a unified filtration system. Capillary pressure is taken as the key factor to drive hydrocarbon expulsion from source rock to reservoir and to prevent hydrocarbon from escaping from the reservoir. The method of migration, accumulation, and escape of hydrocarbon in multiphase and multi-component filtration of oil, gas and water is approved. At last, a method of calculating the loss of oil and gas resources denuded in each tectonic movement under conditions of multi-phase unconformity has developed. This method was also used in calculating the present area extent of oil and gas resources.

A basin description and simulation have been made in Tarim Basin by use of this software system. A wealth of three-dimensional data units has been acquired from structure history to hydrocarbon expulsion in Tarim Basin. The effect of the principal tectonic activities on hy-drocarbon evolution is discussed briefly below.

RELATIONSHIP BETWEEN TECTONIC ACTIVITY AND EVOLUTION OF HYDROCAR-BON

Sedimentary layers with capacity for hydrocarbon generation were deposited from Cambrian to Tertiary in Tarim Basin, and among these are both marine and continental deposits. Principal source layers with marine facies include: Cambrian-Ordovician and Carboniferous-Permian formations, while those with continental facies are mainly Triassic-Jurassic formations. In addition, the dark shale at the bottom of the Silurian in Manjaer Depression is also considered as possible source rocks. Phased generation of oil and gas in each unit of formation and different geological periods are calculated from the model. Figure 3 shows oil-generating rates for different strata in Tarim Basin in different geological periods.

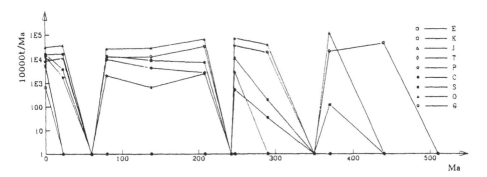

Fig. 3   Phased oil generating rates of Tarim Basin

Four main periods of hydrocarbon generation are distinguished, the early Hercynian uncon-formity, the late Hercynian unconformity and the late Yanshanian unconformity. They were taken as boundaries of the Caledonian Period (initial stage of Hercynian period is also involved), Hercynian Period, India-China Period, Yanshanian Period and Himalayan Period. Hydrocarbon generation ceased at the time of unconformity development, between each two adjacent period of oil generation. Figure 4 shows that the maximum rates of oil generation in the four periods decrease from the old to the new while the layers of source rocks increase gradually.

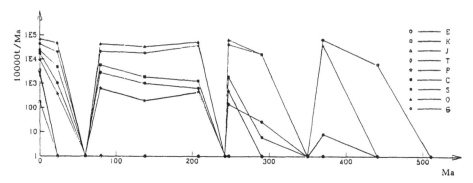

Fig. 4   Phased gas generating rates of Tarim Basin

In the Caledonian period, the amount of oil generated from Cambrian and Ordovician source rocks were dominant and Silurian source rocks started to generate oil. In the Hercynian period, the amount of oil generated from Ordovician and Silurian source rocks were dominant, while the oil-generating rate of Cambrian source rocks decreased by an order of magnitude. Carboniferous formation began to generate oil at that time.

In the India-China and Yanshanian, Ordovician, Silurian, and Carboniferous source rocks were dominant in oil generation. Permian source rocks started to generate oil at the beginning of this period, and reached the peak of oil generation at the end of the Yanshanian. The oil-generating rate of Cambrian source rocks further decreased and that of Ordovician source rocks tended to decrease. The oil-generating rate of Ordovician source rocks reached its peak in the early India-China period, but was less than the rate of Silurian and Carboniferous source rocks in the late Yanshanian.

In the Himalayan period, the oil-generating rate of Cambrian-Permian source rocks gradually increased again. They reached higher values than at the end of the Yanshanian as the uplift of the basin was turned into fast subsidence. Triassic, Jurassic, Cretaceous and Eocene source rocks began to generate oil.

The amount of oil generation from Carboniferous, Silurian, Eocene and Permian source rocks was dominant, whereas the rates of Jurassic and Cretaceous were very low.

The peak of oil generation from Cambrian source rocks occurred at the end of the Ordovician era, while that of Ordovician source rocks occurred at the end of the Devonian era. The maximum amount of oil generation of Silurian source rocks appeared at the end of the Permian era. Carboniferous and Permian source rocks nearly reached the peak of their oil generation at the end of the Cretaceous era. However, the maximum oil generation of Carboniferous,

Permian, Triassic, Jurassic, Cretaceous and Eocene source rocks all occurs at present. It is easily found that the Ordovician source rocks are most important source rocks in the basins. Figure 4 shows generating rate for different source rocks in Tarim Basin in different geological periods. This evolution is similar to that of oil generating formation. Four main periods of gas generation can be distinguished in the same way.

The maximum rate of gas generation from Cambrian source rocks occurred at the end of the Devonian era, while that from Ordovician source rocks occurred at the ends of the Devonian and Permian eras. The rate of gas-generation from Silurian source rocks nearly reached its maximum value at the end of the Permian era and tended to increase gradually therefor, and its maximum amount occurs at present. the rate of gas generation from Carboniferous and Permian source rocks almost reached its peak at the end of the Cretaceous era. The maximum rate of gas generation of Carboniferous, Permian, Triassic, Jurassic, Cretaceous, and Eocene source rocks occurs at present.

RELATIONSHIP BETWEEN TECTONIC ACTIVITY AND EXPULSION OF HYDROCARBON

Parameters of hydrocarbon expulsion were calculated after such mechanisms as hydrocarbon expulsion by compacting, phase transformation of oil and gas and degasification of rising water. Figure 5 shows the rate of cumulative oil expulsion. It can be seen that the rate of oil expulsion from Ordovician and Cambrian source rocks was the highest. They quickly reached their maximum value (that of Cambrian at the latest Ordovician era, and that of Orcovician at the latest Devonian era, both in Caledonian period). The rate of Carboniferous, Silurian and Permian is the second, but tens of times less than that of Cambrian and Ordovician. Not until late Hercynian Period did Carboniferous, Silurian and Permian source rocks reach their peaks of oil expulsion. The rate of Triassic and newer source rocks is still less. They did not start to expel oil in large amounts until the Himalayan period.

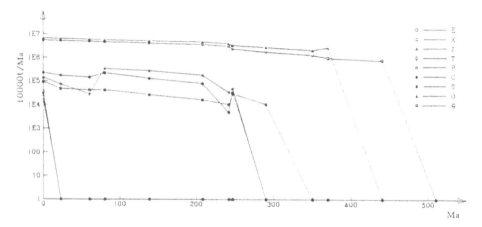

Fig. 5 Cumulative oil expulsion rate of Tarim Basin

The curves of the cumulative oil expulsion rate from Cambrian to Permian source rocks are commonly characterized by a short time interval from beginning to the peak. After the perk, the source rocks underwent several tectonic episodes and sustained strong erosion, and the generation of oil basically ceased when the source rocks were uplifted to sustain erosion. The curves of the rate of cumulative oil expulsion tend to rise gradually, however, which indicates that phased amounts of oil expulsion at later stages were sufficient enough to make up the loss during uplift.

Figure 6 illustrates this phenomenon more clearly. In periods of the basin uplift, the process of oil expulsion did not stop. The rate of oil expulsion decreased only slightly, though new oil generation and further compacting did not occur at the same time. Why did this occur? The volume of gas dissolved in the oil expanded in the period of basin uplift, and thus the saturation of oil increased. So oil could exude from source rocks without an increase in oil amount. From the point of view of dynamics, volume expansion of gas is also a force driving movement of hydrocarbon.

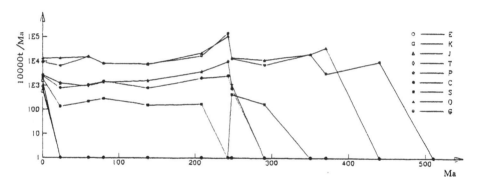

Fig. 6   Phased oil expulsion rates of Tarim Basin

Figure 7 shows the rate of cumulative gas expulsion. Rate of gas expulsion from Cambrian and Ordovician source rocks was the highest, and reached its maximum at the end of the Devonian era. The rate of Silurian, Carboniferous, and Permian source rocks was the second and tens of times less than that of Cambrian and Ordovician source rocks. Not until late Yanshanian period did Silurian, Carboniferous and Permian source rocks reach their peaks of gas expulsion. The rate of Triassic and newer source rocks is still less; they did not start to exclude gas until the Himalayan period.

The curves of cumulative gas expulsion from Cambrian and Ordovician source rocks are similar to those of cumulative oil expulsion. That is, gas expulsion initially rose rapidly to near-peak values and then the curves rise slowly. The rate of cumulative gas expulsion of Silurian, Carboniferous, and Permian source rocks increased slowly to their peaks (Fig. 7).

DENUDED AMOUNT OF CUMULATIVE OIL AND GAS EXPULSION

Amount of cumulative oil and gas expulsion is an important base for calculating the resource

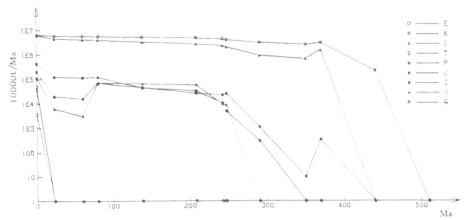

Fig. 7   Cumulative gas expulsion rate of Tarim Basin

extent of oil and gas. The amount of cumulative oil expulsion or gas expulsion in a cell was acquired by accumulating three dimensional data of phased oil and gas expulsion in each time-step through simulation of particle displacement. The feature of cumulative oil and gas expulsion in Tarim Basin was shown in figure 5 and figure 7.

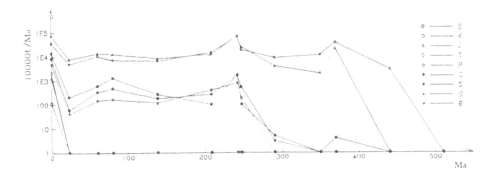

Fig. 8   Phased gas expulsion rates of Tarim Basin

The denuded amount of cumulative oil expulsion and gas expulsion of any formation in any tectonic period is the key problem for the assessment of oil and gas resources in Tarim Basin. The restoration of denuded thickness is the foundation of calculation of them. We took all the cumulative oil or gas in the cells that was denuded in tectonic movements. Further more, all the oil and gas in the cells below surface at a particular depth were taken as denuded. The depth was ascertained by geological study. For oil, it was 1200 m. For gas, it was 2000 m. Figure 9 and figure 10 show the amounts of denuded cumulative oil and gas expulsion of Paleozoic in different tectonic episodes. These figures show that the first period of major loss of oil and gas is the early Hercynian period. The amount of denuded oil expulsion reaches 11,200 million tons, while denuded gas expulsion is 15,100 million tons. The total denuded oil and gas expulsion in Cambrian and Ordovician were more than 30,000 million tons in this move-

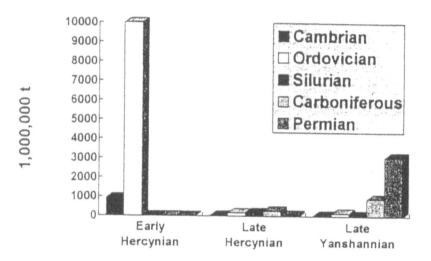

Fig. 9   Denuded volume of Cumulative oil expulsion in Tarim Basin

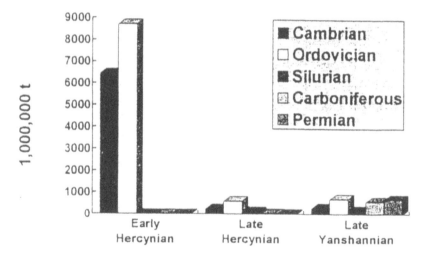

Fig. 10   Denuded molume of Cumulative gas expulsion in Tarim Basin

ment. Fortunately, the flowed movements were not so strong and quite a lot of oil and gas could be remained. The late Yanshanian period is the second major period of loss of oil and gas. Loss of oil and gas caused by erosion in the late Hercynian period is much less than someone said.

ASSESSMENT OF OIL AND GAS RESOURCE IN TARIM BASIN

On the basis of the amount of cumulative hydrocarbon expulsion, the phased resource extent of oil and gas and that denuded in Tarim Basin have been calculated using Monte Carlo calculation.

Figure 11 shows the value of oil and gas resources of each formation in Paleozoic of Tarim Basin with confidence coefficient of 75%. The oil resources were more than 11,000 million tons, while the gas resources were more than 6,000 million tons. The distribution amounts in Mesozoic and Cenozoic was quite small than Paleozoic, less than 10% of the total resources, so they were omitted in the figure.

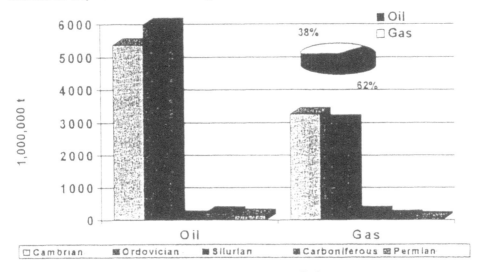

Fig. 11   Speculative oil and gas resources of Paleozoic in Tarim Basin

The simulations indicate that the amount of remaining oil and gas was several times larger than that of denuded of oil and gas though Tarim Basin underwent three strong erosions.

This resulted in part because the residual source rocks still had great potential for hydrocarbon generation after the tectonic episodes. From this point of view, oil and gas resources in Tarim Basin are still very abundant. Oil resource is larger than that of gas in Tarim Basin because of the resource composition in the Basin. The ratio of gas to oil resource is 0.54. The oil and gas resources are concentrated mainly in Cambrian and Ordovician formation. The amount of oil generated from Ordovician source rocks makes up 52.2% of the total oil in the basin, while gas from the Ordovician makes up 44.6% of total gas. Oil from Cambrian source rocks makes up 42.2% of total oil, and gas from the Cambrian is 47.8% of total gas. Oil and gas resources from Cambrian and Ordovician source rocks constitute 62% of the total resource of oil and gas in the basin.

CONCLUSIONS

Three-dimensional simulation modeling of Tarim Basin revealed the following significant facts:

(1) Tarim Basin is a superposed basin with multi cycles, and Cambrian and Ordovician formations that include main source rocks, have the nature of residual basins. Tarim Basin is characterized by multistage hydrocarbon generation, expulsion, migration, accumulation, escape, and deposition.

(2) Oil and gas accumulations in lower Paleozoic formations in the basin underwent multiphase erosion, of which the strongest occurred in the early Hercynian period. The two later tectonic episodes influenced lower Paleozoic formations much less than the first.

(3) In each tectonic episode, the loss oil and gas caused by erosion takes only a small proportion of the total amount of oil and gas generated before the activity.

(4) The main source rocks in the basin, Cambrian and Ordovician, still have great potential for hydrocarbon generation after undergoing the tectonic episodes.

(5) Although oil and gas resources of the basin have been subjected to serious damage, a large resource of oil and gas remains. Thus the basin has good exploration prospects.

(6) Because of multistage modification and migration, distribution of oil and gas is very scattered, some traps show a low degree of filling with oil and gas. Exploration is difficult and great attention must be paid to exploration for subtle traps.

REFERENCE

1. Burnham, A. K. and Braun, R. L.. Development of a detailed model of petroleum formation, destruction, and expulsion from lacustrine marine source rocks, Org. Geochem, Vol. 16,1990.

2. Fu Jiamo and Liu Dehan. Conditions of Gas Migration, Reservoirs and Cap rocks. Science Press, China, 1992.

3. Hao Shisheng et al. Oil and Gas Generation in Carbonate Rocks. Petroleum Industry Press, China, 1993.

4. Lerche, I.. Basin Analysis:Quantitative Methods. Vol. 1, Vol. 2, Academic Press Inc. , 1990.

5. Sachsemhofe, R. F.. Petroleum Generation and Migration in the Styrian Basin (Pannonian Basin System. Austria): An Integrated Geochemical and Numerical Modeling Study, AAPG Bull. , 1994, V. 78,No. 10.

6. Shi Guangren. Numerical Modeling Methods of Petroliferous Basins. Petroleum Industry Press, China,1994.

7. Standvik, E. I. et al.. Primary migration by bulk hydrocarbon flow. Org. Geochem. , 1990, Vol. 16.

8. Ungerer, P. J.. Basin Evaluation by Integrated two-dimensional Modeling of Heat Transfer, Fluid Flow, Hydrocarbon Generation, and Migration. AAPG Bull. , 1990.

PART III

RESERVOIR CHARACTERIZATION

Proc. 30ᵗʰ Int'l. Geol. Congr., Vol. 18, pp. 169~182
Sun Z. C. *et al.* (Eds)
© VSP 1997

The Role of Secondary Mineralization for the Formation of Oil-Gas Reservoirs

E. E. KARNYUSHINA

Geological Faculty, Moscow State University, Moscow , 119899, RUSSIA

Abstract

The specific features of the reservoir secondary transformations of the terrigene Mesozoic formations (in the Daryalyk-Daudan depression of the Karakum basin, the Urengoi and the Talin hydrocarbon fields of the West-Siberian basin) and the volcanogene-sedimentary Oligocene-Miocene formations (in the troughs of the Okhotsk-Kamchatka and the South-Okhotsk basins) have been discussed based on the unproductive, oil-gas containing and hydrothermally transformed deposits. It has been established that progressive catagenesis had resulted in the predominantly dispersed polycomponent secondary mineralization, which caused the gradual lowering of the reservoir porosity. The open porosity decrease gradient is higher in the volcanogene-sedimentary rocks, as compared to terrigene deposits. Relative to this, the secondary reservoirs within the volcanogene-sedimentary formations appear at the early protocatagenetic gradations. The secondary reservoirs are formed in the unproductive terrigene formations in the mesocatagenetic subzone. Superposed hydrothermal processes and phenomena on oil-water contacts are disturb the decrement tendency of changing the porosity and are characterized by contrast transformations of the reservoirs. This is due to the fact that leaching lenses alternate with carbonatized, silicified, kaolinized, zeolitized and others olygo- and monocomponently replaced interlayers which complicate the structure of the reservoirs.

Keywords: secondary mineralization, catagenesis, hydrothermal superposed processes, oil-water contact, terrigene / volcanogene-sedimentary formations, relic-primary / secondary reservoirs, Karakum, West-Siberian, Okhotsk-Kamchatka, South-Okhotsk basins

INTRODUCTION

Catagenesis and superposed processes are responsible for the secondary content and properties of reservoir rocks. These processes depend on the geological development, composition of formations, geothermal conditions and character of fluids of oil-gas-bearing basins. The variety of these factors and their combination form different vertical zoning of the reservoir rock types, which have been studied on samples taken from wells drilled in the basins presented in Fig. 1.

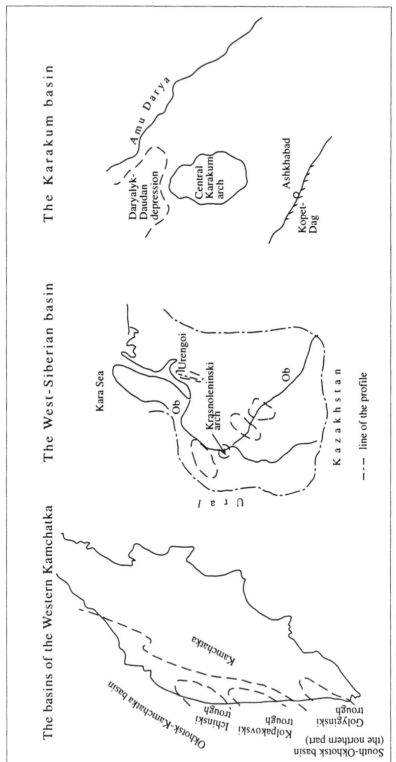

Figure 1. Location of the study areas

The subjects of investigations were the Cenozoic volcanogene-sedimentary formations of the troughs of the Western Kamchatka (the continental units of the Okhotsk-Kamchatka and the South-Okhotsk basins) and the Mesozoic terrigene formations of the Urengoi and the Talin hydrocarbon fields in the West-Siberian basin and the Daryalyk-Daudan depression in the Karakum basin.

More than 500 samples of terrigene rocks were collected from 30 wells at the depths up to 2. 7-5. 5 km. The depths of sampling the volcanogene-sedimentaryrocks ranged from 0. 2 to 3. 7 km (more than1000 samples from 25 wells). All the samples have been described macroscopically and in the thin-sections. The determination of open porosity, mineralogical and bulk density was made for the whole collection too. The scanning electron microscope (SEM) of the BS-311 type was used for the study of morphology and composition of the secondary minerals. X-ray diffractometer (DRON-3) was used for diagnostics clay minerals, zeolites and silica forms which were determined by infrared absorption spectroscopy and different types of chemical analysis [2] . The degree of catagenesis of sedimentary formations was determined by means of the vitrinite reflectance (Ra and $R°$, %) for 500 samples containing coalificated fossil plants. The zone of catagenesis was partitioned into the subzones and gradations according to the Vassoevitch scale[6].

The report are contains the general results of the author's investigations [1, 2, 3]in brief terms. The new data on the change of the terrigene rocks in the zones of oil-water contacts in the West-Siberian basin, are discussed in more detail .

THE CHARACTER OF SECONDARY MINERALIZATION OF THE TERRIGENE FORMATIONS

The unproductive formations of the Northern part of the Karakum basin

In the north of the Karakum basin, in the Daryalyk-Daudan depression the Jurassic humid subcoal-bearing, marine terrigene and semiarid terrigene-carbonaceous formations occur at the depth from 1. 2 to 3 km (temperature 60-130℃) within PC_2-MC_{3-4} gradations. The indicated interval of the Jurassic section is characterized by oil show, but it is commercially ineffective. Recent temperatures in this section are close to maximum values in the geological history of inheritably warping depression[1].

The most remarkable break in the subsidence occurred during Kimeridgian-Tithonian age when a thick salt-bearing formation has been formed in the adjacent areas. The break which in the followed Pre-Neocomian time had resulted in regional seepage of strong salt brine into the most permeable humid formation reservoirs. Salinity of these brines increases down the section of Jurassic deposits[5]. As a result, the secondary minerals of progressive catagenesis (quartz, potash feldspar, clay minerals, siderite, calcite) create paragenesis with halocatagenetic dolomite and sulphates in permeable terrigene rocks of humid formations.

Secondary minerals fill to a different extent the pores in sandstones and aleurolites like dispersed separations and determine the relic-primary character of pore space and decremental (gradually decreasing) capacity of the reservoirs whose median open porosity are decreases from 21 % (gradation PC_3) to 6 % (gradation MC_3). The secondary pore-fissure type of reservoirs appears only in the lowermost part of the section[1].

The productive formations of the West-Siberian basin

Another tendency in the changes of terrigene reservoir rocks is observed in the productive Jurassic and Cretaceous deposits of the West-Siberan basin where the reservoirs of the Urengoi oil-gas-condensate field were studied within PC_3- AC_{1-2} gradations at depths from 2. 2 to 5. 5 km. In this interval recent temperature varies from 60 to 150℃. The maximum paleo-temperature 25-45℃ exceeded the above ones and was associated with Oligocene time [4].

The peculiar feature of this catagenetic zone is the decreasing salinity of reservoir water from the top downwards. At greater depth hydrocarbonate-sodium elisional type is spread with mineralization not exceeding 3-5g/l[7]. The development of superhigh reservoir pressures with an anomaly factor up to 1. 6-1. 8 is also characteristic phenomena[3].

The terrigene essentially marine Lower Cretaceous formations of the Urengoi field contain gas-condensate accumulations and the secondary minerogenesis is governed to a great extent by the processes occurring in the contact zone between reservoir water and hydrocarbon pool. It was established by comparing the secondary transformations in unproductive and productive horizons. In clastic rocks of unproductive part of the section polycomponent dispersed mineralization is spread which appeared as a result of background progressive catagenetic transformations. In this case three cement generations are usually observed: 1 - quartz regeneration, 2 - kaolinite and/or chlorite, 3 - calcite. Concentrated forms of secondary mineralization are observed in productive horizons where a considerable part of original components of terrigene rocks is substituted by carbonate, silica and kaolinite. Sometimes permineralizated rocks alternate with porous lenses of leaching in which primary and secondary components are dissolved as a result of carbon and organic acids action which quantity is higher in the zone of oil-water contact. Leaching is accompanied by destruction of complicated intergranular contacts with pore channels formed there in.

Let us consider a superposed mineralization on the example of $БУ_{10-11}$ productive bed whose gas-condensate accumulation has an oil fringe (Fig. 2). The bed occupies the upper part of mesocatagenetic subzone (depth: 2792-2846 m, MC_1 gradation). At the top of productive bed of the pool roof oligomictic sandstone and aleurolite contain sedimentogene chlorite and illite interstitial cement. Earlier diagenetic concretions are represented by micrite. Background catagenetic minerals are represented by regenerated quartz, more recent (with respect to quartz) kaolinite and sporadic release of calcite in the centre of pores (Fig. 3). Open porosity of such reservoirs depends primarily on grain size distribution of the rock and varies from 12% to 18%.

Within the productive bed observed below the rocks with progressive catagenetic transforma-

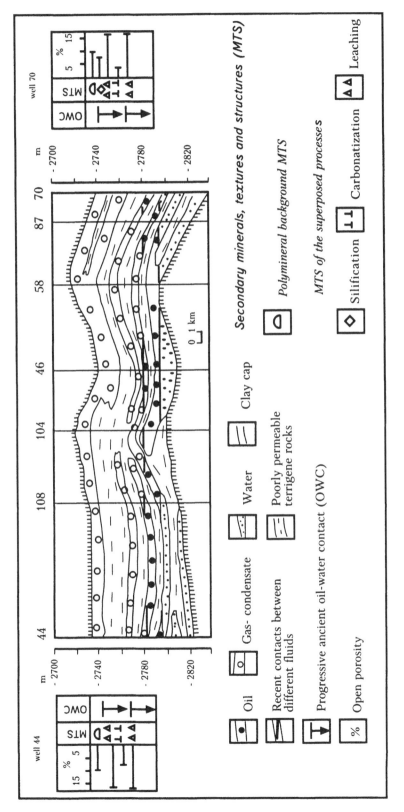

Figure 2. The structure and the secondary change of the reservoirs of the Urengoi oil-gas condensate field (the productive bed Åi 10-11, the Lower Cretaceous).

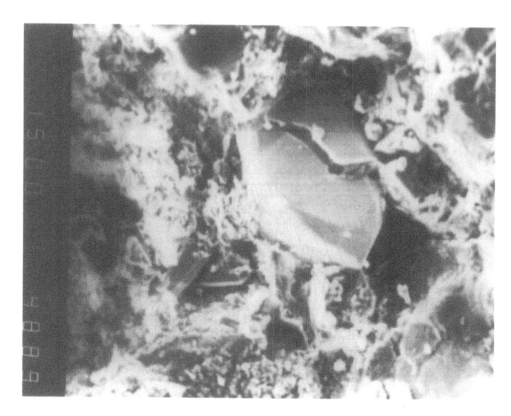

Fig. 3　Background dispersed mineralization of progressive catagenesis: regenerated quartz, polycomponent cement (clay minerals and calcite) and relic-primary pores in the coarse-grained oligomict aleurolite located above the ancient oil-water contact (Urengoi, well 70, depth 2788 m). Scale bar 100 mkm for the Fig. 3 and 4.

tions is the alternation of carbonatized, silicificated, kaolinisated and leached interbeds (Fig. 4 a, b), which mark the level of ancient and recent oil-water contact stabilization (see Fig. 2). Open porosity in silicification reservoirs makes up 10%-15%, in carbonatized reservoirs 2%-6%, in leached varieties reaching 17%-20%. According to empirical data, open porosity increase due to leaching makes up 7%-9% in fine sandstones (3/4 of initial value). The loss of porosity due to partial kaolinization and regenerative silicification does not exceed 2%-3%, carbonitization causes the porosity decrease by 4/5 of initial volume.

Superposed phenomena are traced within the entire productive part of mesocatagenetic subzone, where along side with the primary-relic reservoirs there appear their secondary-pore types. This results in considerable spread of porosity and permeability parameters (permeability coefficient varies from 1 to 130 mD). So, the section of the Lower Cretaceous deposits within MC_1-MC_3 gradations at depths from 2.7 to 3.5 km can be considered as an interval of contrast transformations of reservoir rocks affected by superposed processes due to

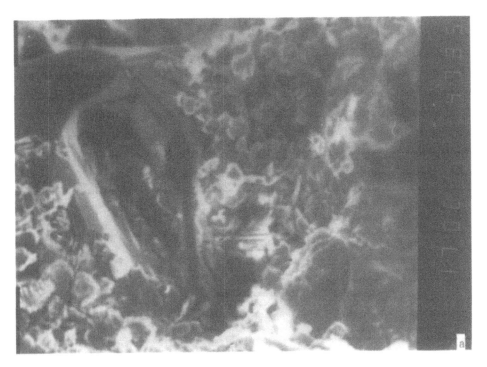

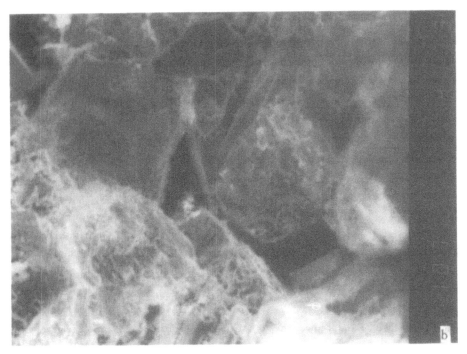

Figure 4. Superposed concentrated mineralization and texture of leaching: a- kaolinization (depth 2865 m); b - solution of different components with generation of secondary pores (depth 2864 m) in the fine-grained polymict sandstone in the zone of recent oil-water contact (Urengoi, well 48)

shifting of oil-water and gas-condensate contacts in the course of formation of the Urengoi field.

Another factor of the superposed processes are the hydrothermal impacts effects being identified in the oil-bearing sandstones of the subcoal continental formation of the Lower Jurassic age occuring at the base of the sedimentary mantle over the Talin oil field of the Krasnoleninsky arch roof, where at the depth of 2. 7 km the recent temperature attains 150℃. Druse-shaped quartzification and kaolinization of the sandstones in the effective beds of this field in this case serve as a positive factor of formation of rock capacitive properties (Table 1).

Table 1   Physical properties of variably changed sandstones of the Talin oil field

well	depth, m; bed	type of sandstone	type of change	open porosity, %	bulk density, g/cm^3
115	2511 IOK$_5$	arcose calcite-clayey	background catagenetic	6. 5	2. 48
105	2707 IOK$_{10}$	quartz kaolinized	superposed hydrothermal	11. 4	2. 34
105	2715 IOK$_{10}$	silicificated with bitumen veins	posthydrothermal on the oil-water contact	8. 7	2. 40

THE CHARACTER OF SECONDARY MINERALIZATION OF THE VOLCANOGENE-SEDIMENTARY FORMATIONS

The hydrothermally changed formations of the South-Okhotsk basin

Contrast transformations of permeable strata were the most intensively developed in the volcanogene-sedimentary formations of the Golyginsky trough of the South-Okhotsk basin under the influence of the superposed hydrothermal processes. Here the Oligocene-Middle Miocene cherty-tuffite-terrigene, tuffite-siliceous, tuffite-diatomaceous and tuffite-sand geosynclinal formations were hydrothermally changed before the sedimentation of the Middle-Late Miocene tuffite coal-bearing molassa (gradation PC$_1$-PC$_3$), which are characterized by the progressive catagenetic change only (Fig. 5). Paleogeothermal reconstructions show that the lower part of the section, now occuring within subzone of mesocatagenesis (gradations MC$_1$-MC$_2$, depth 1. 6-3. 5 km, recent temperature 70-125℃ and thermal gradient decreasing from the top downward the section), was heated to a greater extent before the Late Miocene (up to 100-200℃ and higher), and paleothermal gradient was increasing from the top downward the section.

Lomontite-mordenite and lomontite metasomatite had been formed here under the retrograde-thermal conditions. Zeolitic metasomatites are characterized by low compression strength, which is 4-5 times lower than in the overlying tuffites of the protocatagenetic subzone (gradation PC$_3$), being characterized by the background catagenetic transformations and liberation of smectite and clinoptilolite in the pores. Comparison of samples from these two subzones shows that at the negligible difference of open porosity (25 % - at the depth of

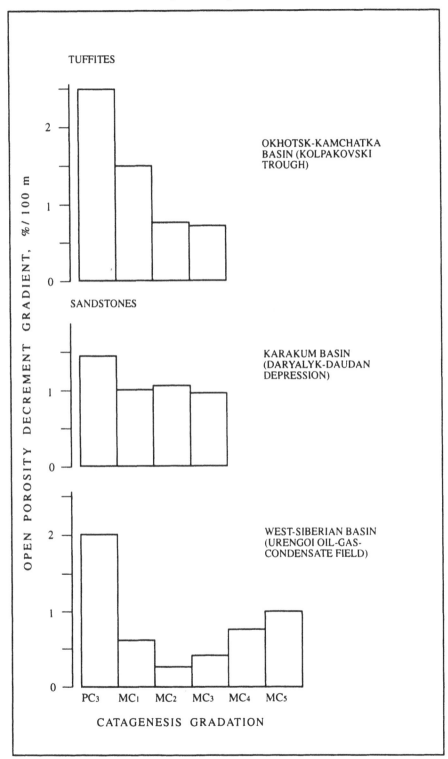

Figure 5. Distribution of porosity decrement gradient in the zone of catagenesis

1230 m and 20 % - at the depth of 3140 m), the permeability makes up 7 and 57 mD respectively with the increase in zeolites due to jointing[2].

The formations of the Okhotsk-Kamchatka basin

Catagenesis of volcanogene-sedimentary formations is characterized by initiation of a special zoning of clay minerals and zeolite, which, on the one hand, considerably decrease the primary porosity and permeability of rock, and, on the other hand, cause the secondary cavernous and fissured reservoirs to appear as early as at the initial transformation stage. It was found in the Oligocene-Upper Miocene formations of the Ichinsky and Kolpakovsky troughs in the Okhotsk-Kamchatka basin where PC_{1-2}- MC_2 gradations are encountered at the depths of 0.3-3.7 km. Recent temperatures in this interval vary in the range of 40-140℃ , the mineralization of reservoir waters is moderate and increases from the top downwards.

In sand-aleurolite tuffites of the Middle-Late Miocene molassa which contain gas pools in the Kolpakovsky trough, authigenic smectite and clinoptilolite is observed. These minerals fill the pores of reservoirs and decrease their open porosity within the protocatagenetic subzone (interval 0.3-1.7 km) from 35% to 26%. Sporadically developed carbonatized interbeds are practically impermeable here. Down the section in the Oligocene-Miocene terrigene-tuffite-siliceous formations within MC_1 and MC_2 gradation (1.7-3.7 km depth, temperature 80-140℃) analcime and stillbiite appear among zeolites. Lomontite is found in the lower part of this succession. Mixed-layered minerals, chlorite and illite are widely represented in cement of tuffites. Open porosity of the reservoirs decreases here to 10%-1% and secondary genesis pore-fissure and fissure types are the main ones here.

DISCUSSION AND CONCLUSIONS

The comparison of the gradient of open porosity decrease (% / 100 m) of the reservoir rock in the zone of catagenesis of the basins under consideration demonstrates that the highest rate of decrease of the pore space is inherent to volcanogene-sedimentary rocks (Fig. 6). Due to superposed processes this rate considerably slows down which is characteristic of productive terrigene formations of the West Siberia. Progressive catagenetic transformations are responsible for gradual decrease of porosity and wide spread of relic-primary reservoirs, as was shown for the terrigene Daryalyk-Daudan formations of the Karakum basin.

In unproductive deposits a wider catagenetic interval of relic-primary reservoir development is observed for terrigene formations. In volcanogene-sedimentary formations there are more intensively developed mixed and secondary reservoirs which appear even at the initial gradations of protocatagenesis. Moreover, in oil-gas-bearing basins hydrothermal superposed processes complicate progressive-catagenetic changing and cause contrast transformations of permeable deposits in the range of any one of catagenetic gradation (Fig. 7).

So, reservoir properties of the rocks are different during subsidence of the sedimentary for-

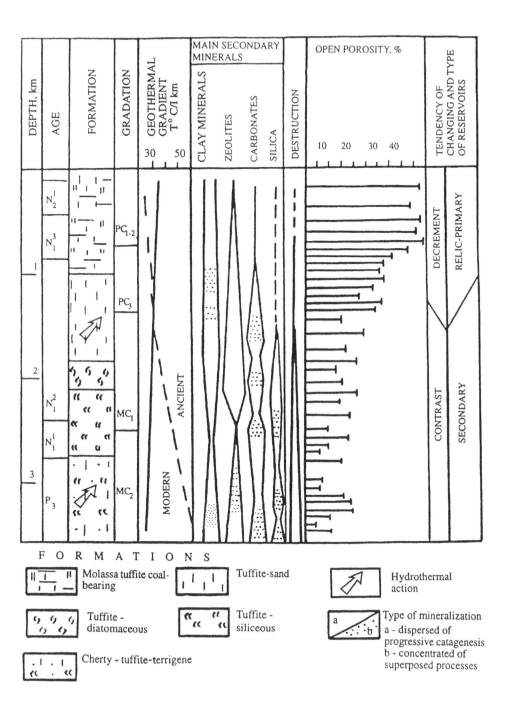

Figure 6. Secondary changing of the formations of the Golyginski trough (the South-Okhotsk basin)

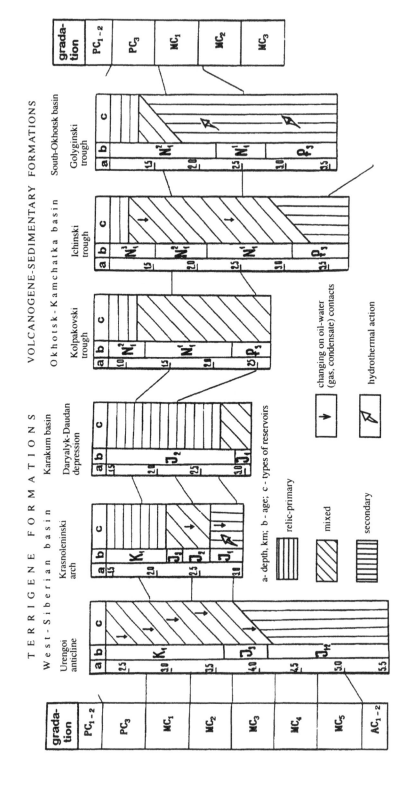

Figure 7. Zoning of reservoir types

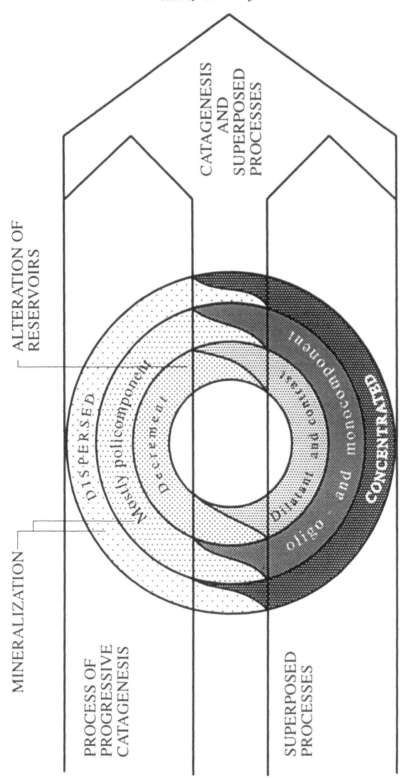

Figure 8. Types of mineralization and reservoirs

mations. Decrement reservoirs with gradually decreasing open porosity are generated due to background catagenetic transformations of rocks marked by dispersed forms of mineralization. Contrast changes in the reservoirs appear due to superposed processes (Fig. 8). Concentrated mineralization, leaching and dilatant phenomena, play an important role in the formation of complicated secondary reservoirs which cause potentiality for oil-gas accumulation over a wide range of the catagenetic zone of sedimentary basins.

REFERENCES

1. V. I. Gorshkov, E. E. Karnyushina. Catagenesis of the terrigene Jurassic deposits of the Northern Turkmenia. Problems of diagnostics of conditions and zones of oil generation. The Institute of Geology and Exploitation of Fossil Fuels. Moscow. 67-74 (1971).

2. E. E. Karnyushina. Volcanogene-sedimentary rocks of the oil-gas bearing basins of the USSR North-East. Moscow State University. Moscow (1988).

3. E. E. Karnyushina, G. N. Leonenko. The reservoir's properties of the West Siberia with the zone of catagenesis. Vestnic of the Moscow University. Ser. 4. Geology. 5, 35-41 (1989).

4. A. R. Kurthikov, B. P. Stavitsky. Geothermal conditions of oil-gas regions of the West Siberia. Nedra. Moscow (1987).

5. O. A. Kuzmina, G. F. Panteleev, I. F. Kuvshinova, V. N. Isaenko. Geology and prospects of the gas-oil bearing potential of the Northern Turkmenia and adjacent areas of the Uzbekistan. Soyuzburgas. 8. Nedra. Moscow (1970).

6. S. G. Nerutchev, N. B. Vassoevitch, N. P. Lopatin. On the catagenetic scale relative the oil-gas generation. Fossil fuels. The problems of geology and geochemistry of the napthides and bituminous rocks 25th IGC. The reports of the Soviet geologists. Moscow 47-62 (1976).

7. M. Ya. Rudkevich, L. I. Ozeranskaya, N. F. Chistyakova et al. Oil-gas bearing formations of the West-Siberian basin. Nedra. Moscow (1988).

Proc. 30ᵗʰ Int' l. Geol. Congr. , Vol. 18, pp. 183~190
Sun Z. C. *et al.* （Eds）
© VSP 1997

A Study of the Pore Structure of Major Sandstone Reservoirs in China

SHEN PINGPING, JIA FENSHU and LI KEVIN
Research Institute of Petroleum Exploration and Development of CNPC,
CHINA

Abstract

Pore structure of reservoir rock is a important factor to describe hydrocarbon reservoir characteristics, it affects oil and gas storage and flowing in porous media directly. About 90% of oil and gas reserves found in China now comes from continental clastic reservoirs which are of some complex properties such as more reservoir types, quick lithologic variation, complex pore structure, high heterogeneity and so on. Hence, it is more important to find a new and better method for describing pore structure and its heterogeneity of reservoirs.

In recent years, fractal geomentry was used to describe pore structure of porous media, usually to study the fractal characteristic of so-called "pore". A new method has been proposed to study the fractal property of "pore-throat" in this paper. Pore structure classification of main sandstone reservoirs in China has been done by this method. It has been proved that the fractal dimension of pore structure calculated by this method can be used to describe the heterogeneity of pore structure quantitatively. The relationship between fractal dimension and some factores of pore structure has been studied in this paper. It has been shown that the smaller the fractal dimension, the smaller the sorting coefficient and pore geometry factor, so the better the pore structure; and vice versa. The effect of fractal dimension of pore structure on oil and water flowing has been also studied in this paper.

Keywords: Pore Structure, Fractal, Core, Oil Recovery

INTRODUCTION

A pore structure is of much influence on the fluid flow in the porous media and also on the oil recovery. There are many parameters proposed to describe the pore structure of reservoir rocks in the past such as pore geometry factor G, threshold pressure Pd[1], puzzier coefficient[2]. The author of this paper proposed microscopic homogeneous coefficient α to describe the heterogeneity of pore structure in 1980[3]. However, it is very difficult to determine which parameter is the best one because of the complexity, the irregularity and the heterogeneity of a pore structure in a reservoir rock.

In recent years, it was found that pore structures of sandstones or other porous media are fractals. Fractal geomentry was usually used to study the fractal characteristic of so-called "pore"[4][5][6][7]. A new method has been developed in this paper to determine the fractal dimension of "pore-throat" in 3-D Euclidean Space by using mercury injection capillary pressure curves. A large number of data has been obtained for many sandstone samples from the oil-fields in China.

THE NEW METHOD

Self-similarity

A fractal is defined as a shape made of parts which are similar to, or repeat the whole in some way. This property is known as self-similarity. Usually, fractal properties can be determined by measuring the length of an interface as a function of the unit of measurement. However, it is too difficult to use this technique to determine fractal dimensions in some cases. Another technique for determining the fractal dimension of an object is as follows:

$$P(r) \infty R^{-D_f}$$
(1)

There $P(r)$ is the probability or the number of the unit whose radius is larger than r. D_f is the fractal dimension of an object. if the plot of $P(r)$ versus r is obtained, the fractal dimension D_f can be calculated from equation (1).

Procedures for Determining D_f

The main procedures of the new method for determining a fractal dimension of a pore structure by using mercurty capillary pressure curves are discribed as follow [8]:

①Select the model (capillary-tube model or improved capillary-rule model). The capilary-tube model is shown is Fig. 1.

②Calculate the value of $P(r)$ according to capillary pressure data.

③Plot the curve of $P(r)$ versus radius r (as Fig. 2).

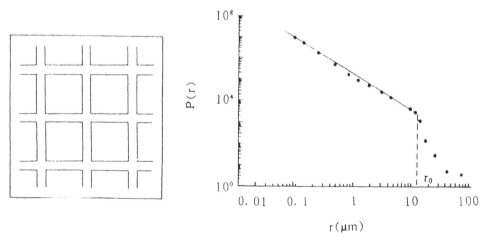

Fig. 1   Capillary-Tute Model   Fig. 2   Fractal Curve

④Calculate the slope of the $P(r)$ vs. r curve from equation (1), the fractal dimension is:

$$D_f = -Slope$$

Results

Fig. 2 indicates that the pore structure of the core sample has self-similarity at some range. When pore-throat radius is smaller than r_0, the pore spaces are fractals. The pore structure can be described by fractal technique. The pore spaces with radius larger than r_0 are not fractals. Pore structures of many other core samples from the major oilfields in China have the similar characteristics shown in Fig. 2.

APPLICATIONS

The relationship between D_f and the heterogeneity of pore structures

Fig. 3 shows mercury injection capillary pressure curves of three core samples from Yidong oilfield. The corresponding fractal curves of these core samples are shown in Fig. 4.

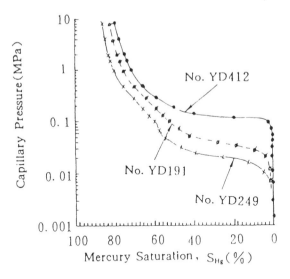

Fig. 3   Capillary Pressure Curves

It is shown in Fig. 3 and Fig. 4 that the more heterogeneous the pore structure of a core sample, the bigger the slope of the fractal curve, in another word, the larger the fractal dimension of the pore structure. So it is possible to quantitatively describe the properties of a pore structure obtained from a mercury injection capillary pressure curve.

There is the same behavior in different oilfields. Fig. 5 shows three mercury-injection capillary pressure curves of three core samples from different oilfields (Shengto, Guangli and Xinglongtai oilfields). The corresponding fractal curves of these core samples are

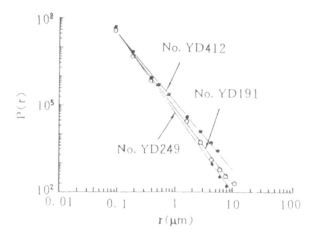

Fig. 4   Fractal Curves (Yidong Oilfield)

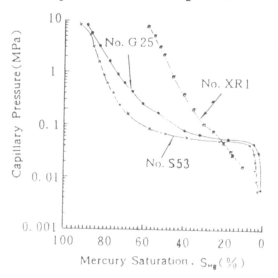

Fig. 5   Capillary Pressure Curves

shown in Fig. 6. It can be seen that the more heterogeneous the pore structure of a core sample, the larger the fractal dimenion of the pore structure.

It is proved that this conclusion is suitable to not only sandstone reservoirs but also sandstones with gravel and pinhole dolomite reservoirs.

The Relationship between D_f and other Parameters

Fig. 7 to Fig. 9 shows the relationship between sorting coefficient, radius sorting coefficient, geometry factor and fractal dimension. In Fig. 7, we can see that the larger the fractal dimension, the larger the sorting coefficient. And so the radius coefficient and geometry factor.

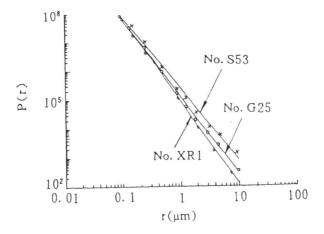

Fig. 6　Fractal Curves

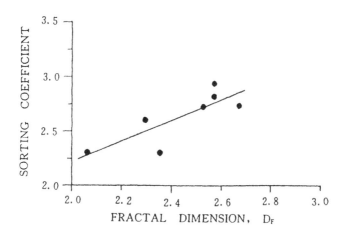

Fig. 7　The Relationship between Sorting Coefficient and D_f

The Influnce on the flowing of oil and water

Fig. 10 shows the relationship between oil recovery of 22 natural core samples at water breakthrough and the fractal dimension. We can see that the larger the fractal dimension, the smaller the oil recovery at water breakthrough. The reason is that the larger fractal dimension implies the stronger heterogeneity of pore structure related to the flowing of oil and water. It is obvious that the stronger the heterogeneity of a pore structure, the more irregular the propagation of the waterflooding front and the smaller the oil recovery at water breakthrough.

Fractal dimensions of core samples from major oilfields in China

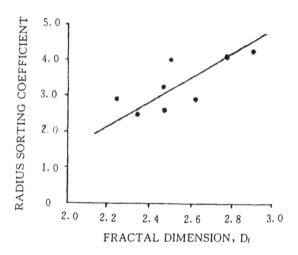

Fig. 8   The Relationship between radius Sorting Coefficient and D_f

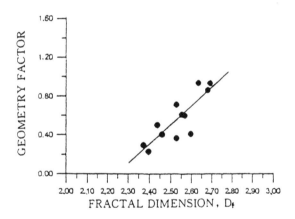

Fig. 9   The Relationship between Geometry Factor and D_f

The fractal dimensions (determined by using the new method from mercury-injection capillary pressure curves) of core samples from major oilfields in China are listed in table 1.
Major China oilfields are classified into 5 types — Ⅰ , Ⅱ , Ⅲ , Ⅳ , Ⅴ . The pore structures in reservoir Ⅰ are the most homogeneous and those in reservoir Ⅴ are the most heterogeneous. It can be seen from table 1 that the better the pore structures of core samples, the smaller the fractal dimensions. The permeabilities in Chunhua and Xinglongtai oilfields are large, but the fractal dimensions are large too because of the strong heterogeneity of the pore structures there. It is obvious that the peremeability cann't be used to describe the heterogeneity of a pore structure in a reservoir rock.

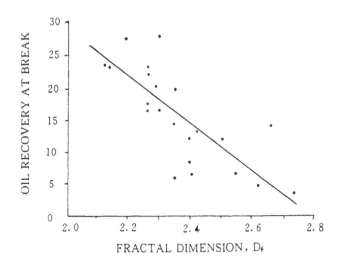

Fig. 10　The Relationship between Oil Recovery at Break and D_f

Table 1　Fractal Dimensions of Core Samples

No.	Sample No.	Name of Oil field	Fractal Dimension	Porosity (%)	Perm.	Category of Pore Structure
1	Sj5318R	Shengtuo	2. 455	33. 2	2000. 0	I
2	XG106R	Shuguang	2. 478	/	892. 0	II
3	NB10R	Binnan	2. 492	30. 0	111. 0	II
4	DQL3R	Lamadian	2. 633	29. 14	416. 9	II
5	GL25R	Guangli	2. 668	26. 9	800. 0	II
6	QH41R	Chunhua	2. 748	24. 3	1825. 0	IV
7	XJ121R	Kelamayi	2. 873	27. 4	41. 7	IV
8	XRT−115R	Xinglongtai	2. 970	22. 0	2071. 0	V
9	GSP−1RT	Gaoshangpu	2. 973	19. 49	3. 74	V

CONCLUSIONS

①The fractal dimensions calculated from mercury injecton capillary pressure curves can be used to characterize a pore structure in a reservoir rock quantitatively. The larger the fractal dimension, the stronger the heterogeneity of pore structure.

②The fractal dimension is ralated to some parameters of pore structure obviously.

③There is a good relationship between the fractal dimension and oil recovery at water break-through. The larger the fractal dimension, the lower the oil recovery at water breakthrough.

REFERENCES

1. Thomeer, J. H.. "Introduction of a Pore Geometry Factor Defined by the Capillary Pressute Curve". JPT, V. 12,73-77,1960.

2. Dulion, F. A. L.. "Is there a Relationship Between Pore Structure and Oil Recovery?" (An Expermental Study), AIME, V. 2,1-11,1970.

3. Shen Pingping, Li Bingzhi and Tu Fuhua. "The Effect of Sandstone Pore Structure on the Oil Recovery by Water Flooding and Its Categories". Presented at the International Technical Meeting of Oil Field Development, Daqing, China,1982.

4. Katz, A. J. and Thompson, A. H.. "Fractal Sandstone Pores Implication for Conductivity and Pore Formation". Phys. Rev. Lett. , V. 54,1325-1328,1985.

5. Krohn, C. E. and Thompson, A. H.. "Fractal Sandstone Pores: Automated Measurements Using Scanning-Electron-Microscope Image". Physical Review B,V. 33,No. 9,6366-6374,1986.

6. Li Kevin, Shen Pingping and Jia Fenshu. "Fractal Property of Sandstones and the Prediction of Oil Recovery". Proceedings of the First Annual Meeting of Chinese Association of Science and Technology, Beijing, P. R. China, April 16-18, 483-486.

7. Jia Fenshu, Shen Pingping and Li Kevin: "Fractal Behavoir of Sandstone Pore Structure and Its Applied Study". Presented at the International Meeting of Petroleum and Chemical Engineering, Beijing, P. R. China, December, 1994.

8. Shen Pingping and Li Kevin: "A New Method for Determining the Fractal Dimension of Pore Structures and Its Application". presented at the 10th Offshore South East Asia Conference and Exhibition, Paper No. OSEA-94092, Singapore, December, 1994.

Proc. 30th Int' l. Geol. Congr. , Vol. 18, pp. 191~204
Sun Z. C. *et al.* (Eds)
© VSP 1997

Reservoir Characterization of the Ekofisk Field: A Giant, Fractured Chalk Reservoir in the Norwegian North Sea-Phase 1, Reservoir Description

BIJAN AGARWAL and SCOTT C. KEY

Philips Petroleum Company Norway, *P. O. Box* 220, 4056 *Tananger*,
NORWAY

Abstract

The Ekofisk reservoir is a high porosity, low matrix pemeability naturally fractured chalk. Fluid flow is largely governed by the distribution. orientation and interconnectivity of the natural fracture system associated with complex structure and reservoir distribution. To facilitate reservoir management decisions, a 3-D reservoir flow model is used as a tool to determine optimum well placement and predict future performance. It is important that the highest degree of heterogeneity be represented in the flow model so preferential flow directions of water or gas from injector sites can be accurately predicted. The current 3-D reservoir flow model captures heterogeneity intrinsic to the chalk, and although this model has done a good job of matching historical performance at Ekofisk on a field wide basis and has produced reliable production forecasts, it does not capture the full geologic complexity known to be present. Water break-though has been observed in areas of the field not consistent with flow model predictions and this further suggests that the reservoir description could be improved.

The Ekofisk Field is currently undergoing a major field re-development in which 45 new wells will be drolled before the end of 1998. This requires that the most comprehensive and detailed reservoir description and geological and fluid flow models be used as the basis for the planning of such a re-development. The situation, as well as new developments in hardware and software and multidisciplinary database and applications integration, led to the dicision in 1994 to completely re-evaluate the reservoir characterization of the field. A major multi-disciplinary effort involving geoscience, petrophysical and reservoir engineering work was initiated through the Ekofish Reservoir Characterization (ERC) project.

The objective of the reservoir characterization project was to improve the existing reservoir description using as available data through the application of new techniques and technology, and to construct and history match a new 3D reservoir fluid flow model using this updated, detailed reservoir description.

Keywords: *reservior characterization*, *3-D reservoir flow model*, *Ekofisk field*

INTRODUCTION

The Ekofisk Field, located in the Norwegian Sector of the North Sea was discovered in 1969. Production started from the Cretaceous-Danian chalks in 1971. Current estimates indicate about 8 billion barrels of oil equivalent originally in place. The field has produced over 1. 2 billion barrels of oil to date from a total of 160 wells. Current production from about 67 devi- ated and horizontal wells is about 240,000 barrels of oil per day and 600 MMCFD of gas. A pilot water injection project was initiated in 1981 in the highly fractured Tor formation[1] and and in the Lower Ekofisk in 1986[2]. Fieldwide water injection began in 1987[3]. Current water injection rates are 840,000 bwpd into 37 active injection wells. A number of additional im- proved oil recovery techniques are being evaluated and currently ongoing is a Water-Alternat- ing-Gas (WAG) pilot in the southern area of the field. Fig. 1 shows a structure map of the Ekofisk Field drawn on the top Ekofisk formation.

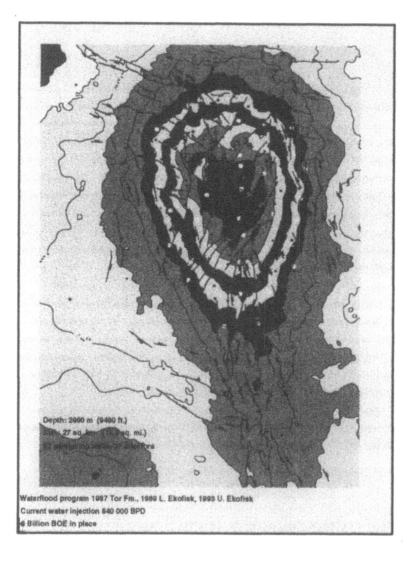

Depth: 2900 m (9480 ft.)

Waterflood program 1987 Tor Fm., 1989 L. Ekofisk, 1993 U. Ekofisk
Current water injection 840 000 BPD
8 Billion BOE in place

Fig. 1   Top Structure Map of the Ekofisk Formation of the Ekofisk Field.

Due to the large reserves remaining in Ekofisk and the potential for increased recovery, Phillips Petroleum Company Norway allocated significant resources to the ERC project. An integrated, multi-disciplinary team consisting of geologists, geophysicists, petrophysicists, engineers and technicians comprised the ERC team and twenty (20) man years of work have been dedicated to the project to date. The reservoir description phase and construction of the 3-D geological model were completed earlier this year, and the process of history matching the 3-D reservoir flow model with 25 years of production data from over 160 wells is now underway.

Numerous strategies were employed during the course of the ERC project, including; 1) reprocessing of seismic and petrophysical data; 2) an integrated team and project plan incorporating the concept of front-end loading; 3) use of state-of-the-art technologies in both software and hardware; 4) a focus on permeability, heterogeneity, and anisotropy; 5) integration of data and tasks; 6) development of new structural/sedimentalogical models; 7) upgrades to the reservoir flow model software to allow refined grid definition and orientation; and 8) an improved link between the geological model and the reservoir flow model.

Reservoir characterization on Ekofisk was directed at gaining a detailed understanding of reservoir hydrocarbon volumes, the architecture of the reservoir and at fully describing the heterogeneity and anisotropy of reservoir parameters. Significant attention was also given to capturing as much heterogeneity as possible, both laterally and vertically, in the new geological model, and subsequently in the new flow model. This would enable better prediction of fluid movement within the reservoir, particularly water movement and possible breakthrough, resulting in improved field wide production forecasts, well placement analysis and special high definition EOR studies.

RESERVOIR DESCRIPTION

The most important aspects in reservoir characterization are understanding the distribution and connectivity of pore volume, as this represents the total energy in the fluid flow system. Once pore volume and additional rock properties such as permeability have been defined at a high resolution, the reservoir description will be upscaled to a coarser resolution 3-D reservoir flow model. To describe the distribution of pore volume an architecture must first be built in three dimensions that constrains the rock-fluid system being described; within this architecture the heterogeneity and anisotropy present in rock properties must be captured; and the individual properties distributed. **Fig. 2** illustrates this process.

The reservoir descriptive phase of the ERC project consisted of detailed mapping and three-dimensional distribution of reservoir properties such as thickness, porosity, water saturation, permeability, petrofacies and quartz in a high resolution geological model. An upscaling technique was used to translate the reservoir description to a lower resolution reservoir fluid flow model such that heterogeneity was preserved. In the following sections we describe highlights of the various tasks associated with this phase.

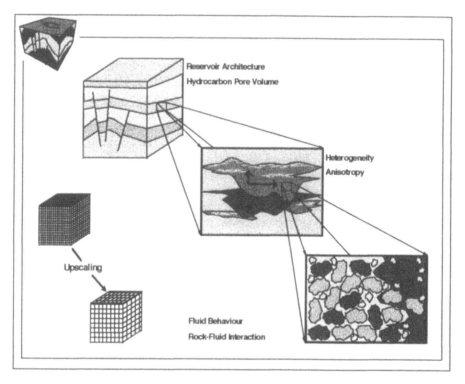

Figure 2. Reservoir Characterization process. To describe the distribution of pore volume, an architecture must first be built in three dimensions that contstrains the rock-fluid system being described.

Data

The first characterization task was the careful editing, integration and analysis of data collected from Ekofisk over 25 years of drilling and production. This included over 160 wellbores with resistivity and porosity data, 125 with well test data, 40 with dipmeter data, 30 with core and 25 with full waveform sonic data. These data were integrated with 3-D seismic data acquired in 1989 and recently reprocessed in 1994.

Architecture

A new structural and sedimentological framework for the field was generated based on an integration of seismic interpretation and wireline log correlation. All wells were carefully re-examined using mineralogy-constrained layer-correlations in the workstation environment. Linear modeling of log responses yielded volume curves for common minerals. Careful use of silica identifier algorithms made it possible to significantly improve previous layer definitions. Synthetic seismograms were generated for over 100 wells outside the gas obscured zone, which guided the seismic interpretation. This produced a consistent reservoir subdivision into 12 layers that honored both well and seismic data.

Heterogeneity

Fault Interpretation

Spatial heterogeneity was preserved in the 3D geological model through detailed layer-based seismic modelling and interpretation of over 300 fault planes in three dimensions. The basis for the fault interpretation was analysis and integration of a number of seismic attributes generated at the major reservoir boundaries. This analysis was linked in three dimensions to provide fully correlated fault planes for each fault. **Fig. 3** compares a model generated in 1991 to the ERC interpretation.

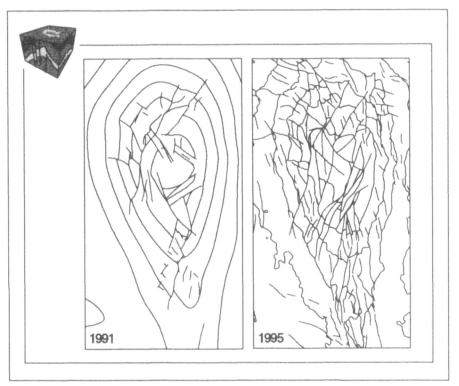

Figure 3. Comparison of the fault model developed through the Ekofisk Reservoir Characterization project to a fault model developed through a previous effort in 1991.

Petrofacies

Production characteristics on Ekofisk are closely related to lithofacies. Chalk petrofacies, with individual flow-characteristics were described and characterized from core analysis. These different petrofacies were then distributed in 3D using cluster analysis based on comparison between core and log data. Lithofacies were described in detail from core data available from 24 wells. The core description was correlated with cluster analysis of log data, combining activity, V-shale & porosity. This generated a distribution of likely facies types where the dominant or "most likely" facies was determined. This process provided a 98% correlation coefficient when compared to foot by foot descriptions from core in target wells. Petrofacies derived for 80 wells were distributed in 3 dimensions generating a continuous 3D facies description. Petrofacies distribution is the basis for the algorithm developed to distribute effective permeability.

Porosity

A petrophysical re-evaluation resulted in more accurate porosity (and water saturation) estimation. The application of new technology allowed seismic inversion techniques to successfully map porosity and isopach values for such seismically resolved layers. A full waveform inversion was carried out for each formation using Phillips proprietary inversion software incorporating simultaneous layer and formation isochron constraints. Forward modeling was carried out with an empirical formula developed for Ekofisk chalks from core and log data with Biot-Gassman used for fluid substitution. Forward models were constructed at 40 wells, generating over 100,000 reservoir models representing the full range of possible petrophysical and geologic parameter variation.

Geostatistical methods were used to simulate porosity distribution within thinner, unresolvable layers. A set of *pseudo*-wells placed across the non-drilled flanks of the field were given synthetic porosity logs, guided from the mapped thickness and porosity and nearby well logs. The synthetic logs ensured a heterogeneous porosity distribution throughout the model. **Fig. 4** is a map of one of the Ekofisk Formation layers, comparing the resulting porosity distribution with the previous characterization based on well control alone. A considerable amount of the intrinsic reservoir heterogeneity is captured as a result of the inversion effort in the well control area and the application of pseudo-wells in the flank areas. Although both datasets match the well data, the location and orientation of significant depositional and diagenetic controls are captured only by characterization of the inter-well space through seismic inversion.

Water Saturation

Accurate distribution of initial water saturation in the geologic and fluid flow models was of extreme importance due to the significant impact it has on hydrocarbon pore volume, hydrocarbon flow characteristics and ultimately on hydrocarbon recovery. The model employed to distribute water saturation was based on a combination of the Bentsen and Anlie correlation[4], the Leverett J-function[5] and the relationship between capillary pressure and height above the free water level. The physical model governing the correlations was a single free water level and water saturation as a function of porosity and height above free water level. All well log data were normalized before standard petrophysical parameters were calculated. A single free water level was established by analyzing pressure data from the first 4 wells drilled on Ekofisk; the intersection of the oil and water gradients defined the free water level. Twenty-four non-linear correlations were generated, representing different geological layers and areal zones, based on fit to

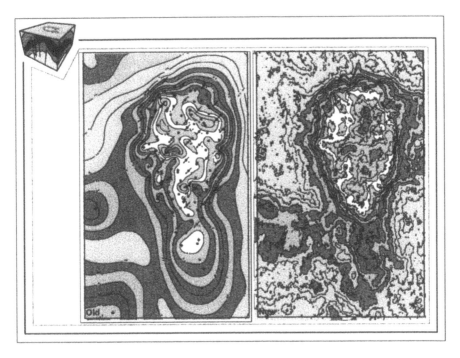

Figure 4 Map of one of the Ekofisk Formation layers comparing the ERC porosity distribution with a previous characterization effort

actual log data. Comparisons to actual log data suggested the correlations were well behaved.

Permeability

Developing a good understanding of the fracture distribution in the chalk is fundamental in predicting the absolute permeability distribution, the hydrocarbon recovery mechanisms and efficiencies for various hydrocarbon recovery techniques. In the Ekofisk Field fracture intensity can be related directly to petrofacies. Five types of fractures, mapped and classified from core data, were identified as healed, stylolite, tectonic, irregular and slump fractures. Each fracture type was then correlated to three distinct facies categories grouped based on general chalk type and analysed with respect to the fracture intensity (number of fractures per foot).

Permeability and its heterogeneity are critical parameters affecting reservoir fluid flow both in terms of volume and direction. Fluid flow characteristics of Ekofisk Field are largely governed by the distribution, orientation and interconnectivity of the natural fracture system. To honor this mechanism, and to capture the intrinsic heterogeneity and complex nature of the field, an algorithm was developed based on the log linear relationship between fracture intensity data from core and well test effective permeability[6]. During development, the basic relationship was modified to incorporate variations associated with: 1) chalk petrofacies; 2) fracture type; 3) porosity; 4) structural location; 5) structural curvature; and 6) silica content.

To calibrate the algorithm, permeability determined from distributing total well test flow capacity (kh) based on production log contribution was used as a tuning parameter. As a final step, geostatistical techniques were applied to tune the algorithm to the well test data. **Fig. 5** represents a flow chart of the effective permeability algorithm. The algorithm is implemented within the geological model and, therefore, provides a three-dimensional distribution of permeability.

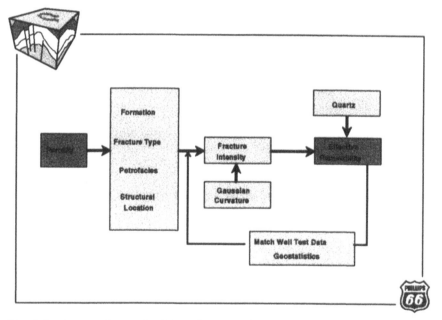

Figure 5. Flow chart representing process for calculating effective permeability within the 3D geological model.

Fig. 6 is an areal view of a stratigraphic slice of an Ekofisk formation taken from the geological model. Immediately obvious from this figure is the heterogeneity, as depicted by the various degrees of shading, present between grid cells; the lighter areas represent higher permeability while the darker areas trend towards matrix permeability.

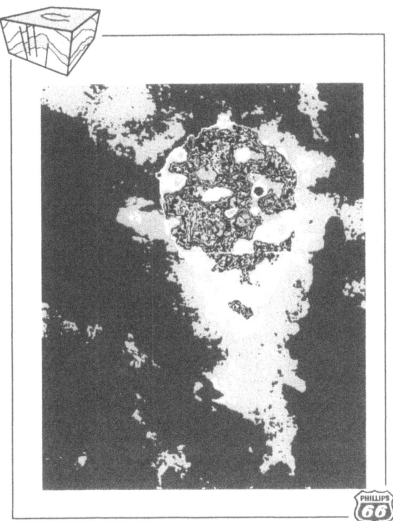

Figure 6. Areal view of a stratigraphic slice of an Ekofisk formation from the geological model.

Relative Permeability

Mechanistic studies were performed to generate pseudo relative permeability curves for the 3D reservoir flow model. Pseudo relative permeability is employed to account for fine scale variations in effective permeability, to reduce the dimension of the model and to control numerical dispersion, when upscaling from the fine scale geological model to the coarser resolution of the flow model. The pseudo curves were developed from 2D fine scale cross-sectional models, extracted directly from the 3D geological model to capture vertical heterogeneity. The number of pseudo relative permeability curves were limited by identifying areas having similar maximum to average flow capacity ratio and vertical heterogeneity. A total of 30 areas were identified with sufficiently diverse characteristics. **Fig. 7** illustrates the process in detail for calculating the pseudo-relative permeability curves.

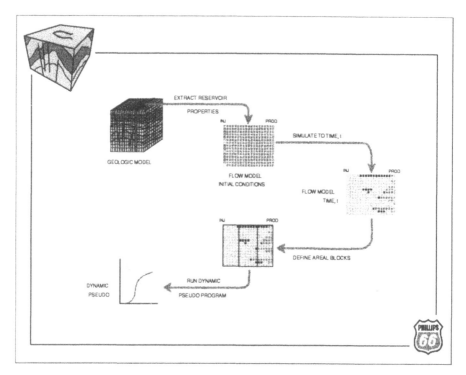

Figure 7. Process for calculating pseudo reslative permeability curves based on a 2D fine scale cross-sectional model extracted directly from the 3D geological model.

Anisotropy

Detailed seismic mapping of fault planes in 3D resulted in the mapping of more than 300 faults. The basis for the fault interpretation was analysis and integration of a number of seismic attributes generated at the major reservoir boundaries. This analysis was linked in three dimensions to provide fully correlated fault planes for each fault. The faults were resolved into three distinct systems, seen in the fault orientation analysis in **Fig. 8**. Here the interpreted Ekofisk faults systems are compared with theoretical data. The excellent agreement between regional data, field data and theoretical models allows the direction of compressional and extensional forces to be established and verifies the orientations of the three fault types; thrust, normal faults and strike slip faults.

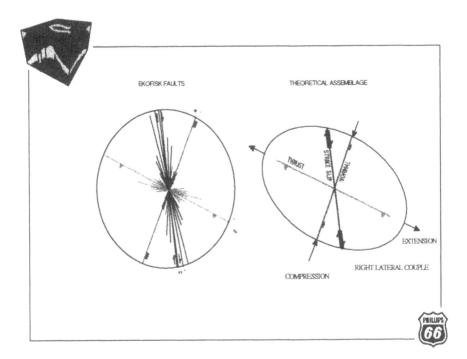

Figure 8. Fault orientation analysis depicting the interpreted Ekofisk fault systems with theoretical assemblage data.

From the orientation of these fault systems and the interpreted stress field it is anticipated that the normal faults and strike-slip faults likely provide higher permeability conduits for fluid flow. The evaluation of well to well tracer tests, interference tests, production/injection well responses and initial history match results suggest these relationships are correct. These data suggest preferential flow NE-SW along open normal faults, secondary flow N-S along partially open strike slip faults and no clear preferential flow NW-SE along closed thrust faults.

The application of a structural curvature correction to permeability and X and Y directional permeability modifiers associated with mapped faults allowed: 1) areas with higher fault density to have higher permeability values, and 2) anisotropy to be captured through the high structural curvature values associated with large normal faults.

3-D MODEL

Field architecture and reservoir attributes were assembled within the 3D geological model. Grid cell size in the model is 36 meter square with a total of 365 rows and 251 columns for each horizon. The following criteria were considered in determining cell size; 1) single wellbore per cell; 2) optimize seismic horizon interpretation; 3) optimize utilization of seismic modeling results; and 4) model fault planes with sufficient detail. Seismic data were based on a 12.5 meter sampling but due to software limitations it was not feasible to define a grid based on this spacing.

Fourteen sequences have been defined in the model representing the geological layers, or flow units, present in the Ekofisk field. The fourteen sequences have been further subdivided into a total of 275 layers which are dependent on the anticipated depositional relationships within each sequence. These are defined by the structural and sedimentological model developed for the field. Optimum layer thickness within each sequence was determined by applying vertical variogram analysis to gamma ray, water saturation and porosity traces to determine the degree of variation in each of the attributes in the vertical direction.

The high resolution definition of the 3D geological model grid in vertical and lateral directions allows for a fairly fine resolution representation of the reservoir description and the ability to capture the intrinsic heterogeneity associated with the field. The final model contains over 20 million data cells. Each model cell is populated by 30 reservoir attributes which include porosity, petrofacies, water saturation, permeability and silica content. The geological model and the detailed 3D picture of the reservoir became the basis for upscaling to a new full field 3D reservoir flow model.

UPSCALING

The geological model has captured the reservoir's heterogeneity at a high degree of resolution. In the upscaling process to the fluid flow model, preservation of heterogeneity is critical. Standard upscaling techniques based on simple averages of permeability do not suffice; for this reason an alternative approach was employed, a technique based on fluid flow, or flux, within the geological model. This technique uses simplified flow simulations for calculating effective properties of complex, 3D reservoir models and addresses several of the drawbacks of standard methods. The method averages flow potential and does not require the user to choose a particular averaging technique. By emphasizing transmissibility over permeability it easily incorporates transmissibility multipliers for modelling flow barriers. In addition it ensures proper volumetric averaging of properties, such as porosity, across no-flow regions. The resulting fluid flow model of 40 thousand cells is defined using the same structural grids as the geological model such that geological units within the two models are identical. Comparisons have shown that the intrinsic heterogeneity of transmissibilities is preserved in the upscaling process.

The direct link between the geological and reservoir models provides new opportunities for multi-disciplinary teams to work with consistent tools at all times and opportunities for multi-disciplinary resolution of well and reservoir performance issues. It also provides a feedback loop between geological and reservoir models using upscaling and downscaling technologies.

CONCLUDING REMARKS

An integrated high resolution geological model and a lower resolution reservoir flow model have been developed for the detailed planning of the redevelopment of the Ekofisk Field. A number of innovative techniques have been applied within the disciplines of geoscience, petrophysics and reservoir engineering resulting in integrated models which enable team-driven decisions to be made regarding reserves optimization (**Fig. 9**). The models have been developed using all data, information and knowledge available during the time span of the project and processes have been and are being put in place to ensure the continuous and rapid updating of the geologoical and fluid flow models as new data, information and knowledge becomes available.

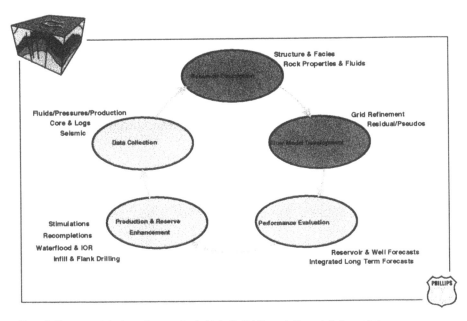

Figure 9. Reserves optimization cycle as associated with the Ekofisk Reservoir Characterization project.

ACKNOWLEGEMENTS

The authors acknowledge permission to publish the above paper from Phillips Petroleum Company Norway and Co-venturers, including Fina Exploration Norway SCA, Norsk Agip A/S, Elf Petroleum Norge AS, Norsk Hydro A.S., Den norske stats oljeselskap a.s., TOTAL Norge A.S. and Saga Petroleum A.S. Special acknowledgement is given to the following individuals for their comments, suggestions and support throughout this work: B. J. Crabtree, L. R. Allen, D. J. Ellen, G. V. Søiland, R. G. Harris, S. Tevik, S. Våge, G. Santellani, C. T. Feazel and G. H. Landa.

REFERENCES

1. Thomas, L. K., Dixon, T. N., Evans, C. E., and Vienot, M. E.: "Ekofisk Waterflood Pilot, " JPT, February, 1987, pp. 221-232.

2. Sylte, J. E., Hallenbeck, L. D., and Thomas, L. K.: Ekofisk Formation Pilot Waterflood, " Paper SPE 18276 presented at the 1988 SPE Annual Technical Conference, Houston, Texas, October 2-5, 1988.

3. Hallenbeck, L. D., Sylte, J. E., Ebbs, D. J., and Thomas, L. K.: "Implementation of the Ekofisk Full Field Waterflood," SPE 19838 presented at the 1989 SPE Annual Technical Conference, San Antonio, Texas, October 8-11, 1989.

4. Bentsen, R. G., and Anli, J.: "Using Parameter Estimation Techniques to Convert Centrifuge Data Into a Capillary-Pressure Curve, SPE 5036, 1975.

5. Leverett, M.C.: "Capillary Behavior in Porous Solids," Trans. AIME, (1941), Vol. 142, 152-169.

6. Agarwal, B., and Allen, L. R.: "Ekofisk Field Reservoir Characterization: Mapping Permeability Through Facies and Fracture Intensity," SPE 35527 presented at the SPE European 3-D Reservoir Modelling Conference, Stavanger, Norway, April 16-17, 1996.

Proc. 30ᵗʰ Int'l. Geol. Congr., Vol. 18, pp. 205~215
Sun Z. C. *et al.* (Eds)
© VSP 1997

On Interwell Prediction for Clastic Reservoir

HUANG SHIYAN

Research institute of Petroleum Exploration and Development, CNPN,
CHINA
WANG KUANGYUEN
Daqing Petroleum Administrative Bureau, CNPC, CHINA

Abstract

This article is about how to use the date from an area of close well pattern rather than from outcrops to work out a methodology to predict the distribution characteristics of interwell sandbodies, which will be applicable in an area of wide well pattern. First, using the data of close well pattern from the reservoir of Pu 1 Unit of the Middle-Northern area of Saertu Oilfield, Daqing, we constructed some single-sandbody framework models and prototype models of straight-low sinuosity channels. Later on, we carried out geo-statistical analysis to reveal the fractal characteristics of petrophysical parameters within a sandbody. Then, we used a fractal method to predict petrophysical parmeters and verify the validity of the method. Finally, we arrived the methodology of interwell sandbody prediction, which will become the basis of constructing a fine and quantitative reservoir predictive model.

Keywords: interwell sandbody prediction, geological model, fractal geometry, petrophysical parameter, Daqing Oilfield.

INTRODUCTION

The Oilfields in Eastern China, such as Daqing, Shengli, and Dagang, etc., have entered into the stage of high water cut. The remaining oil more being decentralized, the more difficult it is to produce oil in a water drive. Besides this, the reservoirs in Eastern China are mainly continental sedimentary facies with intense heterogeneity, which leads to oil recovery is very low in water flooding. Unfortunately, the high water cut stage is and still will be an important phase in the future. In order to improve development effect and enhance the oil recovery, the distribution of subsurface reservoirs and remaining oil must be known clearly, so a methodology to predict the interwell distribution features of sandbodies is needed badly.

The interwell sandbody prediction consists of two aspects: the first is the prediction of the sandbody distribution, including the distribution of depositional genetic units and architectures (flow units and flow barriers) in a sandbody; the second is that of the distribution of some reservoir properties (such as permeability). The essential work is to work out a methodology to effectively predict the continuously changing features of a subsurface reservoir.

There are quite few references on this topic. The conventional method is outcrop analogue studying. This is, to apply outrop reservoir characterizations to a similar subsurface reservoir. First, an outcrop that is similar to a reservoir is chosen. Then, the geological knowledge of the outcrop can be obtained through the detailed description and architecture element analysis, and a prototype model of petrophysical properties can be worked up through geostatistical analysis and stochastic simulations. Finally, these methods can only be applied to the oilfield modeling on a premise that the outcrop is the same as or similar to the subsurface reservoir.

There are no outcrop appearances around the oil bearing basins of Eastern China. As a result, it is impossible to understand the reservoir properties through detailed study of outcrops. However, the data of close well pattern are very rich, and the well spacing is so close that it is less than 50m and even as close as 20m on some occasions because of twice infills. That makes it neccessary and possible for us to take advantage of the abundant subsurface data to work out some prediction methods of interwell sandbodies.

PROCEDURE OF THE INTERWELL PREDICTION

The procedures are carried out as follows:

Step 1: Choosing a typical area of close well pattern

In order to study a suitable target zone, we must first investigate its geological settings, which include startigraphy, sedimentary environments, even structural backgrounds, as well as its process of exploration and development. Then we can choose a typical area to study, in which the well spacing is very close; data are rich; especially, the facies types are abundant and complete.

Step 2:Constructing the single sandbody frame models through detailed microfacies analysis

Associating the core observations and descriptions with corresponding well logging curves, it is possible to determine the relationship of the lithofacies and the log response. The different kinds of microfacies can be distinguished by different logging curve shapes. Through log correlating and mapping on the plane, frame models of single sandbodies can be constructed.

Step 3: Studing the reservoir heterogeneity

In this step, we study shapes and the scales of the sandbodies, the different level boundaries and architecture elements within a sandbody, and the in-layer heterogeneity of a sandbody. The aim is to summarize the geological knowledge of different kinds of sandbodies. All the knowledge can be applied to the prediction of the distribution of sandbodies and their heterogeneity in a wide spacing area or an exploration area.

Step 4:Evaluating reservoir with mullet-well logging curves

After normalization, cored wells can be used to study the relationships among four features:

lithology, petrophysical properties, log responses and oil-bearing. After that, we will acquire a log interpretation model that can be used to calculate the petrophysical parameters dot by dot in a given interval. All kinds of parameters of a single well can be obtained through multi-well log processing and interpreting.

Step 5: Constructing a section prototype model of petrophysical parameters

In order to accurately predict the petrophysical properties in a grid size of 50. 0m0. 5m, at first we must study the prediction method for an existing model that always has the indentical predictive accuracy. It is very important to construct some prototype models through the study of an area of close well pattern. This can be done by multi-well assembling under the control of some sedimentary theories and rules. In the paper, we will construct some permeability prototype section models of straight-low sinuosity channels.

Step 6: Studying the fractal parameters of predicting the interwell permeability distribution

We can determine the fractal parameters that can characterize the permeability distribution by analyzing the data from the prototype models with a fractal method. These parameters include: Hurst index (H), fractal changing range (V_H), the short axis angle of the anistropic ellipse O, and the ratio of the long axis to the short axis (a/b). All these parameters can be applied to the interwell prediction of the same kinds of reservoirs which have the identical depositional environments and sandbody scales in the same oilfield.

Step 7: Using the predicting method of the interwell permeability distribution and verifying its validity

The trend distribution can be generated by the Kriging method under the control of some geological conditions. A series of equi-probable fractal realizations can be gained through the fractal field interpolating method with fractal parameters. The validity can be verified by taking out some wells from the prototype models. After taking out some well data, a model is near to a wide well spacing case. Then we generate a series of equi-probable realization with the above-mentioned method and parameters and calculate the average, the standard deviation and histogram of them. We can say the method is valid, if these statistical indexes are similar after comparing them with that of the prototype models.

A CASE STUDY OF STRAIGHT-LOW SINUOSITY CHANNEL SANDBODIES OF PU 1 UNIT IN DAQING OILFIELD

1. Choosing a typical area of close well pattern

Daqing Oilfield is located in the Northern Songliao Basin. The pay zones belong to the Cretaceous System. The sedimentary environment is fluvial-delta facies. The main reservoirs of Pu_1 are straight to low sinuosity sandbodies.

Daqing Oilfield has been developed for more than 30 years since the early 1960s. Now the well spacing is very close in some places after twice systemic infill drillings. We chose three blocks of close well pattern to study, in which the primary well spacings are less than 100m, most of them within the range of 40m to 60m due to infills for development tests. There are quite a few double wells in site or triple wells in site, which are very beneficial to our study on the interwell prediction.

2. Constructing single-sandbody frame models

One of the blocks we studied is located in the north part of the Saertu Oilfield. It is a development testing area with thick pay zones. After observing the cores of seven wells and their logging curves in this block, we set up the relationships of different level boundaries with logging responses. Then a detailed microfacies analysis of the three typical blocks was carried out under the guidance of sedimentological theories.

To construct a frame model, the first thing to do is to discriminate between channel and non-channel sandbodies, which can be done by recognizing the difference of the features of lithology, texture, and thickness of the sandstone, as well as logging curve features and the rhythm of the sandstone.

The comparative altitude change to the mark layer, the change of logging curve shapes and the horizontal thickness variation in the different parts of a channel sandbody; and the difference of the production performances between two neighboring wells; can be used to distinguish a single sandbody from a complex channel belt.

The accurate locations and the shapes of the boundaries can be determined according to the sedimentological theories and the channel types. It is not accurate enough just to put a sandbody boundary in the midpoint between two wells. The local change of a boundary should conform the sedimentological rules of a sandbody deposition and distribution. However, the accuracy of the boundary locations is still dependent upon the well pattern density.

The well spacing in the chosen area is very close, so the errors of the boundaries are less than 50 meters and few exceed 100 meters. The possible relative error is in range of 10% to 20% of the channel sandbody width. It can be believed the constructed single-sandbody frame models are determinate (Fig. 1).

3. Reservoir Characteristic of the Pu1 Unit

The reservoir in the Pu1 consists of straight to low-sinuosity channel sandbodies. From a statistical analysis of 85 horizontal sections, we found there are two kinds of width scales: 550m and 300m.

The channel sandbody thickness varies from 2.0m to 6.0m, in average, 5.0m or so. The lithofacies is fine-medium sandstone with medium grain size of 0.13-o.18mm.

The porosity and permeability in Pu 1 Unit are medium to high. The porosity varies from 20% to 30%, and the permeability from $200 \times 10^{-3} \mu m^2$ to $2000 \times 10^{-3} \mu m^2$. The variation factor of the permeability is greater than 0.7. It can show there is intense in-layer heterogeneity in the Pu1 Unit. The permeability is the highest in the lower part of a channel sandbody.

The muddy barriers of low permeability consist of gray or white-gray silty shale or shaly siltstone. The thickness of a single barrier is about 0.1m to 0.2m. Its air permeability is less than $30 \times 10^{-3} \mu m^2$. The barrier appearing frequency (barrier number per meter) is about 0.4

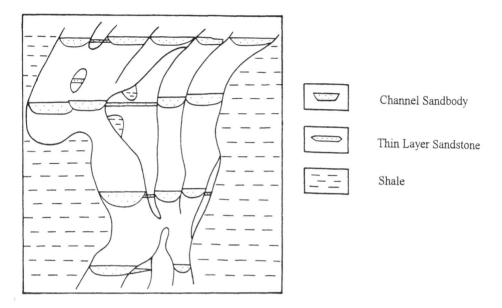

Fig. 1 The single sandbody framework model of Blockwhich there are 275 wells in 9. 2 km². There are 30 wells in each square kilometer.

and the density is 8% to 10% to unit sandstone thickness, so a barrier is very hard to trace in a horizontal direction.

The change of some parameters in the area of close well spacing is shown in Figure 2.

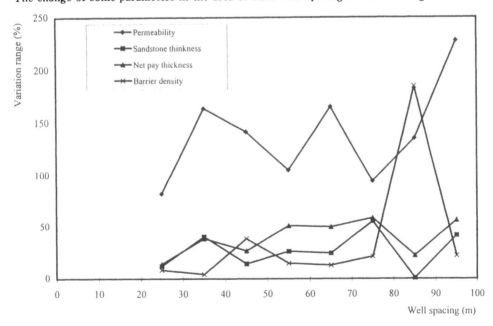

Fig. 2 The relationship of the variation range of some reservoir parameters versus well spacing in the channel sandbody.

Form the figure we can find the permeability varies more rapidly than other parameters. In the work, the permeability fluctuates intensely between wells. Therefore the applications of linear interpolation and other determinate interpolation to the prediction of petrophysical property distribution are not very suitable to the actual case. It is essential to find a good method of stochastic simulation to predict the distribution of interwell petrophysical parameters.

4. Constructing permeability prototype section models

According to sedimentological rules, the sandbodies belonging to the identical environment, identical channel type and scale have the same or similar reservoir characteristics. The procedure to construct a model is as follows:

Choosing basic sections: the horizontal sections in same scale in a channel are chosen as basic sections.

Dividing the basic sections into intervals of equidistance according to the scale of a channel sandbody: in order to construct prototype models of 50. 0-0. 5m in the horizontal direction, a section of 300m width must be divided into 6 equidistances, and a 550m wide section be divided into 11 equidistances. Each interval has a number, the first from the left is number 1, the second is number 2, so any well in a section has its number.

Constructing the prototype models of section: One of the sections that has the maximum numbers of wells is chosen as a key section. The well-absent interval can be replaced by an interval with the same number from the neighboring section reasonably. Therefore, we get an ideal section for a prototype model.

Endowing the prototype models with permeability values: Each well in the ideal section is endowed with permeability values from logging interpretation in 0. 5m interval. Finally, we get a prototype model with an accuracy of 50. 0m ∼ 0. 5m.

All the models in the three chosen blocks are constructed in the above-mentioned steps.

5. Fractal analysis of the prototype models

Fractal geometry is a means used to describe the shapes or phenomena that are unstable, irregular, with self-similarity but a characteristic length. Some fractal parameters and formula can be used to describe the complex fractal phenomena in nature.

Fractal parameters and their geological meanings

When the fractal means is used to study the spatial change of the reservoir permeability, the features of the pemeability distribution are characterized by the parameters as follows:

HHurst index, the topological dimensions minus the fractal dimensions, refers to the varying frequency of a fractal phenomenon. It can be calculated from the varigram, R/S and FFT analysis.

V_H variable range of a fractal phenomenon.

a/b ratio of the anistropic ellipse long axis to the short axis. It refers to the heterogeneity in the different directions.

The short axis angle of the anistropic ellipse. It refers to the direciton of changing most rapidly.

The fractal parameters from the data of each cored well are calculated with the varigram. R/S and FFT methods. The results are shown in Table 1.

Table 1 The fractal parameters of the single well cored interval

channel scale	well name	sample number	sample interval	average permeability	standard deviation	H_R	H_V	H_F	variation factor
300m	H332	55	0.10	1184.0	1124.0	0.73	0.71	0.80	0.95
	H333	55	0.10	656.0	456.0	0.67	0.65	0.74	0.69
	H334	53	0.10	891.0	982.0	0.66	0.71	0.78	1.10
	H344	49	0.10	584.0	956.0	0.71	0.69	0.70	1.64
	J335	40	0.10	1112.0	1266.0	0.70	0.60	0.77	1.14
550m	H335	45	0.10	1259.0	1203.0	0.64	0.65	0.68	0.95
	ZG2-8	90	0.05	1217.0	1096.0	0.63	0.56	0.77	0.90
	ZJ3-7	67	0.10	2180.0	1240.0	0.73	0.72	0.77	0.57
	ZJ4-8	70	0.10	1881.0	1475.0	0.69	0.67	0.72	0.78

From Table 1, it can be found the permeability of the channel sandbodies presents fractal feature with H varying from 0.6 to 0.8 in the vertical direction. It indicates that the distribution of permeability can be characterized by the fractal means.

Fractal analyzing to the prototype models of cross section.

The fractal parameters of the cross section permeability prototype models are calculated with varigram method (Table 2).

Table 2 The fractal parameters of the cross section prototype

channel scale	block name	data number	H	a/b		average permeability	standard deviation	variation factor
	block 1	51	0.68	3	90	293	167	0.56
300m	block 2	49	0.57	3	90	481	435	0.90
	block 3	51	0.63	3	90	382	375	0.98
	average		0.62	3	90			
	block 1	93	0.83	7	90	418	284	0.67
550m	block 2	109	0.86	7	90	623	431	0.69
	block 3	88	0.82	7	90	411	213	0.71
	average		0.83	7	90			

The results show that the variation of the permeability in the vertical direction is greater than that in the horizontal. Hurst index (H) and the ratio a/b are in direct proportion to the channel sandbody scales. All of these conform the geological rules. These paramaters can be used to predict the distribution of the permeability in the straight to low-sinuosity channel sandstone bodies.

6. Predicting of the interwell permeability distribution

In order to verify the methods of predicting interwell permeability distribution in an area of wide spacing, we take out some wells from a prototype model and make the section similar to the well spacing of 200-300 meters. The trend distribution can be generated by the Kriging method under the control of some geological conditions. Considering this, a series of equi-probable fractal realizations have been produced by the application of the fractal field interpolating method with the fractal parameters (Figure 3). We calculated the averages, standard deviations, and the histograms of these realizations, the Kriging field and the prototype model data (Table 3, Figures 4).

Comparing these indexes with these of different type models, it can be seem that the statistical results from the fractal realizations are similar to those of prototype models. In Kriging field the heterogeneity of the permeability is smoothed. The fractal stochastic generated from taking out some wells from the 550m width scale prototype model of Block. The black dots in the figures present the sample location.

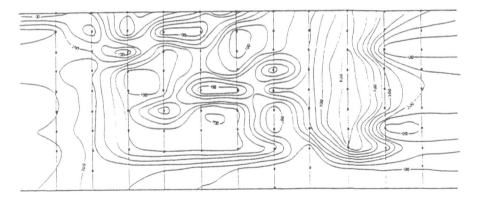

3(A) The cross section permeability prototype model

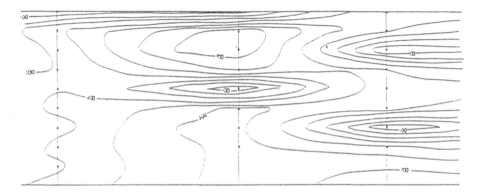

3(B) The Kriging field after taking out 8 wells from the prototype model

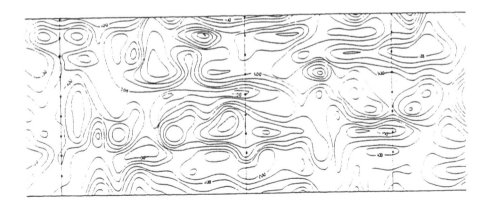

3(C) The fractal realization 1, H=0. 83, a/b=7, V_H=150

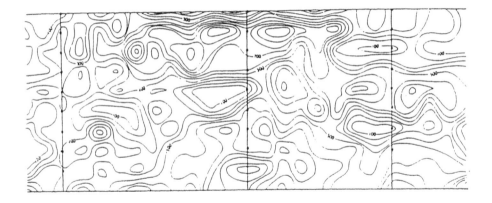

3(D) The fractal realization 2, H=0. 83, a/b=7, V_H=150

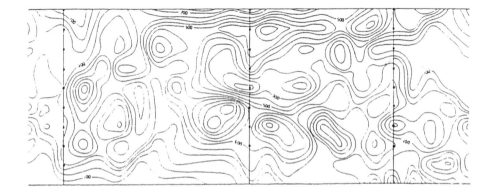

3(E) The fractal realization 3, H=0. 83, a/b=7, V_H=150

Figure 3   The cross section permeability prototype model and different models

Table 3 Statistical characteristics of the different permeability distribution models of the channel sandstone bodies of Block

channel width scale	Model	Grid size		Average permeability	Standard deviation	Fractal parameters
	Prototype model	50	0.5	293.00	167.50	
	Wide spacing section	200	0.5	313.30	132.80	H=0.68
	Kriging field	50	0.5	263.30	140.00	a/b=3.0
300m	Fractal realization 1	50	0.5	221.30	193.40	V_H=150.0
	Fractal realization 2	50	0.5	278.90	182.70	
	Fractal realization 3	50	0.5	259.30	196.80	
	Prototype model	50	0.5	418.00	284.00	
	Wide spacing section	300	0.5	408.40	221.00	H=0.83
	Kriging field	50	0.5	408.50	209.20	a/b=7.0
550m	Fractal realization 1	50	0.5	313.40	270.80	V_H=150.0
	Fractal realization 2	50	0.5	520.30	378.00	
	Fractal realization 3	50	0.5	441.70	284.90	

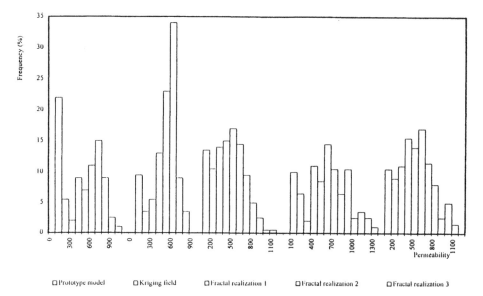

□ Prototype model   □ Kriging field   □ Fractal realization 1   □ Fractal realization 2   □ Fractal realization 3

Figure 4   The histogram of the different permeability distribution models in 550m width scale sandbody Block.

Realizations generated under the control of the fractal parameters not only maintain the changing trend of permeability districution but also reflect the permeability unpredictability. All of these verify that the fractal interpolating method and the fractal parameters are valid to predict the permeability distribution of the straight and low-sinuosity channel sandstone bodies in a wide well spacing case.

CONCLUSIONS

When the oilfield development enter into the high water cut stage, we can work out the methodology of interwell prediction by using the subsurface data of a close well pattern area. Through detailed microfacies analysis we can construct some single-sandbody frame models and analyze their architecture elements. Then we can summarize the geological knowledge of the different types of sandstone bodies that can be used to predict the distribution of the sandstone bodies in another wide spacing case.

The changing of permeability is fractal. So permeability distribution can be predicted by fractal means with some fractal parameters under the control of the geological rules.

REFERENCES

1. T. A. Hewett. Fractal distribution of reservoir heterogeneity and their influence on fluid transport. SPE 15386,1986
2. Qiu Yinnan. Reservoir geological model. ACTA Petrolei Sinica, Volume 12,1991

PART IV

MARINE CARBONATE SOURCE BEDS AND RESERVOIR

Proc. 30ᵗʰ Int' l. Geol. Congr., Vol. 18, pp. 219~223
Sun Z. C. *et al.* (Eds)
© VSP 1997

The Geochemical Characteristics of Marine Oils in China

HUANG DIFAN and ZHAO MENGJUN

Research Institute of Petroleum Exploration and Development, *Beijing* 100083,
CHINA

INTRODUCTION

Now, the oil and gas exploration mainly related to Palaeozoic marine strata has been in progress in the Tarim Basin, Sichuan Basin, North China Basin and Ordos Basin of China. All Palaeozoic marine source rocks in China generally have a low abundance of organic matter, high maturity and good types. In the article, we discuss the geochimical features of marine oils in China, mainly based on the marine oils of the Tarim Basin. And all of us know that the important paybeds of these oil fields are Ordovician carbonates. Carboniferous and Triassic sandstones and their source strata are mainly Cambrian-Lower Ordovician carbonates and Middle-Upper Ordovician mudstones.

RESULTS AND DISCUSSION

The characteristics of marine oils with middle wax and low sulfur

Marine oils generally have a specific gravity from 0. 80 to 0. 92 in the Tarim Basin, of which 52 percent are low wax oils (wax, <5. 0%), 38 percent are middle wax oils (wax, 5. %-10. 0%) and 10 percent are high wax oil (wax, >10. 0%). According to our study, part of the higher wax oils are formed because of the following factors.

(1) Specific source   Bacteria could generate certain amounts of waxy hydrocarbons;

(2) High maturity   A simulation test showed that the kerogen composed of bacteria and algae could generate waxy hydrocarbons in high maturity (t, >375 ℃).

There are 88 percent of low-sulfur oils (sulfur, <0. 5%), 10 percent middle-sulfur oils (sulfur, 0. 5%-1. 0%) and 2 percent high-sulfur oils (sulfur, >1. 0%) in all marine oil samples of the Tarim Basin, and the high-sulfur oils are heavy oils subjected to biodegradation. There are two reasons to explain the marine oils with low sulfur: open platform and basin facies sedimentary environment for the source strata; deep preservation condition for the oils.

The carbon isotopic ratios of marine oils

The carbon isotopic ratios of marine oils are much lower and have a narrow variation from

−30. 45‰ to −33. 83‰ in the Tarim Basin. We considered that the marine oils were derived
from the kerogens of the Cambrian and Ordovician whose carbon isotopic ratios are from −
27. 48‰-−31. 28‰ which had a good correspondance with the ratios of the marine oils. And
the carbon isotopic ratios of the marine oils are consistent with those of Lower Palaeozoic oils
and rocks in the world. The marine oils have an obvious difference from the oils derived from
lacutstrine strata and coal measures in carbon isotopic ratios of whole oils and their compositions (Fig. 1).

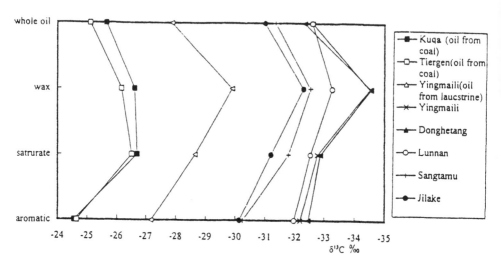

Fig. 1   The carbon isotopic ratios of the whole oils and their compositions in the Tarim Basin.

In addition, the whole oil carbon isotopic ratios in oil fields of the eastern Tarim are different
from the ones of the western Tarim. The mean carbon isotopic ratios of whole oils are
−31. 0‰ to −32. 66‰ in the eastern oil fields and −32. 40‰ to −32. 63‰ in the western
oil fields (Table 1). And the ratios of their compositions in the western oil fields are also
lower than the ones in the eastern oil fields. Different maturities of the oils are one of the
causes responsible for the carbon isotopic ratio difference between the oils in the western
fields and the oils in the eastern fields.

The biomarkers of marine oils

The mean ratios of pristane to phytane of the marine oils in the Tarim Basin are from 0. 88 to
1. 17, which have a narrow change, and the odd carbon number predominance of the normal
alkanes is not obvious, as contrasted with the Ordovician oils in mid-continent whose ratios
of pristane to phytane are from 0. 63 to 1. 82. The marine oils of the Tarim Basin contain
more isoprenoid hydrocarbons, in which Pr/nC_{17} only are from 0. 04 to 0. 28. But the content
of the long-chain isoprenoid hydrocarbons whose carbon number range is from C_{21} to C_{45} is
much lower than the one of the short carbon number isoprenoid hydrocarbons such as pristane and phytane in the Tarim Basin. Generally, the long-chain carbon number isoprenoid
hydrocarbons were considered to be generated by bacterium source.

Table 1 The geochemical parameters of marine oils in the Tarim Basin

Oil fields		Samp. No.	Main peak carbon No.	Pr/Ph	Pr/nC$_{17}$	Ph/nC$_{18}$	CPI	$\delta^{13}C_{whole\ oil}$ Mean‰	tri./hop.[a]	R°(%)[b]	$\dfrac{V}{V+Ni}$
W. area	Yingmaili	5	15-19	0.88	0.42	0.51	1.13	−31.87−−33.40 (−32.63)	0.60-0.95	0.89	0.64
W. area	Donghe-tang	8	18,19	0.93	0.35	0.40	1.10	−31.69−−32.80 (−32.35)	0.60-1.00	0.78-0.88	0.80-0.84
W. area	Lunnan	6	17,19	0.92	0.40	0.51	1.15	−32.00−−33.83 (−32.61)	0.90-1.50	0.91	0.87
W. area	Tazhong 10	1		1.17	0.33	0.33	1.13	−32.20−−32.60 (−32.40)	2.60	0.85	0.60
E. area	Sangta-mu	6	13-17	1.01	0.27	0.31	1.05	−30.45−−32.10 (−31.37)	1.50-2.50	1.02	0.10-0.54
E. area	Jilake	3	15,17	1.05	0.22	0.27	1.01	−30.82−−31.20 (−31.01)	1.50-3.50	1.07	0.10-0.29
E. area	Tazhong 4	6	15-19	0.88	0.26	0.29	1.09	−32.10−−33.28 (−32.66)	1.50-3.00	0.99	0.69-0.80
E. area	Tazhong 1	5	14-16	1.14	0.35	0.40	1.07	−31.21−−33.30 (−31.97)	1.50-3.00	1.02	0.39

a. tri./hop. presents the ratio of tricyclic terpanes to hopanes; b. R°(%) is obtained by calculating using methyl phenanthrine parameters (Radke, 1982,1984).

The relative contents of C_{27}, C_{28} and C_{29} regular steranes respectively are 25%-40%, 15%-30% and 40%-60%, and the regular steranes are much higher than other steranes (Fig. 2). The high concentration of C_{29} steranes is an important index to distinguish the Ordovician marine oils from the marine oils of Mesozoic and Cenozoic, and also states the importance for cyanobacteria contributing to the oils. The rearranged steranes with the carbon number from C_{27} to C_{29} are rich, similar with the oils in the mid-continent. However, the typical features of the marine oils in China are much abundant pregnanes which reflect the high maturation of these oils. From Fig. 2, we could see that the highest peak of the C_{21} and C_{22} pregnanes is as tall as or taller than the peaks of C_{29} steranes.

C_{30} hopane is the highest peak among C_{29}-C_{35} hopanes whose mother source is tetrahydroxy-bacteriohopane. The marine oils in the Tarim Basin contain more tricyclic terpanes than other Palaeozoic marine oils in the world, and the ratios of tricyclic terpanes to hopanes can reach to 3.5 (Table 1), which reflect the high maturation of the oils, too. We conclude that the marine oils with the ratios of tricyclic terpanes to hopanes from 0.60 to 1.5 are generally normal specific gravity and mature, and have an obvious C_{23} peak preponderance of the tricyclic terpanes in the western oil fields of the Tarim Basin; the marine oils with the ratios

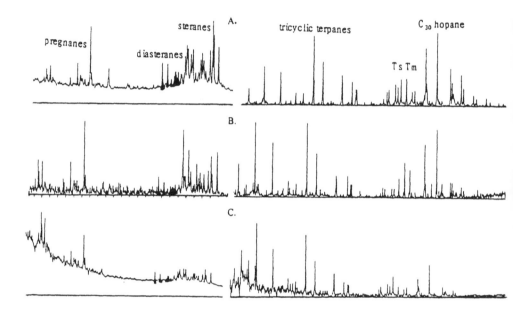

Fig. 2   The distribution of the steranes and terpanes in the marine oils in the Tarim Basin
A. Lunnan oilfield; B. Sangtamu oilfields; C. Jilake oilfields

from 1. 5 to 3. 5 mainly belong to condensates with a high maturity, and have the preponderance of both C_{21} and C_{23} peaks of the tricyclic terpanes in the eastern oil fields of the Tarim
Basin. Palacas et al. (1984) concluded that the crude oils derived from carbonates generally
had C_{23} peak preponderance of the tricyclic terpanes. So, in addition to the effect of the oil's
maturity, the lithology of the source rocks is another possible factor to make them different
for western and eastern oils of the Tarim Basin. The western area is composed mainly of carbonates of platform facies and the eastern area mainly mudstones of basin facies.

The aromatic features and maturity of marine oils

The compounds of thiophenes are rich in the marine oils although the sulfur content is very
low. The ratios of dibenzothiophenes to phenanthrene are more than 10, and the ratios of the
oils in Tazhong 4 are in the range from 122 to 379. It was very interesting to find retene in
the marine oils, which was considered only to be generated by the terrestrial plants before,
and the fact testified that it could also be generated by the source of bacteria and algae. Based
on the equation calculating oil's maturity by advanced aromatic parameters, we got the maturation values $R°$ of the marine oils in the Tarim Basin (Table 1). The marine oils are mature
in the western oil fields ($R°$ 0. 85%-0. 91%), and mature to highly mature in the eastern oil
fields ($R°$ 0. 99%-1. 07%).

The V-Ni fractions of marine oils

The V-Ni fractions (V/V+Ni) generally are from 0. 10 to 0. 90 in the marine oils of the

Tarim Basin. According to the fractions combined with their low sulfur, we could conclude that all source rocks of the marine oils were deposited in the environment of weak reduction to reduction, under which nickleous cations and vanadyl cations were available for metallation with vanadyl cations being hindered in part by the formation of hydroxides and nickleous cations being hindered in part of meatastable sulfide ions. It is noteworthy that the V-Ni fractions of the marine oils in the eastern oil fields are generally less than 0.50, and more than 0.50 in the western oil fields. The marine oils of the eastern oil fields were possible mainly expelled from mudstones, in which iron was rich and sulfur was poor, so there was abundant iron to react with sulfurous ions forming metastable iron sulfide, nickelous cations were available for bonding in petroleum, and the proportion of vanadium to nickel would be low ($V/V+Ni < 0.5$). The marine oils of the western oil fields were possibly mainly expelled from carbonates, in which iron was relatively poor and sulfur was relatively rich, so nickelous cations were hindered in part of metastable sulfide ions, and the proportion of vanadium to nickel would be high ($V/V+Ni > 0.5$).

CONCLUSIONS

(1) The features of marine oils in China, such as abundant normal alkanes, low sulfur, light carbon isotopic ratios, preponderance of C_{29} steranes among regular steranes and abundant rearranged steranes, are similar to those of the Ordovician in other basins of the world. The similarity of Ordovician oils in different basins states that the prokaryotic biota has a great contribution to the oil generation.

(2) The marine oils in China have some special features such as higher wax, weak odd carbon number predominance, abundant pregnanes and tricyclic terpanes and much lighter carbon isotopic ratios. This may be caused by the oil's high maturity, special depositional environments of the source rocks and special species of the prokaryotic biota in the Tarim basin.

(3) The marine oils have different maturities from the eastern to the western oil fields. In addition, the distribution of the tricyclic terpanes and the V-Ni fractions in the marine oils in the Tarim Basin may account for the difference of the lithology of the source rocks in the eastern area from that in the western area.

Proc. 30ᵗʰ Int' l. Geol. Congr. , Vol. 18, pp. 225~237
Sun Z. C. *et al.* (Eds)

Stratigraphic Zonation of Bangestan Reservoir, Binak Field Persian Gulf Coast, SW Iran

B. HABIBNIA and D. JAVANBAKHT

Faculty of Petroleum Eng. , The University of Petroleum

Ahwaz Fax. 0216153926 *IRAN*

Abstract

The present work deals with stratigraphic study of Binak (BK I-IV) oil filed (Bangestan Reservoir), located at Persian Gulf coast, about 22 km NW of Genaveh port SW Iran. More than seven wells have been drilied during these years, in which five of these are from Bangestan group. Two formations, Illam and Sarvak of Bangestan group are subdivided, based on reservoir rocks into five zones and one subzone. Zone I , with a low porosity carbonate, 350 feet thickness consisting mainly of grey to brownish fine limestone, characterized by the high content of argillaccous material. The subzone I which is overlain by zone I is about 260 feet thick, composed of dark grey to dark brown shale with subordinate thin beds of fine grained chalky limestone which contains no matrix oil. Zone I and subzone I are equivalent to Illam and Surgah formations (Cenomanian-Coniacian) in Lurestan state, which developed locally in Binak field. The top of Sarvak formation was placed at 10612 feet at the top of limestone bed, contained the Middle Cretaceous faun, *Valvulammina* sp. , *Nummuloculina* sp. and *Dicyclina* sp. The upper 1000 feet of Sarvak in BK- I is rich in *Oilgosteginids* associated with pelagic foraminfera, the most common species include *Stomoisphaera sphaerica* and *Calciosphaerula immominata*. The zone IV logged in BK- IV and recognized by a high gamma ray and very low neutron log response. The stratigraphic sections of all the zones petrophysical properties have been studied.

Keywords: Persian Gulf, Zonation, Bangestan, Cenomanian, Stratigraphy

INTRODUCTION

Petroleum occurs on all the continents of the world, although some continents are much richer in petroleum than others and it occurs in all the geologic system from Precambrian to Reccent, though some system are notably more profilic than others. Binak is a relatively small highly under saturated field, located on the Persian Gulf shore, about 22 km NW of Genaveh port South Iran.

On the surface, the Binak structure is a low relief feature expressed in highly eroded Aghajari Formation. Based on seismic and well data, the Bangestan structure is a symmetrical an-

ticline, about 22 km long and 10 km wide, as measured at the oil-water contact of 13/780
feet subset on the top of Zone (I) (equivalent to Sarvak Formation). Figure 1 shows the
location of the Binak Field. The vertical closure on the top of Khami level is about 7/500 ft.
The entire succession is believed to be conformable and no structure complications have been
noted. The Binak field was discovered in 1960, upon completion of BK-1 and commercial
production commenced in December 1967, after BK-2 was completed in Bangestan. BK-2
was drilled to a total depth of 12/462 ft, approximately 100 ft above the base of the oil bearing Sarvak formation. Figure 2 shows the fence diagram of Bangestan group in Rag-e-Safid,
Binak, Kilure, Karim, Gulkhari and Kharg fields.

ZONATION OF BANGESTAN FORMATION

A detailed geological study by previous workers (G. N. I. O. C), in the field showed that the
Bangestan formation could be sub-divided, based on reservoir rock, characteristic in to five
zones. The present authors on lithological bases have described the above zones. Figure 3
shows structure map on the top of the Sarvak formation.

ZONE- I

This is a low porosity carbonate 350 ft thick, consisting mainly of grey and grey-brown very
finely granular limestone, mudstone-wackstone characterized by the high content of argillaceous material and associated thin beds of marl.

Subzone(I)

ZONE- I and subzone- I are equivalent to the Ilam and Surgah Formations in the Lurestan
which developed locally in the Binak Field. These zones are completely missing, probably by
non-deposition, possibly by truncation as in the Rag-e-Safid and Kharg fields and are considerably thinner towards the Kilur Karim and Gulkhari Fields.

ZONE- I

ZONE- I is a carbonate zone, with about 190 ft, thickness , consists mainly of very line to
medium grained limestone, wackstone to mudstone, containing the Middle Cretaceous fauna.
Argillaceous material increases towards the base. Figure 4 shows the structure map on top of
zone I of the Bangestan group. The lithology and petrophysical properties of the Bk-2 in
zone I are listed in the table 1.

ZONE I

This zone has an average thickness of 320 ft, and composed primarily of interbedding wackestone to packestone, rich in pelagic foraminifera. This changes downwards into a thick

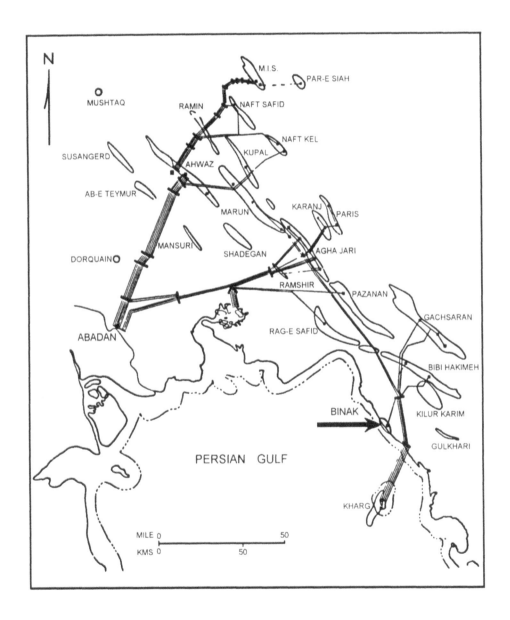

Fig. 1 Oil field location map.

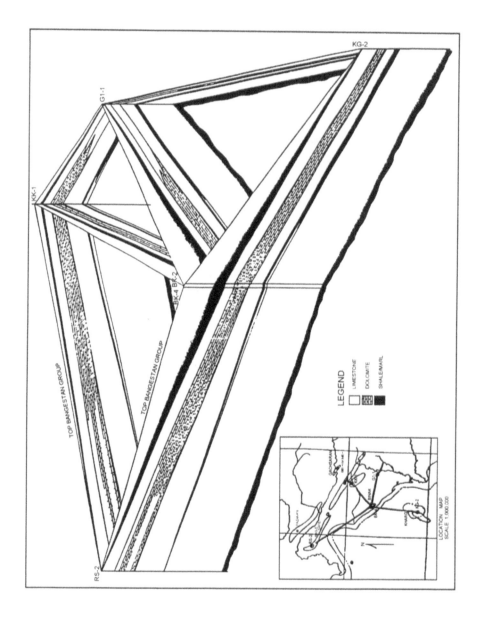

Fig. 2   Fence diagram of Bangestan group in Reg-E Safid, Binak, Kilur Karim, Gulkhari and Khard fields.

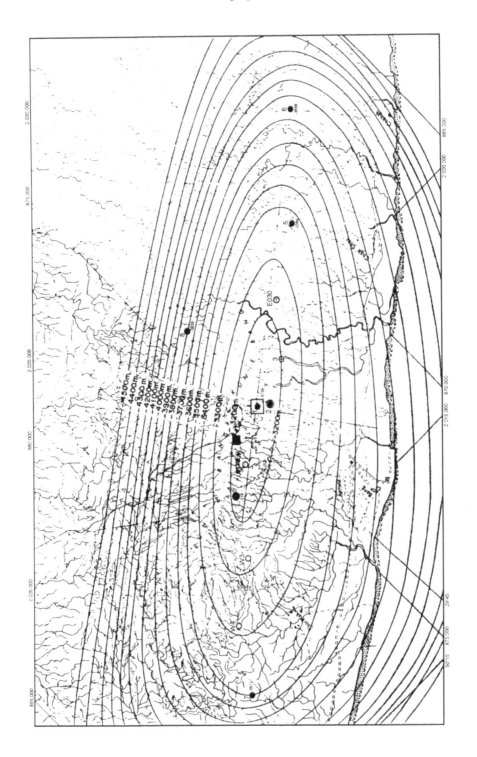

Fig. 3   Structure map on the top the Sarvak formation.

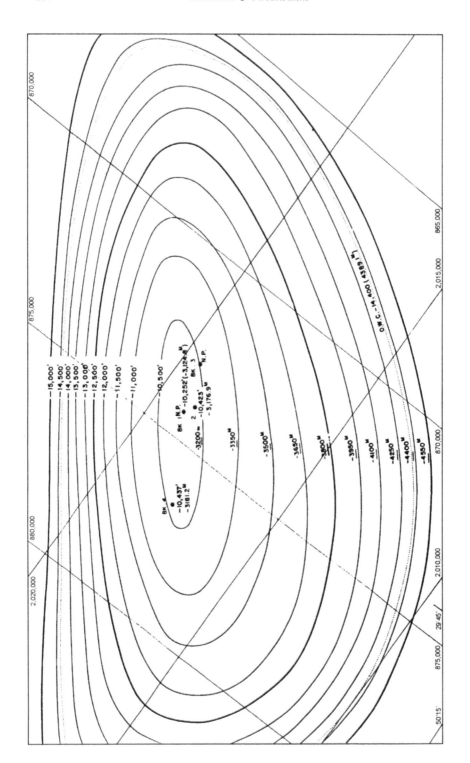

Fig. 4   Structure map on top zone **I** of the Bangestan

Table 1   The lithology and petrophysical properties of zone I in BK-2

Well	:BK-2
Zone	: I
Formation	:Sarvak
Top	:10614 ft.
Bottom	:10814 ft.

Depth ft	Lithology	Thickness ft	Porosity &	SW &
10624	LST	7	10. 1	30
10634	LST	10	8. 0	34
10640	LST	6	6. 7	39
10643	DST	3	7. 5	40
10663	DOL	8	9. 7	38
10668	LST	5	7. 4	44
10695	LST	7	6. 5	38
10717	LST	11	7. 3	34
10721	LST	4	7. 3	34
10750	DOL	12	25. 9	21
10769	DOL	2	17. 1	27
10775	DOL	3	14. 6	27
10798	DOL	12	7. 9	29

dolomite, Rudist facies, the major porous interval of the zone. This porous section which thins somewhat NW wards to BK-4 can be easily traced across the Binak Field. Figure 5 shows the structure map on top of zone Ⅱ. The average porosity, water saturation and net to gross ratio of Bk-2 in this zone are 9. 42%, 28. 17% and 0. 388 respectively. Figure 6 shows the ratio of limestone to dolomite for zone Ⅱ.

The lithology and petrophysical properties of BK-2 in zone Ⅱ are listed in table 2.

ZONE- Ⅳ

It is a low porosity carbonate zone, about 350 ft. thick, characterized by a very finely granular, sometimes chalky limestone with high content of argillaceous material, passing downwards into a 30-40 feet thick bed of greenish-grey shale at the base of the zone. Figure 7 shows the struture map on top of zone Ⅳ. The average net to gross ratio, porosity and water saturation of well BK-2 in this zone are about 0. 280, 8. 55% and 21% respectively. Table 3 shows the lithology and petrophysical properties of BK-2 in this zone.

Table 2   The lithology and petrophysical properties of zone Ⅱ in BK-2

Well	:BK-2
Zone	: Ⅱ
Formation	:Sarvak
Top	:10814 ft.
Bottom	:11149 ft.

Depth ft	Lithology	Thickness ft	Porosity &	SW &
10817	LST	1	10.6	38
10818	LST	1	10.6	38
10825	LST	4	9.3	38
10836	LST	11	11.6	43
10845	LST	7	11.1	42
10872	LST	6	6.5	50
10885	DOL	5	16.5	38
10937	LST	3	7.6	31
10986	LST	15	5.0	37
11012	LST	26	10.1	43
11017	LST	5	10.2	42
11018	LST	1	10.2	42
11026	LST	8	11.0	44
11033	LST	7	8.2	38
11041	DOL	8	11.2	22
11046	DOL	5	6.4	41
11064	DOL	18	9.6	29
11105	DOL	5	6.5	25
11116	DOL	11	5.2	42
11128	DOL	4	7.1	28
11142	DOL	5	4.9	46

ZONE V

Zone V is essentially a low porosity argillaceous limestone and shale interval. The limestone is grey, fine grained with a minor amount of movable oil in a few areas. The zone was logged in BK-4 and is easily recognized by a high gamma ray and very low neutron log responses. The average porosity, water saturation and net to gross ratio of this zone in well B-4 are about 5.95%, 44.96% and 0.018% respectively.

Table 4 shows the petrophysical properities of BK-4

Table 3   The lithology and petrophysical properties of zone Ⅳ in BK-2

Well	:BK-2
Zone	: Ⅳ
Formation	:Sarvak
Top	:11149 ft.
Bottom	:1147 ft.

Depth ft	Lithology	Thickness ft	Porosity &	SW &
11264	LST	9	5. 1	11
11272	DOL	4	9. 2	6
11288	LST	7	4. 5	20
11302	LST	14	6. 3	6
11313	DOL	7	7. 4	14
11331	LST	6	6. 6	13
11388	LST	4	5. 1	38
11403	LST	5	4. 7	48
11407	LST	4	10. 0	30
11415	LST	5	18. 4	20
11417	LST	2	9. 1	32
11422	LST	5	15. 2	21
11425	DOL	3	22. 4	19
11429	LST	4	15. 9	22
11433	LST	4	5. 7	47
11439	LST	6	9. 1	36
11447	LST	3	4. 7	39

Table 4   The petrophysical properties zone Ⅴ in BK-4

Well	:BK-4
Zone	: Ⅴ
Top	:11549 ft.
Bottom	:12661 ft.

Depth ft	Thickness ft	Porosity &	SW &
11890	10	6. 4	41. 7
11900	10	4. 5	49. 6

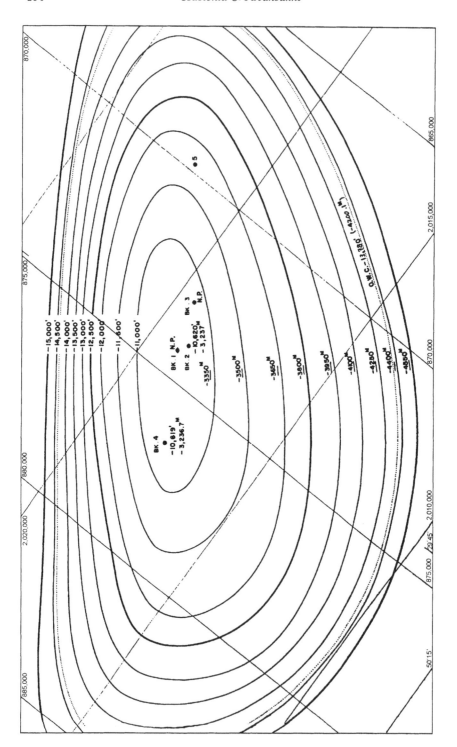

Fig. 5   Structure map on top of Zone II

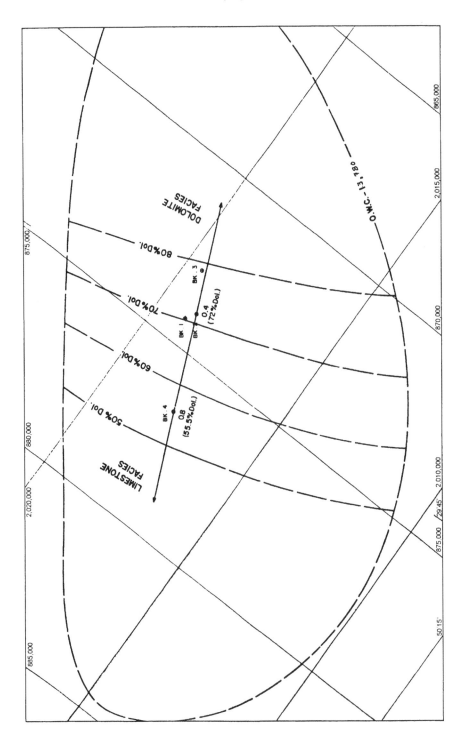

Fig. 6   The ratio of limestone to dolomite for Zone **II**

BANGESTAN GROUP (9. 782 ft-112462 ft.)

In BK-2 well, Bangestan Reservoir is divided in two formations that is described as follows:

ILAM FORMATION (9/872 ft. 10,363 ft.)

The top of the Ilam formation was placed at 9/782 feet from Gamma-ray and neutron log correlation with BK-1. This formation consisted of limestone Archie type 1, fine grained, in part cores grained.

From 10/100 ft an Archie type I / I and type I , coarse grained, partially dolomitized limestone was encountered and persisted to 10/363 ft. Where it passed into a glauconitic limestone Sarvak formation (10612 ft-12462 ft.). The top of Sarvak Formation was placed at 10/612 feet, at the top of a limestone bed which contained a middle Cretaceous fauna *Valvulammina* sp. , *Nummuloculina* sp. , *Dicyclina* sp. , The Sarvak Formation in BK-2 primarily consisted on Archie type IA, and type I limestone with several thick sections of type I / I limestone and dolomite. The upper 1/1000 ft. of the Sarvak Formation in BK-2 is rich in *Oligosteginids* associated with pelagic foraminiferea, the most common species includes *Stomoisphaera sphaerica* and *Calciosphaerula immominata*.

The Binak-4 well, (BK-4) was drilled to test the Khami Group (Lower Cretaceous and Jurassic) in a large anticline known to contain hydrocarbons in the Asmari Formation (Miocene) and in the Bangestan Group, (Cretaceous). BK-4 (original hole) was drilled to 13/095 ft. in Bangestan Group and was plugged back when oil and gas flows were encountered. It is completed as a Bangestan group oil well the Khami Group was abundant because of indication of a collapsed 5″ liner and liner leaks below 11064 ft.

CONCLUSIONS

In the present work the authors have reviewed the stratigraphic zonation of Bangestan group in Binak field, Persian Gulf Coast. Two formations, Ilam and Sarvak are subdivided, base on reservoir rocks into five zones and one subzone. All the zones lithologically discussed. Besides, different fauna of two formation and their ages introduced. The lithology, thickness and petrophysical properties of the zones have shown.

ACKOWLEDGEMENTS

The authors are grateful to the head of department of faculty of Petroleum Eng. Ahwaz for the providing different (Laboratry) facilities in the Institute. We are highly indebted, head of Petroleum Eng. N. I. O. C on shore field Ahwaz, who kindllly providing the information and data of Binak oil field, Bangestan reservoir and his detailed instruction in the matter.

REFERENCE

1. A. Ansari and Shirmohammadi. Geological and Reservoir Engineering study of the Binak field. Ahwaz-Petr. Eng. (1975).

2. B. C. Craft, M. Hawkins and etc. , Applied Petroleum Reservoir Eng. Englewood Cliffs, New Jersey (1991).

3. A. I. Levorsen. Geology of Petroleum, second edition, 724 p. (1967).

4. K. H. Nazem Roaya. Technical note 3192, P. V. T. Analysis of Binak well BK-2, Production Chem. Dept. Ahwaz. Iran (1967).

Proc. 30th Int' l. Geol. Congr. , Vol. 18, pp. 239~253
Sun Z. C. *et al.* (Eds)
© VSP 1997

Reservoirs of Chalkified Miocene Reef-banks on Dongsha Platform in South China Sea

WU XICHUN and LI PEIHUA

Department of Petroleum, Chengdu Institute of Technology, Chengdu, Sichuan, 610059, *CHINA*

HU PINGZHONG

China Offshore Oil Naihai East Corp. , *Guangzhou,* 510240, *CHINA*

Abstract

Chalkified reservoir rock is a kind of white, loose, friable and porous limestones. Habitat of chalkified porosity, in narrow sense, is in chalkified wackestones and mudstones (Moshier, 1989), but in broad sense, is not only in mud-supported carbonates, but in chalkified grainstones, packstones, dolomites of dedolomites (Sun, 1995, per. com.), framestones and baffllestones. Early near surface diagenetic chalkification of Early Miocene reefs and corrosion, is the main cause of reservoir generation. Much evidence of early exposure of the reef-bank facies has been found by the study of microscopy, SEM, cath. lum. , x-ray, etc. analyses. Based on the study of paleo-hydrodynamics and paleo-hydrogeochemistry, we have found that chalkification of the reef-banks was enhanced by multiphase compaction-released groundwater in geohistory.

The study of core columns of tens holes demonstrates that the reefs and banks on Dongsha platform were not immediately drowned by transgression (Erlich et al, 1990,1993), but were exposed for a certain time before drowning, unless chalkification episode of reef-band reservoirs could not occur.

Chalkification is restricted in the reef-bank facies of paleorises on the platform, constructing the early Li-uhua and the late Dongsha horseshoe reef belts. The lagoonal and basinal facies are not a bit chalkified.

Keywords: *chalkified carbonate reservoirs, Early Miocene reef-banks, South China Sea*

INTROCUCTION

Significance of Chalkified Carbonate Reservoirs
The authors consider that all carbonate reservoirs in the world through geologic time can be summarised into 4 essential types: dolomitized porous reservoirs, chalkified porous reservoirs, paleokarstic porous-vuggy reservoirs and fissured reservoirs, of which the chalkified reservoirs amount to 4. 5% of giant oil fields in the world (Wilson, 1980). Chalkified carbonates of chalkified reservoirs are common in Tertiary and Mesozoic sequences in south East

Asia (Friedman, 1983; Moshier, 1989), the Middle East (Wilson, 1975; Moshier, 1989), North Africa (Bebout et al, 1975), the North American Gulf Coast (Loucks, 1987, Perkins, 1989), Texas, U. S. A. (Ahr, 1989; Dravis, 1989) and Marshall islands in Pacific (Saller et al. , 1989). Some Middle Devonian reefal oil fields in Canada also contain chalkified reservoirs (Kaldi, 1989). Hence the chalkified reservoirs merit attention for oil and gas explorers.

Understanding of Chalkification

Chalkified reservoir rocks are a kind of white, loose, friable and porous limestones, which are the product of diagenesis and must be distinguished from true chalks, being of deep-ocean accumulations of calcareous nannofossil tests. There are two viewpoints of understanding of "chalkification" or "chalkified/chalky limestones".

(1)In narrow sense:

The chalky limestones are caused by the diagenesis of only pure micritic mudstones, originated from limemud in platformal lagoons. Intercrystal micropores in carbonate matrix are typically 5-10 μm. Such micropores are commonly responsible for $>20\%$ porosity and >1 md permeability (Moshier, 1989).

(2)In broad sense:

The diagenesis of chalkification is not sedimentary-facies-selective and not rock-texture-selective. All of the following limestone textures can be chalkified, caused by certain diagenesis:

(a) Limemud-supported matrix;

(b) Grain-supported particles, such as bioclasts, oolites, oncolites or other intraclasts;

(c) Texture of organic bafflestones, bindstones of framestones;

(d) Particles of dolomites of de-dolomites (calcified dolomites) (S. Q. Sum, 1995, per. com.).

Hence, every kinds of carbonate sedimentary facies can be chalkified, such as reefs, mounds, banks, lagoons, platforms, deeper water facies, caused by appropriate diagenesis. Chalky limestones contain not only intercrystal microporosity, but also intergranular, intragranular, moldic, sheltered, skeletal and texture-unselective porosities.

GEOLOGIC SETTING

Dongsha uplift is located near the eastern coast of Guangdong province in South China Sea (Fig. 1). It is a nose-type uplift, trending and dipping from northeast to southwest, 370 km in length and 65-130 km in width. The incipient uplift was originated from Mesozoic, while most part of the uplift was exposed with much less deposits.

Up to Zhuhai stage of Oligocene a transgressive sand sequence, called Zhuhai formation. 0—585 m in thickness, deposited on the uplift. Comparable with the periphery basinal deposits.

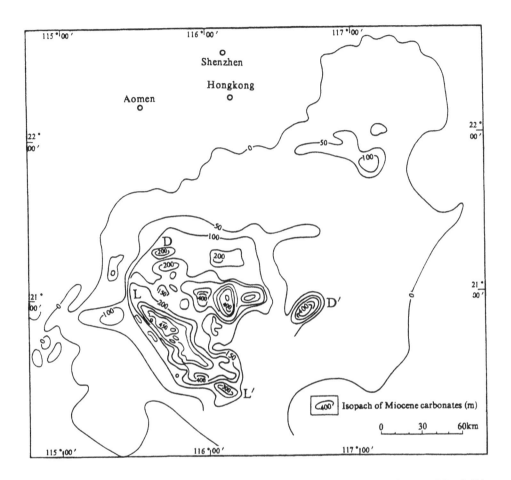

Fig. 1   Early Miocene carbonate isopach map of Dongsha uplift in Pearl River Mouth basin of South China
Sea.

The early Miocene Zhujiang formation overlies the Zhuhai formation. Zhujiang formation
contains a little sandstones intercalated by claystones at bottom. The lower to middle part of
the formation consists of a carbonate sequence, composed of greyish white coral-red algae
framestones, rhodolith and foraminifer packstones and wackestones, corresponding to reef,
bank and lagoon facies. The thickness of carbonates on the platform is 0-500 m or more. The
upper part of the formation consists mainly of grey claystones intercalated by fine sandstones, corresponding to shallow marine shelf facies. The total thickness of Zhujiang formation is 373-1080 m.

Hence, the Dongsha carbonate platform developed during the early to middle Zhujiang stage.
The area of Dongsha platform corresponds to the Dongsha uplift (Fig. 1) (Chen et al.,
1989; Turner et al., 1991).

There are 1 horseshoe reef belts on the platform. The one is Liuhua reef belt (Fig. 1, L-L'), and the other is Dongsha reef belt (Fig. 1, D-D'). along with the transgression or drowning from the windward (from the southeastern side of the platform), the Liuhua reef belt developed earlier and the Dongsha reef belt developed later. According to seismic facies anlysis, the development of Dongsha platfrom can be differentiated into 5 stages (Fig. 2).

Stage Ⅰ : A narrow and small bioclastic bank facies developed on the Dongsha uplift, locally with gentle and small algal-bryozoan mound facies found, drilled by LF 15-1-1 well.

Stage Ⅱ : A broad bioclastic bank facies developed along the southwestern part of the platform, with an incipient marginal reef belt, drilled by LH4-1-1 well and others, called LH4-1 reef.

Stage Ⅲ : The LH4-1 marginal reef belt developed further, with pinnacle reefs (such as HZ33-1 reef) in front, and with a large patch reef at back. called LH11-1 reef, drilled by LH11-1-1A,LH11-1-3, LH11-1-4, etc. wells. The reefs were built by corals. The reef flat, reef back bank and lagoon facies also deveolped.

Stage Ⅳ : The reef-bank complex developed fully. The reef-builders turned from corals into coraline algae.

Stage Ⅴ : The reef-bank complex was exposed for a certain time in the low stand system tract, and was covered by shelf clay and sand, caused by intensive drowning and transgression. Only the LH4-1 marginal reef locally developed further.

RESERVOIR ROCK-TYPES OF LIUHUA OIL FIELD

The Liuhua reef-bank complex contains very good quality of reservoir rocks, which can be sorted into 3 types:

1) Chalkified coral-coralline algae framestone with intercrystal and skeletal porosity: Such reservoir rocks are massive coral or algae framestones, within the LH4-1 marginal reef and the LH11-1 patch reef. For the better chalkified intervals, the reservoir rocks have the porosity of 15. 5%-33. 3% and the permeability of 26-9329 md. And for the moderately chalkified intervals, the porosity is 6. 1%-27. 8% and permeability is 0. 01-1663 md.

2) Chalkified rhodolith-bearing packstones of rhodolith rocks with intercrystal intergranular, intragranular or moldic porosity: They are mostly rhodolilth enrichment sediment within reef flat facies belt, behind the LH4-1 marginal reef ridge (Fig. 2). Taking the rhodolith-bearing interval of LH11-1-4 well for example, the porosity is 6. 7%-36. 0%, and the permeability is 0. 01-5609 md.

3) Chalkified algae-forminifer-bioclastic grainstones and packstones with intercrystal, intergranular, intragranular and moldic porosity: They are seen mainly in the reef back bank facies, such as interval 1233-1236 m of LH11-1-2 well, interval 1239-1248 m of LH11-1-1A well, etc. The porosity is 6. 5%-31. 6% and permeability is 0. 01-3540 md.

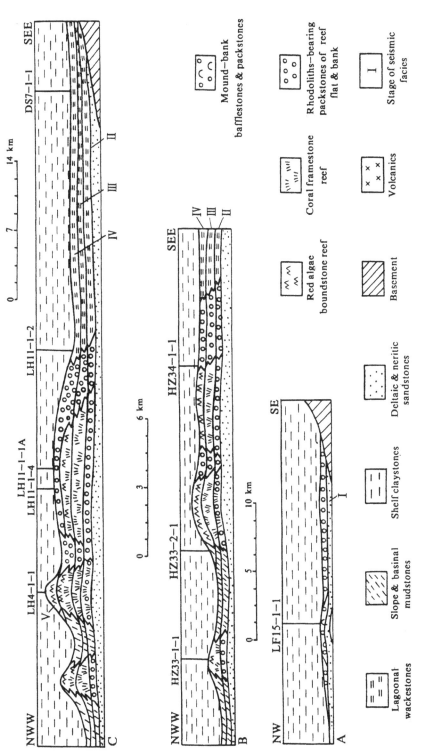

Fig. 2   Sedimentary patterns of different seismic facies stages on Early Miocene Dongsha carbonate platform in South China Sea.

MECHANISM OF THE CHALKIFICATION

Observation of Drilling Core

At first we observed that all of the coring intervals under oil water contact of the wells in Liuhua oil field, such as these in LH4-1-1, LH11-1-4, LH11-1-1A, LH11-2-1,etc. wells, represent typical chalkified texture. The cores of such intervals are mostly of corals, pack-stones or rhodolith rocks. It seems that chalkification was caused by the bottom water of the oil field (Fig. 3). But while we extracted the oil-bearing core samples with chloroform, we found that most of the oil-bearing coring intervals of the wells in Liuhua oil field are also chalkified to different degree. The cores along the intervals of high quality oil-bearing reser-voirs are loose and fragmentary, with low coring percent, for the core rocks have undergone intensive chalkification. The observation of drilling cores shows that the porosity, perme-ability and oil abundance are in direct ratio to the intensity of chalkification.

Fig. 3   Chalkified porous coral framestone of coring interval (1331. 79-1334. 35 m) near oil water contact of LH11-1-4 well.

Most of the coring intervals of wells drilled into the LH4-1 marginal reef, the LH11-1 patch reef, reef flat and reef back bank facies are mediately or intensively chalkified, with mediate or high abundance of oil.

But all of the coring intervals of wells drilled into lagoonal mudstone or wackestone facies are not a bit chalkified. The fact of this area is just different from that reported by Moshier (1989). The coring intervals of HZ33-1-1 well drilled into the HZ33-1 incipient pinnacle reef in the basinal facies are also not a bit chalkified. The reef rocks are tight in light-grey colour, representing normal reefal limestones. Along the coring intervals of the well only fis-sured reservoirs have been found with certain amount of oil.

Observation Under Microscope and Electronic Scanning

The orginal structure and texture of chalkified rock of this area are well preserved, without deformation of abrasion, but become blurred (Fgi. 4). We call such texture retrogressive texture, i. e. , all of the lime matrix and organic tissues, including these of coraline algae and foraminifers, which are diagenesis-resistant, have undergone micritization and recrystallization. The size of calcite particles of the matrix and tissue has been reduce while recrystallizing and then corroded to some degree. So that the limestones become loose, soft and friable, with a large amount of corroded intercrystal micropores (Fig. 5).

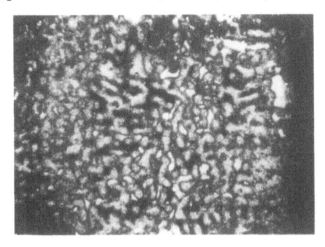

Fig. 4   Microscopy of chalkified coral, still maintaining skeletal structure with a lot of intra-skeletall micropores, coring at 1312. 95 m of LH4-1-1 well.

Under microsope, the megapores created in meteoric fresh water vadose zone during early diagenetic stage can be observed. At the top of the Liuhua reef-bank complex or within the complex, some corrosion surfaces or exposures can be seem, with microkarstic pockets or caves, vadose slit, geopetals, gravity cement, etc. , containing megapores, caused by immediately successive dissolution.

Some evidence of pores, created in the fresh water phreatic zone during early diagenetic stage also can be seen, such as residue of rimmed crystsls of scalenohedrons, crystals coarsen toward center of pores, syntaxial overgrowth on echinoderms, etc. , mostly

Fig. 5   Intensively chalkified foraminifer tissue in packstones of reef back bank facies with corroded calcite crystals and intercrystal pores, coring at 1278. 3 m of LH11-2-1 well.

dissoloved by immediately successive corrosion. It seems that the diagenetic horizon was oscillatory between the active saturated zone and unsaturated zone (Longman, 1980; Flugel, 1982).

The images of electronic scanning of the chalky rock show that the microcrystals of calcite on the pore walls are commonly scalenohedrons, with a lot of solution pores and holes, and always without sharp edge angles, caused by intensive corrosion. The scalenohedrons indicate that the chalkification occurred initially in the phreatic zone during the early diagenetic stage (Fgig. 6,7) (Flugel, 1982).

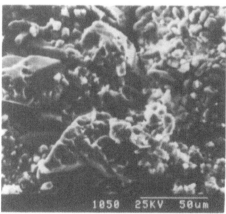 

Fig. 6   Many corroded holes or pores on scalenohedron crystals of chalkified packstone, coring at 1304. 8 m of LH11-2-1 well.

Fig. 7   Corroded scalenohedron crystals on the wall of an intergranular pore in the packstone, coring at 1286. 93 m of LH11-2-1 well.

Some corroded pores and vugs contemporaneous with pre-stylolites, syn-stylolites or post-stylolites, during the middle to late diagenetic stages can be also observed. Hence the chalkification has undergone multiphase dissolution in multistages through geologic time.

But the unchalkified HZ33-1 incipient pinnacle reef shows that its primary fossil skeletons are well preserved (Fig. 8,9). Intergranular pores of the reef were fully filled up by mica-sheet like calcite spar, created in the fresh water phreatic zone during the early diagenetic stage, while the sediment became tight, and successive sulution fluids could not go through (Fig. 10).

Geochemical Characteristics

(1) Stable isotopic values

Table 1 shows the comparison of the average values of $\delta^{13}C$ and $\delta^{18}O$ in ‰PDB. The samples are from some drilling wells of the inhomogeneously chalkified reef-bank complex of Liuhua oil field. The result shows that the average isotopic value of intensively chalkified samples is heavier than the average value of samples in total. The fact demonstrates that the chalkifica-

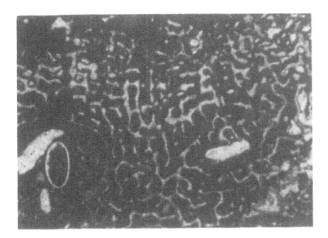

Fig. 8   Microscopy of unchalkified coral framstone, coring at 2026. 25 m of HZ33-1-1 well.

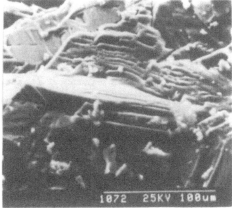

Fig. 9   Columnar calcite crystals of unchalkified coralline algae. , coring at 1999. 34 m of
HZ33-1-1 well.

Fig. 10   Mica-sheet like calcite spar filling up a
pore in the reef rock, coring at 2029. 73 m of
HZ33-1-1 well.

Table 1

Drilling Well	Rock Character	Average Value of Chalkified Samples		Average Value of Total Samples	
		$\delta^{13}C‰$	$\delta^{18}O‰$	$\delta^{13}C‰$	$\delta^{18}O‰$
LH4-1-1	Coral-algae framestone	0. 82	$-3. 81$	0. 46	$-4. 27$
LH11-1-4	Rhodostone Coral-algae framestone	0. 70	$-3. 42$	$-0. 39$	$-4. 36$
LH11-1-1A	Algae rudstone	0. 98	$-2. 89$	$-2. 05$	$-5. 11$
LH11-2-1	Bioclast packstone	0. 83	$-3. 32$	$-0. 23$	$-3. 52$

tion is caused merely by leaching, which results in the loss of impurity from the intergranular space. So that the heavier isotope composition expresses the original isotopic value of grains and skeletons.

(2) Trace elements

Taking the reef matrix mudstone of the area as reference material, Fig. 11 shows that the chalkified reef-bank rock is rich in calcium, which indicates the loss of impurity caused by intensive leaching, and that it is rich in manganese, which indicates that the leaching occurred in an open system during early diagenetic stage (Flugel, 1982). The Sr content slightly increases comparable to that in the reefal matrix, indicating that Sr is concentrated in the grains and skeleton, protected from leaching. The Zn content is also richer than that of the reefal matrix, indicating that chalkification continued during late diagenetic stage (Flugel, 1982).

(3) X-ray diffraction

The X-ray diffraction shows that the chalkified rock is rich in calcite and ankerite, indicating that the intensive leaching has resulted in the enrichment of calcium, and that the sediment had undergone the leaching of compaction fluid under shallow burial (Schroeder,1986).

MACROSCOPIC GEOLOGIC CONTROL OF THE CHALKIFICATION

Paleorise on the platform

The chalkified carbonate sequence of Zhujiang formation is only located in the Liuhua horseshoe reef-bank complex found, where has been a paleorise on the Dongsha carbonate platform (Fig. 1, L-L'), during the early time of the Early Miocene. Until now there have been no dilling wells on the Dongsha horseshoe reef-bank complex (Fig. 1, D-D'). As above mentioned the Liuhua reef-bank complex on the paleorise of the platform includes the LH4-1 marginal reef, the LH11-1 patch reef behind, the reef flat and the reef back bank facies. The wells drilled on the paleorise can meet with the chalkified interval with good reservoirs and oil potential, such as LH4-1-1, LH11-1-1A, LH11-1-4, LH110201 wells.

The HZ33-1 incipient pinnacle reef is not located on the paleorise of the platform, so that it has not been chalkified and just contains some fissured carbonate reservoirs. The lagoonal facies is also not located on the paleorise of the platform, so that it has not been chalkified and contains no oil, such as DS7-1-1 well, LH11-1-2 well, etc. .

Sequence Stratigraphy

The chalkified interval in the Liuhua reef-rank complex is just located beneath the sequence stratigraphy boundary of 16. 5 Ma. The boundary corresponds to the top surface of the Ⅳ seismic facies belt (Fig. 12). The chalkified interval is about 100 m more in thickness within the Ⅳ seismic facies belt.

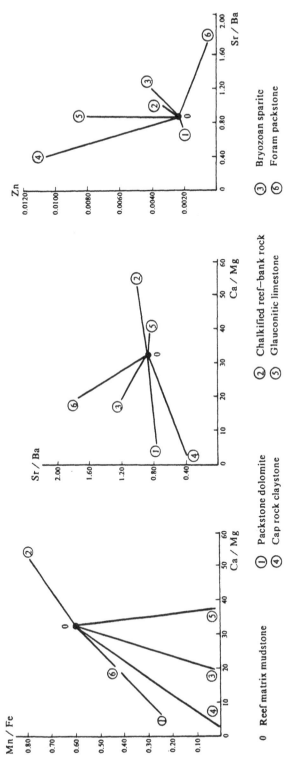

Fig. 11 Correlation diagram of trace elements in core samples of Liuhua reef complex.

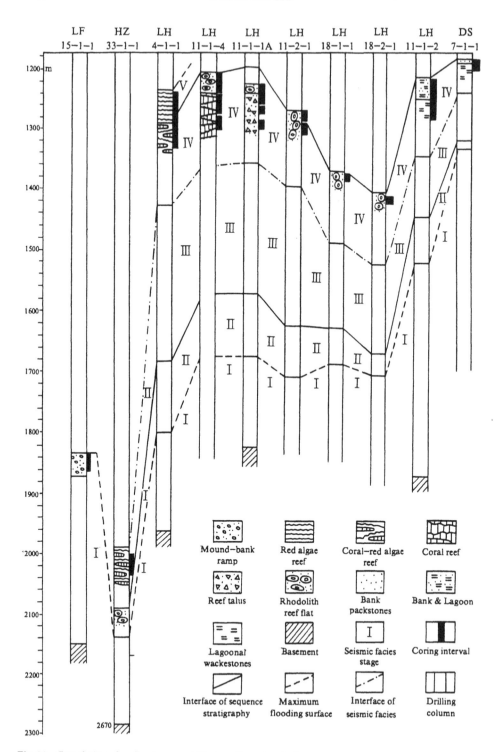

Fig. 12   Correlation of well columns, drilled on Dongsha platform.

The LH15-1 mound-bank facies drilled by LH15-1-1 well within the I seismic facies belt, and the HZ33-1 pinnacle reef drilled by HZ33-1-1 well within the I seismic facies belt are not beneath any sequence stratigraphy boundary. So that they have not been chalkified.

Hence the chalkification of this area is firstly related to the early diagenetic stage, for it is closely related to the sedimentological sequence stratigraphy. Erlich et al (1990,1993) explained the event of 16. 5 Ma. sequence stratigraphy boundary as simply an immediate drowning succession. But the authors think that the boundary was exposed for a certain time before drowning, unless the important first chalkification episode of the reef-bank reservoirs could not occur.

It is evident that the first stage of chalkification occurred contemporaneously to the lowstand system tract caused by the exposure of 16. 5 Ma. boundary.

Hydrogeology and Hydrodynamic Field

A detailed study of hydrogeology and hydrodynamic field in a much broader area of the Zhujiankou (The Pearl River Mouth) basin has been completed. It is found that the chalkified interval in Liuhua reef-bank complex or on Liuhua paleorise has been a groundwater catchment area with continuous low hydrodynamic potential from the beginning of the Early Miocene up to the Recent. It means that the chalkification of Liuhua reef-bank complex originated from the early diagenetic stage and was enhanced by multiphasic compaction-released groundwater through geologic time. The chalkified area has been a continuous dissolution zone but not a precipitaition zone from the Early Miocene up to the Recent.

CONCLUSION

Chalkification of this area is the result of repeated leaching, flushing and corrosion process through diagenetic stage. Accrding to the macroscopic and microscopic study of the area, the chalkification of this area has passed many stages, summarised into 4 major stages, corresponding to diagenetic stages:

1) The meteoric diagenetic stage

The early exposure of Liuhua reef-bank complex at the 16. 5 Ma. sequence boundary is the essential step of chalkification. The first step of chalkification occured in the unsaturated zone of fresh water phreatic zone. The chalkified rocks of this area contain a large amount of texture-selective pores, such as intergranular, intragranular, moldic and skeletal pores. The dog tooth-shaped calcite crystals often rim pore walls. The greater part of the primary and secondary pores in the Liuhua reef-bank complex has kept open, without sparry, cement, to be favourable for further dissolution.

The HZ33-1 incipient pinnacle reef is located in the basinal facies belt. There are many early exposure marks found, such as geopetals, vadose silt, gravity cement, etc. , indicating the fresh water leaching into the reef. But it is not a bit chalkified, with porosity only up to

5.3%, for the reef bank into the CaCO₃ precipitate zone quickly, and most of the primary pores were filled with spar. The reef already became too impermeable to let in further any corrosion fluid, after the early diagenetic stage.

No exposure marks (geopetals, vadose silt, etc.) have been found in any drilling well in the lagoonal facies belt. It means that the facies has not undergone the meteoric diagenetic action. There are much more stylolites in the lagoonal facies than in the reef-bank complex, caused by the early compaction. It became impermeable quickly, and unfavourable to let in the further corrosion fluid. So that it maintains un-chalkified.

2) The shallow burial diagenetic stage

The porous reef-bank complex has undergone further corrosion by the compaction fluid, which is a kind of compaction-released groundwater with great ionic saturation deficit, capable of corrosion. The chalkification mechanism described by Moshier (1989) took place in this stage, while the intercrystal micro-porosity occurred, caused by lime particle recrystallization in the sediments.

3) The deep burial diagenetic stage

The deep buried reef-bank complex was not exposed and affected by the Dongsha movement, occurred at the end of the Middle Miocene. The further chalkification of the complex is caused by the action of acidic groundwater originated from the degradation of petroleum, or from the dissolution of CO_2 from molten rocks. The non-texture-selective pores and vugs in the chalkified rocks were formed in this stage.

4) The bottom water immersing stage

The part of the reef-bank complex immersed under the oil-bottom water contact of the Liuhua oil field becomes white and more soft (Fig. 3).

ACKNOWLEDGEMENTS

The study project is supported financially by China Offshore Oil Nanhai East Corp. and by the State Key Laboratory of Oil/Gas Reservoir Geology and Development. We are grateful to Dr. Sizhong Chen, Dr. Changmin Chen, Dr. Zesong Li, Dr. Shice Xu, Dr. Jinzhong Wang, Dr. Jiamin Li, Dr. Xiaozeng Liu, Dr. Huanrong Song, Dr. Shangyu Huang, Dr. Yan Shi, Dr. Lilan Duan and Dr. Daming Gong, Who have encouraged and helped us in the study.

REFERENCES

1. W. M. Ahr. Early diagenetic microporosity in the Cotton Valley Limestone of East Texas. Sediment. Geol. 63, 275-292 (1989).

2. C. G. Bebout and C. Pendexter. Secondary carbonate porosity as related to early Tertiary, Zelten field, Libya. A. A. P. G. Bull. 59, 665-593 (1975).

3. Chen Sizhong and Hu Pingzhong. Tertiary reef complex in the Zhujiangkou (Pearl River Mount) basin

and their significance for hydrocarbon exploration. China Earth Sciences, 1:1,43-58 (1989).

4. J. J. Dravis. Deep-burial microporosity in Upper Jurassic Haynesville oolitic grainstones, East Texas. Sediment. Geol, 63,325-341(1989).

5. R. N. Erlich et al. . Seismic and geologic characteristics of drowning events on carbonate platforms. A. A. P. G. Bull. ,74,1523-1537 (1990).

6. R. N. Erlich et al. Response of carbonate platform margins to drowning: evidence of environmental collapse. A. A. P. G. Vem. ,57,241-266 (1993).

7. E. Flugel. Microfacies analysis of Limestones. Springer-Verlg, Berlin, Heidelberg. New York, 1-631 (1982).

8. G. M. Friedman. Reefs and porosity: examples from the Indonesian Archipelago. SEAPEX Proc. , 6, 35-40 (1983).

9. J. Kaldi. Diagenetic microporosity (chalky porosity), Middle Devonian Kee Scarp reef complex, Norman Wells, Northwest Territories. Canada. Sediment. Geol. , 63,241-252 (1989).

10. M. W. Longman. Carbonate diagenetic textures from nearshore diagenetic environments. A. A. P. G. Bull. , 64:4, 461-487 (1980).

11. R. G. Loucks and P. A. Sullivan. Microrhombic calcite diagenesis and associated microporosity in deeply buried Lower Cretaceous shelf-margin limestones (abstract). Soc. Econ. Paleomtal. Mineral. Annu. Midyear Meet. Abstr. , 4,49-50 (1987).

12. S. O. Moshier. Microporosity in micritic limestones: a review. Sediment, Geol. , 63, 191-213 (1989).

13. R. D. Perkins. Origin of micro-rhombic calcite matrix within Cretaceous reservoir rock, west Stuart city, Trend, Texas. Sediment Geol. , 63,313-321 (1989).

14. A. H. Saller and C. H. Moore. Meteoric diagenesis, marine diagenesis, and microporosity in Pleistocene and Oligocene limestones, Enewetak Atoll, Marshall islands. Sediment. Geol. , 63,253-272 (1989).

15. J. H. Schroeder. Diagenetic diversity in Paleocene coral knobs from the Bir Abu El-Husein area, S Egypt. In: Reef Diagenesis. J. H. Schroeder and B. H. Purser (Eds). Springer-Verlag, Berlin, Heidelberg, New York, 132-159 (1986).

16. N. L. Turner and Hu Pingzhong. The lower Miocene Liuhua carbonate reservoir, Pearl River Mouth basin, offshore Peoples's Republic of China. Offshore Technology Conference, Society of Petroleum Engineers, 97-107 (1991).

17. J. L. Wilson. Carbonate facies in geologic history. Springer-Verlag. New York, N. Y. 1-471 (1975).

18. J. L. Wilson. Limestones and dolomite reservoirs. In: G. D. Hobson (Ed). Petrol. Geol. 2, Applied Science Publishers LTD. Essex, 1-51 (1980).

Proc. 30ᵗʰ Int' l. Geol. Congr. , Vol. 18, pp. 255~261
Sun Z. C. *et al.* (Eds)
© VSP 1997

Deformational Characteristics and Oil/Gas Accumulation in Tarim Basin of China

YANG KEMING   CHEN ZHAOGUO

Institute under Southwest Bureau of Petroleum Geology, MGMR, P. R. China

GONG MING   WU YAJUN

Comprehensive Research Institute of Petroleum Geology, MGMR, P. R. China

Abstract

The structural styles in Tarim basin can be divided into compressional structure, externsional structure, strike-slip structure and complex structure based on the mechanical property. The compresional structure style is the main deformational one in Tarim basin, which is dominated by the basement involved back-thrust structures within the basin, and dominated by the double overthrust and imbricate overthrust structures in the basin's edge area. Different structure styles have different ability to control the oil and gas accumulation, among which the buried hill and drape formed by the back-thrusting are favourable for oil and gas accumulation.

Keywords: Tarim basin, deformation, structural style, oil and gas accumulation

INTRODUCTION

Tarim basin which is surrounded by Tianshan, west Kunlunshan and Aerjinshan (Fig. 1) is the important part of basin-and-range terrain in the west China.

In Tarim basin with the area of $560 \times 10^3 km^3$, Phanerozoic strata with the thickness of 8000-15000m is very developed. The resource extent of the three source rocks in Cambrian-Ordivian, Carboniferous-Permian and Triassic-Jurassic reaches $(20-50) \times 10^9 t$, having plenty of oil resource.

Tarim basin is the large scale of complex baisn stacked by different period and style basins. Since phanerozoic period, the basin has undergone at least 13 structural movements and has a plenty of structure deformation. They have an important effect to control oil/gas migration and accumulation.

The structural style is the geometric expression to structural deformation in Tarim basin. According to the mechanics, the structural styles can be divided into compressional, extensional, strike-slip and complex structures.

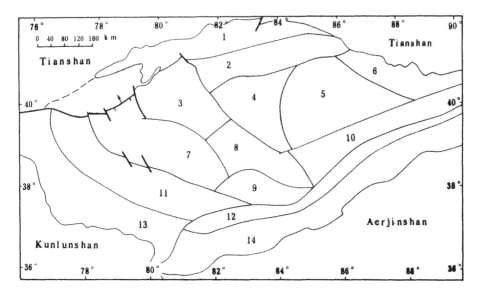

Fig. 1   Structural zonation map of Tarim basin
1. Kuche depression; 2. Shaya uplift; 3. Afanti fault depression; 4. Shuntuoguole uplift; 5. Manjiaer depression; 6. Kongquehe slope; 7. Bachu uplift; 8. Katake uplift; 9. Tanggubasi depression; 10. Guchengxu uplift; 11. Maigaiti slope; 12. Qiemo uplift; 13. Yecheng depression; 14. Beiminfeng fault depression

COMPRESSIONAL STRUCTURAL STYLE

This style is an important deformational one in Tarim basin. There are two kinds of different deformational characters in the basin and the basin edges.

Within the basin, the deformation existed under early or are Hercynian orogeny (Fig. 2) and was dominated by the basement involved back-thrust structure, which can be classified into two orders: the first order style and second order style. The first order back-thrust structure was composed of the double thrust in the both sides of the large scale positive structure zone, such as Shanya, Katake and Qiemo uplifts (Fig. 2). The second order back-thrust structure, which is less in scale, was composed of double thrusts in the both sides of local structure, such as Katake No. 1 and No. 2 back-thrust structures (Fig. 3).

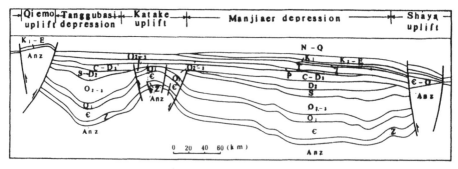

Fig. 2   North-south seismic interpretation section in Tarim basin

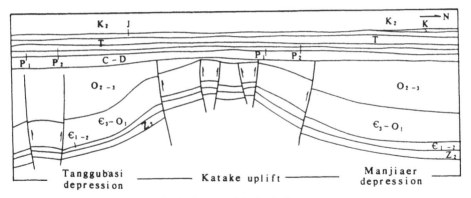

Fig. 3　Seismic interpretation section in the central Tarim basin

The back-thrust structures in Tarim basin have some basic characteristics as following:

① These back-thrust structures in northeast or northwest strike were formed by early or late Hercynian, which are identical with regional maximun horizontal compressional stress at that time.

② The first order structure located in the source rock depression, created the background for oil/gas accumulation and controlled the second order structure's development.

③ Due to uplifting, the back-thrust structure often resulted in a great amount of strata erosion. forming several buried hills with different range. The low range drapes were developed in the overlain strata due to the buried hill effect. At present, most oil/gas pools are discovered in these kinds of buried hill-drape association in the basin, and have good oil/gas accumulation condition.

In the basin edge area, the main deformational styles are thin-skinned detachment structures. There are some deformational differences between the north and the south edges in the basin.

The deformation in Kuche depression is most active and strong in the north edge area of the basin. The seismic geological cross section is shown in Fig. 4, which has the following characteristics.

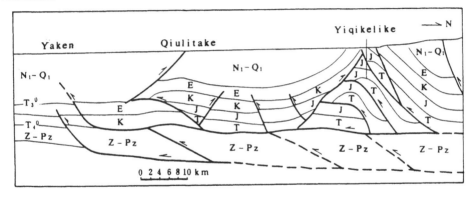

Fig. 4　Seismic interpretation section in Kuche depression

① Double thrust structures exist in Palaeozoic, and thrust structures in Mesozoic and Geno-
zoic.

② About 49 fold structures associated with thrusting are developed in Kuche depression. According to the formational mechanics, they can be divided into three structure zones, such as
Yiqikelike imbricate pile, Qiulitage fault-bend fold and Yaken fault-propagation fold from the
south to north. The upper structure may not coordinate with lower structure due to the multiple detachment.

③ The shortenning ratio is 23%-38% based on the calculation of regional balanced section,
which is similar to that in some foreland basins abroad and at home.

④ The imbricate pile structures might be transformed by faulting and weathering, resulting
in oil and gas leakage. So oil and gas shows are commonly distributed along the fracture of
fold axis. The preservation condition may become better southward with the great depth or
weak structure movement. So Qiulitage fault-bend fold structure and Yaken fault-propagation fold structure may be favourable for oil/gas accumulation.

Yecheng depression in the southwest edge area of the basin was characterized by the imbricate thrust structures composed of several faults and folds (Fig. 5), which possess upper
steep and lower gentle characters. These faults extended downward into Kalawuyi formation
mudstones or coal beds of Carboniferous, resulting in detachment along the layers, and possibly merged with the main detachment of Sinian in the front of Tiekelishan southward.
Based on the general deformational characteristics, we speculate that thrust sheets were
formed from early in the south to late in the north, showing the propagation style. Compared with Kuche depression, it has the moderate fold, complete shape and structure highs i-
dentical in the upper and lower structures, so the favourable oil and gas accumulation exists
in the area.

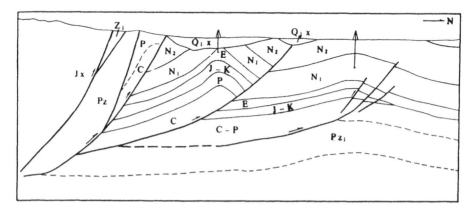

Fig. 5   Seismic interpretation section in Yecheng depression

EXTENSIONAL STRUCTURE STYLE

The extensional structure style is difficult to recognize in Tarim basin due to the strong compression durig Mesozoic and Cenozoic. Through the geophyscial study in recent yerars, we find some extensional structures formed by the local pull-apart, which are most horst association and synthetic rotational fault association.

Horst structure association:

Several normal faults in north-south strike with 2 km long externsion and 15-35 m fault displacement, are developed in Akekulei structure during Triassic from west to east (Fig. 6), breaking Triassic up into 8 independent traps such as faulting folds, faulting noses and faulting blocks, which were the horst structure association in the cross section. Exploratary practise shows that this horst structure has favourable oil and gas condition.

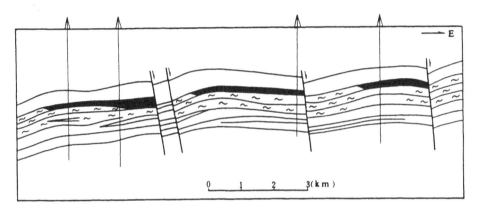

Fig. 6   Horst structure section of Triassic in Akekule area

Synthetic rotational fault association:

A series of normal faults in north dip developed in Shaya uplift were formed during early to middle Himalays movement (Fig. 7), resulting in thickening northward in Tertiary. Because the rotational direction of fault block is identical to fault inclination, so it is called as syntheric fault structure of skeleton structure.

STRIKE-SLIP STRUCTURE STYLE

Most strike-slip structure styles in Tarim basin belonging to basement involved structures, were characterized by the large scale and strong deformation, and distributed in Qiemo fault uplift, Selibuya fracture and Aqia fracture. Selibuya fracture is taken as the example (Fig. 8), having following characteristics.

① Fracture zone, 60km long and 3km wide, is characterized by narrow zone, showing strike-slip fracture system.

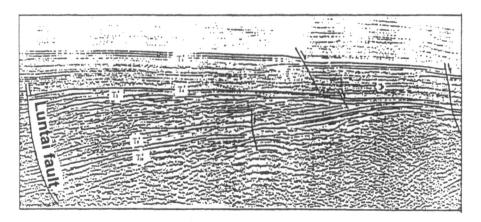

Fig. 7   Synthetic rotational fault blocks

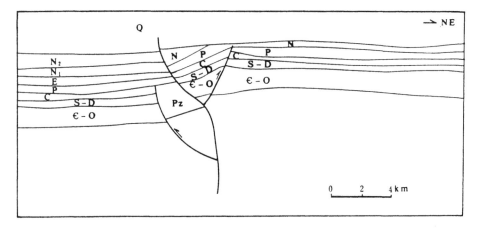

Fig. 8   Flower structure schematic diagram in Yasangdi area

② Above the top of Sinian, Yasangdi flower structure is formed by three fractures; under the top of Sinian, the three faults were emerged into one and vertically got into basement, showing the basement involved structure.

③ Positive flower structure was formed in the end of Miocene, controlling Pliocene deposit. Exploration practise shows that the flower structures have good oil and gas condition.

COMPLEX STRUCTURE STYLE

In geologic history, Tarim basin has undergone several extension and compression, and resulted in the mechanical alternation of primary structure, forming a series of complex structures which include positive reverse structure and negative reverse structure.

Positive reverse structure：

Based on the seismic interpretation. Pihakexun fault in southwest Awati depression was the normal fault before Carboniferous, and was alted as reverse fault during Himalays period due to the regional compression, so showing the reverse fault in the upper section and the normal fault in the lower section.

Negative reverse structure:

Luntai fault, Yanan fault and Yao No. 4 fault in Shaya uplift were the reverse faults before Tertiary, and were alted as propagation reverse faults during Himalaya period due to local extension, so resulting in the normal fault in the upper section and the reverse fault in lower section. Taking the top of Palaeozoic as zero position, the structure with 80m reverse range belongs to slight negative reverse structure.

CONCLUSION

1. The structure styles may be divided into the compressional, extensional, strike-slip and complex structures based on the mechanical property, among them, the compressional structure true style might take the main position.

2. Tarim basin has undergone several deformation, among which Hercynian and Himalaya movements were an importance to the deformation.

3. Within the basin, the deformation was located under the Hercynian uncomformity, being basement involved back-thrust structure.

4. In the basin edges areas, the deformation was located within Mesozoic and Cenozoic, being the double thrust and imbricate thrust by the caprock detachment, showing the propagation fault, with the shortenning ratio of 23%-38%.

5. Back-thrust structure formed early and sealed better have a good condition for oil/gas accumulation.

REFERENCES

1. Sun Zhaocai. Evolution of collision orogens and foreland basin. Northwest University Press, Xian (1993).

2. Zhu Zhicheng and Song Honglin. Structural Geology. China University of Geosciences Press, Wuhan (1990).

3. Chan Wenlang and Song Honglin. Theoretic, method and practise analysis of structural deformation. China University of Geosciences Press, Wuhan (1991).

4. Wang Xiepei, Fei Qi and Zhang Jiahua. Structual analysis of petroleum exploration. China University of Geosciences Press, Wuhan (1991).

5. Yang Keming. The formation and evolution of the western Kunlun continental margin. Geological Review, Vol. 40, No. 1, 19-81 (1994).

6. Yang Keming, Gong Ming, Duan Tiejun and Wu Hong. Carrying and sealing properties of faults in Tarim basin. Oil & Gas Geology, Vol. 17, No. 2, 123-127 (1996).

7. Wang Xingwen adn Chen Fajing. Deformational characteristics of Kuche depression in northern Tarim basin. Oil & Gas Geology, Vol. 16, No. 1 (1995).

PART V

OIL/GAS BASIN SYSTEM AND OTHERS

Proc. 30th Int' l. Geol. Congr. , Vol. 18, pp. 265～281
Sun Z. C. *et al.* (Eds)
© VSP 1997

Oil and Gas Basin Systerm

ZHAO ZHONGYUAN

Institute of Oil and Gas Basin, Northwest University, Xian, PRC

JIN JIUQIANG

Research Institute of Petroleum Exploration & Development, CNPC, PRC

Abstract

Sedimentary basins are the fundamental elements of hydrocarbon generation, migration and accumulation. The formation of oil and gas as well as pools were the results of interactions between various geological factors and agents in geological time. The inherent geological laws of hydrocarbon distribution can be revealed only by regarding the basin as an integrated system, viewing the basin as a whole and studying it dynamically and comprehensively. Therefore, integrated, dynamic and comprehensive analyses are considered as the three general principles of the basin research.

There are three subsystems in the system: (1) basin geology subsystem, which gives the all-round studies of the basin; (2) pool-forming subsystem, which studies the processes of pool formation and uses it to reveal the geological rules of pool distribution; (3) resource assessment subsystem, which evaluates and predicts the hydrocarbon resource of different scales and classes.

The system can be applied to the various stages of oil and gas exploration, but the research level will be advanced along with the accumulation of data.

Keywords: Oil and gas basin system, Integrated dynamic, comprehensive analyses, pool-forming, resource assessment

Oil and gas basin system takes sedimentary basins as the fundamental elements of hydrocarbon generation, migration and accumulation. The formation of oil and gas as well as pools in sedimentary basins were the result of interactions between various geological factors and agents in geological time. The inherent geological laws of hydrocarbon distribution can be revealed only by regarding the basin as an integrated system, viewing the basin as a whole and studying it dynamically and comprehensively.

RESEARCH CONTENTS OF OIL AND GAS BASIN SYSTEM

Integrated, dynamic and comprehensive are the three basic principles of oil and gas basin system [16] , in which there are three subsystems: basin geology, oil and gas pool-forming, and hydrocarbon resource assessment (Fig. 1).

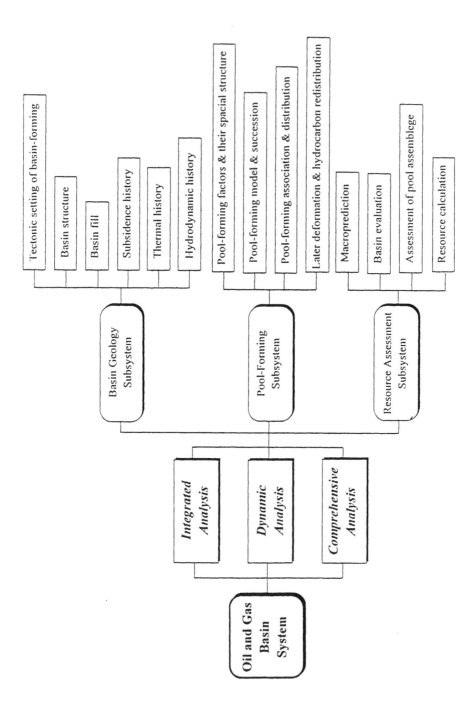

Fig. 1 Flow diagram of oil and gas basin system. T-S—Time-space.

GENERAL PRINCIPLES

Integrated analysis
Sedimentary basins are the fundamental elements of hydrocarbon generation, migration and accumulation. Subsidence in sedimentary basins causes the formation of source rocks, reservoir rocks, topseals and traps. Moreover, hydrocarbon is generated due to burying and may accumulate as pools under favorable conditions. Therefore, the reasonable judgment that hydrocarbon exists or not, how much it is, and where it can be found may be achieved only by the integrated analysis of the overall formation and evolution of the basin as well as hydrocarbon.

Dynamic analysis
Hydrocarbon is a kind of mobile mineral. Moreover, during its accumulation into pools, its parent source rock as well as its objects of migration pathway, reservoir rocks, topseals and traps which it passes through and deposits in continuously change in state and characteristics. In geological time scale, the present state of hydrocarbon distribution in basins is only an instantaneous situation. Therefore, only through the dynamic analysis of pool-forming process can the inevitable laws of oil and gas distribution be discovered.

Comprehensive analysis
The formation of hydrocarbon reservoirs results from the synthetic interaction between a series of geological factors and agents. These factors and agents control the formaiton and distribution of hydrocarbon reservoirs. The restrict and compensate each other during the pool formation. The mistake of one-sidedness can be avoided only by a comprehensive study of all these factors and agents and their relationship in time-space matching.

BASIN GEOLOGY SUBSYSTEM

Tectonic setting of basin formation
The sedimentary basin is a kind of lithospheric features. Its origin and evolution are controlled by lithospheric geodynamics, especially by geothermal dynamics. This phenomenon is obvious in the basins located at the margin of lithospheric plate, but not so evident in intraplate or intracontinental basins. The main reason is that the crust in the inner part of plates and continents is relative thick and the deep-rooted geodynamics hardly exerts a considerable influence on the basins.

The interior of China continent is far away from active plate margins. Generally, people have to attribute the formation of inland basins in Mesozoic and Cenozoic to the distant plate tectonic movements. Although this explanation is valid in kinematics, it is not enough for the thermal dynamics of the basin in such tectonic setting.

The geotectonics of China generally consists of landmasses with various sizes and orogenic

zones that joined the landmasses together. Before the amalgamation, i. e. , before the Indosinian tectonic epoch, marine sediments of different scales developed in the margin or interior of these landmasses. After they combined into a unified continent, these landmasses
were distant from the marginal areas of tectonic activity. Therefore, most geologists think
that the origin of the inland downwarped basins formed in West China during hte Indosinian
and Yanshanian was related to the opening and closure of the Tethys Occean remote to the
south, and that the fault-depressed basins in East China were related to the sinistral wrench
motion of compreso-shearing between Asian continent and Kula-Pacific Ocean Plate that was
connected with Tethys Plate by transform faults orienting in northeast-north direction [11].
The piedmont basin, or known as post-orogenic foreland basins, formed in West China during Himalayan epoch were obviously the results of the collision between Indian Plate and
Asian continent and the later northward progradation of the former, which elevated the
mountains and Qinghai-Tibet Plateau to the north. In East China, graben basins were associated with the dextral transtensional stress caused by the southward movement of the eastern part of China in respect to the northward approaching of Indian Plate. In addition, the
back-arc extension in western Pacific Ocean was also responsible for this tectonic feature[14,15].

However, in addition to the above, further studies must be conducted to answer the question
that how is the geodynamics, especially the thermal dynamics of the continent interior itself.
The reason is that this is the key to understand the formation and evolution features of the
basins in the area of thick crust, as well as that of the hydrocarbon generation in the basins.

Basin structure

Basin structure includes two aspects:basin geometry and structural styles.

Basin geometry reflects the basin type and the regional tectonic setting in which the basin exists and the geometry changes along with the basin evolution. Thus, the study of basin geometry includes not only the present geometry, but also the changing geometry during basin
evolution, which is even more important.

There are two methods to show the basin geometry. The first is to use the isopath map of
depositional sequences in different times (considering the top and bottom surfaces as
isochrononous) to represent the floor geometry in the time when the sequence was deposited.
The second is to use the thickness of each sequence to stack one by one from old to young to
show the structural feature of the basin in different times, and can be considered as a kind of
palaeostructural maps. When combined with the study of sedimentary facies, it can be used
to determine the locations of source rocks, reservoirs and topseals as well as stratigraphic
pinch-outs and possible eroded zones in the basin. In the second method, in addition to
showing progressive changes of basement geometry during the basin subsidence, the isopth
map made up of all sequences represents the present structural configuration of the basin
basement.

Sturctural style are referred to the structural configuration caused by the stratigraphic deformation in a basin. Different types of basins have different features of structural styles due to
their different plate location [3] .

The purpose for the studying of structural styles is to determine the association and distribution of structural traps characterized by different types of basin, and to combine with pool-
forming models to improve the prediction accuracy of pools.

Basins associated with convergent plate mainly have the structural style of compressive
blocks and basement thrust faults. Nappe structures thrusting from orogenic belts toward
basins, or anticline and syncline zones in lines with equal spacing commonly develop in basin
margin. On cross-section, the altitude of these structural zones, the intensity of folds or
thrusts and nappes, and the asymmetry of anticline folds all decrease toward the interior
basin.

Basins in association with divergent plate are predominated by fault blocks in structural
styles. For instance, rollover anticlines generally distribute along basement formal faults and
listric normal faults, and drape structures commonly occur on the top of titled-blocks or
horsts.

Basins related with transform faults normally have the structural style of transcurrent faults.
The anticline and syncline folds that distribute along the faults show en echolon arrangement, in which the ends of folds are in offset or connect to each other and most of them are
in equal spacing.

The downwarped basins in cratonic interiors are characteristic of basement uplifts and depressions in structural style. Anticlines with short and long axes commonly develop during
differential uplifting and subsiding.

It must be noted that there is generally not only one type of structural style in one type of
basin. The reason is that the forming mechanism of a basin is generally not simple, but complicated. Besides, later deformation of the basin can not only modify the previous structural
styles, but also create new structural styles.

Basin fill

During subsidence, sediments fill in basin, which not only contain hydrocarbon source rocks,
reservoirs and caprocks, but also record and preserve variety of information about the process of basin evolution, such as paleogeography, paleoclimate , paleoecology, paleotecton-
ics, paleo-geotemprature and etc.. Therefore, the research on the sediments in a basin provides the basis for basin studies.

The sequence framework, depositional types and general features of sediments in basins had
been comprehensively discussed by Miall [5] and Allen [1] in their books. Wu Chongyun and
Xue Shuhao [13] conducted a systematic study on the sedimentology of Mesozoic and Cenozoic
basins in China, most of which are inland basins.

Compared with most basins on the world that are predominated by marine sediments, the de-

position of inland basin, in addition to tectonic controls, shows more evident influence of paleoclimate zones, paleotopography, and paleodrainage patterns as well as the marine environment in offshore basins. Therefore, for the study on the sediment of inland basin, in addition
to obtaining the tectonic and thermal data of basin evolution, emphasis must be placed on the
research of lacustrine depositional systems controlled by various paleoclimate zones and tectonic settings. The reason is that nearly all of the hydrocarbon source rocks in inland basins
come from lacustrine sediments. In the research of reservoir, it is necessary to study fluvial
and deltaic depositional systems. The reason is that based on the known data, the clasitc
reservoirs of these two systems make up respectively 46% and 35. 4% of the total oil reserve
of inland basins in China (Fig. 2).

Subsidence history

The subsidence history of basins is controlled by two factors: tectonism and the
isostatic adjustment caused by sediment
loading. The subsidence curve obtained
from geohistorical analysis [10] represents
the composite result of the above two
factors. Backstripping technique [12] can
be used to separate these two and to
study their roles in basin subsidence respectively as well as their increase or decrease state inversely proportional to
each other.

Subsidence analysis is an important tool
for the study of basin evolution. Since
basin subsidence is one of the response of
regional tectonic evolution, subsidence
curves can be used to deduce the happened regional tectonic events or to veri-

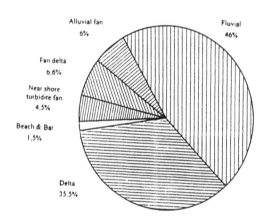

Fig. 2   Ratio of oil reserves in various clastic reservoirs of the Mesozoic and Cenozoic basins of China.
The fluvial reservoir contains the distributary sands of
delta plain, and the delta reservoir only consists of
front sands (after Wu et al. , 1992).

fy each other with the studies on regional tectonic setting. Different type of basins has its
own particular shape of subsidence curve because the type of basin is controlled by its regional tectonic setting.

The compensation relationship between the basin subsidence and deposition can be determined through the comparison between depositional rate and the corresponding subsidence
rate calculated by subsidence curve.

Meanwhile , the subsidence history of basins reflects the process of the sediments in progressive thickening, compaction, temperature increasing by burying. The combination of burial
and thermal histories can be used for the further studies on the maturity of source rocks, the

time of hydrocarbon expulsion, and the porosity change of reservoirs during compaction and diagenesis.

Based on the one-dimensional subsidence curves at different locations in basin, it is possible to construct the two-dimensional subsidence section of the basin in different times as well as paleosubsidence maps.

Thermal history

Geothermal mainly comes from the earth interior and meanwhile also from the decay of the radioactive elements contained in the sediments of basins. Therefore, along with basin subsidence, formation temperature will gradually increase, that is, the temperature increasing caused by burying and subsidence.

Besides, there is still another temperature increasing caused by tectonics, i. e. , the compressive friction of strata to create heat or the temperature increasing of burial or magmatic activity caused by tectonic movements. This kind of temperature increasing is quite common in the inland basins on China, which are related with polycyclic orogenesis. For example, the tectonic-thermal event from the end of the Jurassic to the Early Cretaceous (the Middle Yanshanian) in the North China is an important temperature increasing of tectonism (Fig. 3). This event enabled the hydrocarbon resource rock of the Paleozoic in most of the North China to get into the phase of dry gas generation, and the Upper Triassic source rock of Ordos Basin into the phase of oil generation. The temperature increasing of tectonics may be one of the important thermal mechanisms for inland basin.

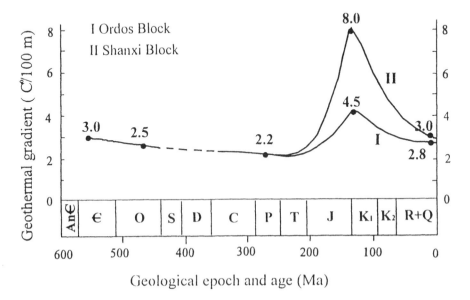

Fig. 3 Geothermal evolution profile of North China, showing an important tectonic-thermal event in the Late Jurassic-Early Cretaceous (the Middle Yanshanian).

Nowadays there are varieties of methods for the measurement of paleotemperature. There-fore, the profile of thermal evolution should be built up as soon as possible in the study of oil and gas basins. In this study, the special attention must be paid to the last tectonic-thermal event of the basin and its surrounding area. This is because it would cancel most geological records of temperature indicator if its temperature exceeds the highest one in the history. In this case, it is difficult to determine that to which level hydrocarbon had evolved before the event occurred.

Hydrodynamic history

Hydrodynamic history is used to study the role of hydrodynamics in hydrocarbon generation, migration and accumulation, and to determine its influence on oil and gas accumulation as well as later preservation.

The key in the study of hydrodynamic history is to analyze the fluid potential energy in differ-ent geological times. In the study of potential energy, the first is to determine the major fac-tor creating potential energy-paleoaltitude. The second is to analyze the abnormal fluid pres-sure caused by the differential compaction of strata, increasing pressure of aquatherm, the alteration of clay minerals, and hydrocarbon generating. The last is to study the potential difference of oil, gas and water caused by density difference at a specific height point.

Hydrodynamic history can be divided into two stages: before and after pool formation. For the first stage, studies are concentrated on the role of hydrodynamics in the process of pool-forming. For this purpose, a special paleostructural map can be drawn to determine the flow direction of water. This kind of paleostructural map is represented by an isopach map. It is constructed by using the top surface of the formation corresponding to the different phases of source rock maturity to the bottom or top surface of a carrier bed nearest to the source rock. Based on this map, the potential map of abnormal fluid pressure can be made and further study is carried out for the potential difference of oil-gas-water at a specific height point in order to determine the flowing direction of oil and gas.

The hydrodynamic study in the second stage places emphasis on the present hydrodynamic state and the destruction in pools and in the dynamic balance of underground fluids caused by tectonism after pool formation. In the case of relatively stable tectonic movement, there is little destroy of pools caused by hydrodynamics, and water-sealed pools can even form under particular geological condition, in which the balance between water and oil and gas is achieved. If intensive tectonic movement occurs and oil pools are destroyed, ground water will adjust together with oil and gas to determine if and to which level the pool can be pre-served and whether new secondary pools can form.

POOL-FORMING SUBSYSTEM

Pool-forming elements and their time-space structure

Pool-forming elements include source rock, carrier bed, regional topseal, and trap unit that consist of trap, direct topseal and reservoir . The relationship of pool-forming elements and their time-space structure and will decide if the hydrocarbon expelled from source rocks can migrate to trap units to accumulate as pools in the most economical and reasonable way.

Generally, the simple and reasonable spatial structure of the elements is that the source rock is located in the lower part of the basin interior and directly connects to the updip carrier bed at its top and bottom. The carrier bed is covered by the regional topseal and connects with the trap units in a short distance (Fig. 4). This arrangement will decrease the loss of hydrocarbon in migration to the minimum.

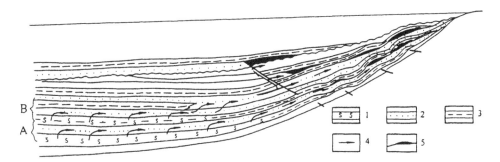

Fig. 4　A sketch profile showing the spatial structure of pool-forming elements. A. The source and carrier beds, and regional topseals extend laterally and connect directly with the traps. B. The topseal pinches out, and the oil may migrate along the pinch-out line from the lower carrier bed to the upper one. Then it may 1) onward pass through the fault to accumulate in the frontal traps, 2) go along the fault to enter the pinch-out sand in the hangingwall of the fault, and 3) go along the fault continuously to the unconformity surface and accumulate in the onlap sand.

1. source bed; 2. carrier and reservoir bed; 3. regional topseal; 4. migration path; 5. oil pool.

However, In most cases the spatial structure of the elements is relatively complex. This complexity is mainly caused by discontinuous carrier beds, lack of regional topseals or well-developed faults. In this case, hydrocarbon will pass through several intermediate pathways till it can finally arrive in the trap units.

The study on the spatial structure of pool-forming elements should not only concentrate on the present structure. It must also dynamically analyze the spatial structure during different maturity phases of source rocks in order to reconsturct the process of petroleum expulsion, migration and charge into reservoirs.

In the basin with high maturity of source rocks, the spatial structure maps of pool-forming elements have to be drawn at least for three periods: the periods of oil, wet gas, and dry gas generating thresholds. This map is based on the stratigraphic isopach map which is from the top of the strata corresponding to the time of getting into the threshold to the top or bottom of the carrier bed nearest to the source rock (i.e., paleostructural map). Then the extent of the source rock matured in this phase is delineated on the map. The last step is to draw out

on the map the carrier bed and the structural and nonstructural traps just above the bed in
that time, as well as their connection patterns.

The reason for drawing the spatial structure map of pool-forming elements in different time
is to fully consider the changes in structural framework which may be caused by tectonic deformation and basin differential subsidence in the corresponding time. Its purpose is to correctly determine the efficiency of the spatial structure in catching hydrocarbon.

Pool-forming model and succession

According to the spatial structure patterns of pool-forming elements which are commonly
seen at present, generally two pool-forming models can be identified. One is the model of
backward catchment basin which is predominated by lateral migration, and the another is
pot-flower model in which vertical migration is dominant.

The Model of backward catchment basin is similar to the catchment basin in catchment process, but contrast in the flowing direction of fluids (Fig. 5). The water source (Rs) of the
catchment basin is located in the basin margin, whereas oil and gas come from the basin interior (Sb). In the former case, water is transported by the rivers within their drainage system
from high to low position to accumulate in reservoir lake (Rl). However in the later model,
oil and gas are transported by the drainage system composed of carrier beds (Cb) from low to
high position to accumulate as pools in trap units (Tu). The drainage system of oil and gas is
similar to that of rivers, but they flow along the migration pathway (Mp) constituted by the
anticline-type fold axes of carrier beds. This model is commonly observed in the basin which
is lack of faults and with well-developed regional topseals.

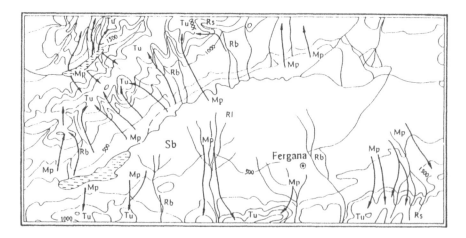

Fig. 5   A topographical map of the catchment basin in Fergana valley, showing its relationship with the
pool-forming model of the backward catchment basin if the map is considered as a structural contoural
map.

Rl—reservoir lake; Sb—source bed; Rb—river baisn; Rs—river source; Tu—trap unit.

The structure of pool-forming elements in pot-flower model is similar to the situation of flowers in pot, and the flowing directions of fluids in both patterns are the same. Both obtain the fluids—oil and gas, or water and nutrition needed by flowers from the bottom. Then the fluids are transported to the above traps or flowers and leaves through faults or stems. This model is common in the basin lack of regional topseals but with well-developed faults (Fig. 6).

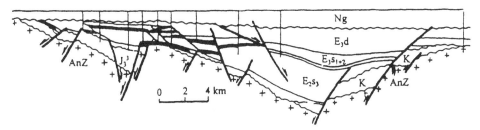

Fig. 6 Profile of hydrocarbon accumulation in the western part of Liaohe Depression, Bohai Gulf Basin, showing the pot-flower model of pool-forming. The hydrocarbon came from the source bed, E_2s^3 in the lower part of the basin, passed through faults, and accumulated in the traps of upper horizons. AnZ—pre-Sinian; K—Cretaceous; Es^3—3rd member of the Shahejie Formaiton of the Eocene; E_3s^{1+2} —1st and 2nd members of the Shahejie Formation of the Oligocene; Ng—the Guantao Formation of the Necocene (after Liaohe Oil Field).

Pool-forming model is a kind of summary about the forming pattern and state of pools based on the spatial structure of pool-forming elements. They combine with each other to draw the prediction map of pool-forming in the different periods of source rock maturity.

Pool-forming succession refers to the accumulation order of oil and gas pools which formed in the time corresponding to the different phases of source rock maturity. For example, in the early phase of source rock maturity ($R° < 0.6\%$), gas accumulation will occur in the trap u-nit at the low position if there is gas generated by biochemical action (Fig. 7A). In the oil generating phase ($R° = 0.6\% \sim 1.2\%$), oil will pass over the formed gas pool to accumulate as oil or oil and gas pool in the trap unit at a higher position (Fig. 7B). If the maturity of source rock is in the phase of wet gas generating ($R° < 1.2\% \sim 2.0\%$), the subsequent gas will replace the oil in the trap at the higher position and drive it to the trap at an even higher position (Fgi. 7C). In the phase of dry gas generating ($R° < 2.0\% \sim 4.0\%$), if there is no fa-vorable trap unit at the even higher position, there will be no oil but only gas pools in the basin. Otherwise, oil will accumulate in the trap at the highest position (Fig. 7D).

The above succession of pool-forming can only occur in the case when every factor and agent is in a favorable combination. However, it is still important in identifying the hydrocarbon types of pools generated in the different phases of hydrocarbon generation.

Pool association and distribution

The study on pool assemblage and distribution is to analyze the plane and sectional or three-

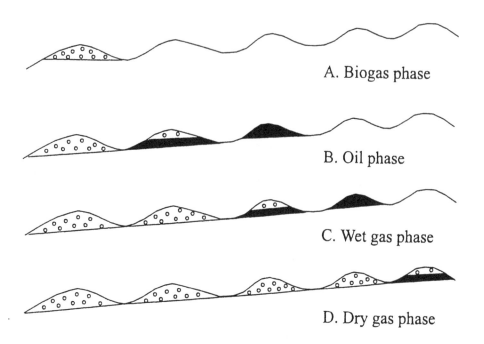

Fig. 7   Skecth profile of pool-forming succession (see the text for the explanation of the order from A to D).

dimensional assemblage and distribution pattern of trap units that are capable of catching hydrocarbon.

The assemblage pattern of structural pools is related to the structural styles which are unique for the different type of basin. The distribution of the former is controlled by the style and pattern of the latter. Nonsturctural pools are associated with lithofacies change zones, the slope of paleouplifts, paleochannels, paleoweathering crust, unconformity surfaces, reefs and halite, etc. They are generally characteristic of zoning distribution.

Not all traps have the apportunity to catch hydrocarbon to become pools, no matter how is the assemblage pattern of structural and nonstructural traps. Therefore, in the study of pool assemblage and distribution, the first step is to analyze the traps which are possible to form pools, and then to determine their position and distribution in the pattern of structural or nonstructural assemblage. For this purpose, it should be at first to find out the conditions for these traps to form trap unit, that is, the trap elements (vertical closure, closing area, and spill point), and the match of reservoir with the direct caprock. The second work is to determine which trap units, in a specific model of pool-forming, are at the favorable positions in the structure of pool-forming elements to catch oil and gas.

Not all the above conditions are very clear in exploration, but this does not impede to make a reasonable judgment based on the available data. Then the previous judgment can be modified or corrected along with the data accumulation and knowledge improvement.

Later deformation and hydrocarbon redistribution

After the formation of pools, it is very common for the formed hydrocarbon accumulation to be deformed by the later geological action of tectonic movement because the evolutional process of basin and lithospheric tectonics do not cease.

In the case that no evident tectonic deformation occurs in the basin, pools may not undergo a significant modification. But during the subsidence or uplift of the basin, the amplitude and spill point of the previous traps will considerably change to led to the overflow of oil and gas because the dip angle of strata may vary in a regional scale. In this case, a comparison between the present structural map of the previous carrier bed and the paleostructural maps of the formation in different phases of hydrocarbon generation can be made. The purpose is to determine to which level the reservoir was modified or destroyed, if the secondary reservoirs formed, and where they may distribute.

The Mesozoic and Cenozoic inland basins in China were influenced by the multicyclic tectonic movements of the neighboring orogenic belts, and thus underwent a sinnificant later deformation. This modification not only caused the deformation of the previous basin, but commonly destroyed and separated the previous basin, or even created a new generation of basins on the base of the old. This deformation can be generally divided into two cases: before and after hydrocarbon generation.

As stated above, the Mesozoic inland downwarped basins widely developed after the Indosinian epoch in the western part of China. The Turpen-Humi and Junggar basin, which are at present located to the south and north of the Bogeda mountains, was a unified basin in the Early-Middle Jurassic (Fig. 8). The Lower-Middle Jurassic source rocks of lacustrine-swamp

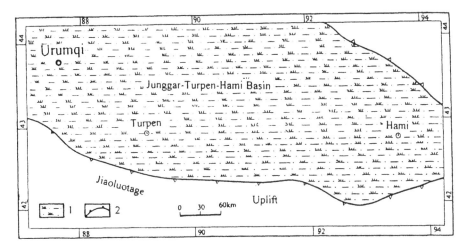

Fig. 8   The inferred distribution of the Junggar-Turpen-Hami Basin in the Early-Middle Jurassic.
1. fluvial swamp deposits; 2. inferred basin boundary (simplified after Zhou Lifa et al. , 1995).

facies deposited in the basin. The Middle Yanshanian Movement in the end of the Jurassic caused the Bogeda area to rise as mountains, which separated the Junggar Basin to the north

and the Turpen-Hami Basin to the south (Fig. 9), during the Cretaceous and Eocene, wide
depositional basins developed in the both sides of the low-gentle Bogeda mountains. But until
the Neogene and due to the Himalayan Movement, the Neogene foreland basins formed in the
two sides of the mountains. During the Yanshanian tectonic-thermal event, coal-derived hydrocarbon might have generated from the source rocks of the Lower-Middle Jurassic, and
then the shale source rock went into the threshold since the Paleogene. The hydrocarbon
richness of the Mesozoic in the two basins will be controlled by the quantity and quality of
source rocks they obtained from the separation. However, both had been deformed by the
Himalayan Movement after the pool formation.

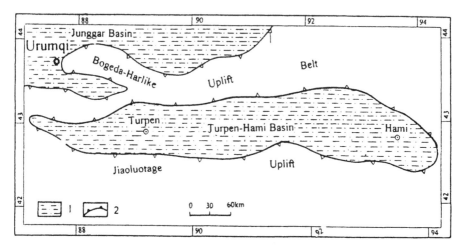

Fig. 9   The inferred boundary of the Junggar and Turpen-Hami baisns in the late of the Middle Jurassic
and the Late Jurassic. The Bogeda area rose as mountains and separated the Junggar Basin to the north
and the Turpen-Hami Basin to the south.

1. shallow lacustrine deposits; 2. inferred basin boundary (simplified after Zhou Lifa et al., 1995).

The similar case exists in the Jiuquan Basin and the Hauhai-Jinta Basin. These two basins
were a unified basin in the Mesozoic and shared the Cretaceous source rock of lacustrine facies. Their present separation was formed in the end of the Eocene (Fig. 10). And in the
same time, the source rocks in both basins were matured. Rich hydrocarbon accumulation
had been discovered in the Jiuqiuan Basin in the thirties of this century. However, until
nowadays there have been no oil and gas discoveries in the Nuahai-Jinta Basin. One of the
reasons is that the quantity and quality of the source rocks obtained by the Hauhai-Jinta
Basin from the separation are not so good as that of the Jiuquan Basin. Another reason is
that the structural tops formed since the Himalayan movement were not in good match with
the major phase of hydrocarbon generation, which might result in the loss of huge amount of
oil and gas or some remained only in the underlying Mesozoic strata [8].

The Dianquangui Basin is an example of deformation after hydrocarbon generation. The
basin is situated in Yunnna, Guizhou and Guangxi provinces of the southwestern China, with

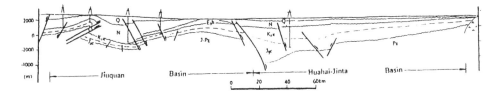

Fig. 10 Structural cross-section of the Jiuquan and Huahai-Jinta Basin, showing that they were a unified basin in the Mesozoic, and separated into two basins by Himalayan Movement in the end of the Eocene. Pz—Paleozoic; J-Pz—Paleozoic to Jurassic; J_{3c}—Chijinbo Formation of the Upper Jurassic; K_1x—Xinminbo Formation of the Lower Cretaceous; E_2h—Huihuibao Formation of the Oligocene; N—Neogene; Q—Quaternary (modified from Li Guoyu et al., 1988).

an area of 410,000 km². The marine strata deposited in the Paleozoic is over 5000 m thick, and multilayers of source rocks of the Lower Cambrian, Ordovician and Silurian developed in the basin, which were in mature and overmature phase before of during the Indosinian. The basin underwent many times of deformation during the Indosinian, Yanshanian and Himalayan. The formed oil and gas pools suffered a heavy destruction. Until now there is no commercial oil and gas reservoirs discovered in this basin.

RESOURCE EVALUATION SUBSYSTEM

Macro-prediction

The major basis for the macro-prediction of hydrocarbon resource in a basin is the tectonic setting analysis of regional geology, surface and subsurface geological survey and research, and geophysical and geochemical exploration as well as the corresponding mapping. The purpose is to determine the tectonic type and evolution of the basin, the development of source rocks, reservoirs, carries beds regional topseals and their quality and type, and the assemblage and distribution of traps and trap units. Based on the above evaluation and integrated analysis of all the basic geological conditions for hydrocarbon reservoir formation, the judgment about the possible process of hydrocarbon generating and the pool-forming and preservation condition can be achieved. Meanwhile, geological analogue with the similar mature basins has to be made and then the macro-prediction can be obtained, which answers the questions that if there are conditions for hydrocarbon accumulation and how the possible reservoir richness is.

Macro-prdiction provides the basis for determining if the basin is worth of further exploration.

Basin evaluation

The basis for basin evaluation is the prediction map of hydrocarbon accumulation in the basin. This map uses the paleostructural map of carrier beds in the time of hydrocarbon generation as a basis to outline the distribution of the matured source rock and the spatial structure of traps (Shown in plane projected position supplemented with cross-section) as well as their connection relationship with carrier beds through faults and other migrating pathways. Then the model of pool-forming is determined by analyzing the structural relationship between its source beds, drainage area of hydrocarbon, migration pathways and traps.

Pool-forming model is the basis of the prediction for hydrocarbon accumulation. Constrained by the specific model, hydrocarbon accumulation zones of different classes can be distinguished based on the connect relationship between traps with the primary and secondary migration pathway and the drainage area of oil and gas.

Assessment of zones or areas

Assessment of hydrocarbon accumulation zones or areas takes the zones or areas as objects to draw the distribution map of traps and possible oil and gas pools. Emphasis is placed on the pools to make the comparative assessment of trap units, in which in addition to trap units themselves, the major work is through the comparision to determine if they are at the favorable positions for catching oil and gas. Finally it provides targets for drilling.

Resource calculation

Resource calculation is made for different zones or areas of hydrocarbon accumulation according to the assessment class of potential pools. Finally, the hydrocarbon resource of different classes as well as the total resource in the entire basin are calculated.

CONCLUSION

Oil and gas basin system provides the theory and method for the systematic study of oil and gas basins. It is different from the general analysis of sedimentary basins [5,1], and also from petroleum system or petroleum play [6,7,4,9,2]. However, the general analysis of sedimentary basins is contained in the research contents of oil and gas basin system and provides the basis for the latter. Oil and gas basin system also includes the theory and method of petroleum system and play.

The principle and method provided by oil and gas basin system are applicable to the different stages of oil and gas basin studies and exploration, and there should be different missions for different stages. therefore, it is not the simple summary of research and exploration results in each stage, but an understanding in advance of the petroleum geological conditions and exploration potential in a baisn at a new start point. Only in this way can it serves as the guidance for the research and exploration of oil and gas basins.

REFERENCES

1. P. A. Allen and J. R. Allen. Basin analysis — principles and applications. Blackwell Scientific Publica-

tion, Oxford, London(1990).

2. G. Demaison and B. J. Huizinga. Genetic classification of petroleum system, Bull. Amer. Assoc. Petro. Geol. 75,1626-1643 (1991).

3. T. P. Harding and J. D. Lowell. Structural styles, their plate tectonic habitats, and hydrocarbon traps in petroleum provinces. Bull. Amer. Assoc. Petrol. Geol. 63,1016-1058(1979).

4. L. B. Magoon. The petroleum system: a classification for research ,exploration and source assessment. In: Petroleum system of the United States. L. B. Magoon (ed). pp. 2-15. U. S. G. S. Bull. 1987(1988).

5. A. D. Maill. Principles of sedimentary basin analysis. Springer-Verlag, New York(1984).

6. A. Perrodon. Geodynamique petrolire. Masson, Paris(1980).

7. A. Perrodon. Geodynamics of oil and gas accumulation. Elf Aquitaine, Pau(1983).

8. Ren Zhanli, Zhang Xiaohui and Liu Chiyang. Measurement of paleo-geotemperature of source rocks may point out the exploration direction of the Huahai-Jinta Basin, China. Chinese Science Bulletin, 40, 921-923(1995).

9. J. T. Smith. The petroleum system as an exploration tool in a frontier setting. Bull. Amer. Assoc. Petrol. Geol. 75,673(1991).

10. J. E. Van Hinte. Geohistory analysis, application of micropaleontology in exploration geology. Bull Amer. Assoc. Petrol. Geol. 62,201-222 (1978).

11. Wang Hongzhen, Liu Benpei and Liu Sitian. Geotectonic units and tectonic development of China and adjacent regions. In: Tectonopaleogeography and paleobiogeography of China and adjacent region, Wang Hongzhen et al (eds). pp. 3-34. China University of Geoscience Press, Beijing (1990).

12. A. B. Watts and W. B. F. Ryan. Flexure of the lithosphere and continental margin basins. Tectonophysics, 36,25-44(1976).

13. Wu Chongyun and Xue shuhao. Petroliferous basin sedimentology of China. Petroleum Industry Press, Beijing(1992).

14. Zhao Zhongyuan and Liu Chiyang. Meso-Cenozoic tectonics and petroliferous basins of North China craton and their formation and evolution. In: Selected Works of Symposium for the 45th Anniversary of the Founding of the Geology Department of Northwest University. pp. 399-406. Shaanxi Science and Technology Press, Xi' an (1984).

15. Zhao Zhongyuan. Tectonic setting for the formation and development of the Meso-Cenozoic petroliferous basins in East China. In: The formation and evolution of the sedimentary basins and their hydrocarbon occurrence in the North China craton. Zhao Zhongyuan and Liu Chiyang (eds). pp. 1-9, Northwest University Press, Xi' an (1990).

16. Zhao Zhongyuan. On integrated, dynamic and comprehensive analysis of the oil and gas basin. In: Developments in oil and gas basin geology. Zhao Zhongyuan, Liu Chiyang and Yao Yuan (eds). pp. 3-18. Northwest University Press, Xi' an (1993).

Proc. 30ᵗʰ Int' l. Geol. Congr., Vol. 18, pp. 283~297
Sun Z. C. *et al.* (Eds)
© VSP 1997

The Evolution of Geotectonic System and Formation of Hydrocarbon Generation in China

WANG JINGQI

Southwest Bureau of Petroleum Geology, Ministry of Geology and Mineral Resources, Chengdu, Sichuan 610081, *CHINA*

Abstract

The paleo-Chinese continent has passed both convergent and divergent periods. Passive continental marginal sediments that is favorable for hydrocarbon generation are widely developed in the separated continental plate. A variety of marine-continental transitional source rock deposited during the merging of the plate. After Indosinian movement, Chinese continent basically became an integrated body. There is an obvious macroscopic regulation that Tethys, Paleo-Asia ocean and Pacific ocean are associated each other in one side, and play their major roles in different stages respectively on the other side, and control the forming, developing of the continental basin, and the source rock deposit successively.

Keywords: geotectonic system, formation of hydrocarbon generation, foreland basin, extensional rift-depression, giant hydrocabon-bearing zone

INTRODUCTION

The geotectonics in China varies frequently, and the late history of the ancient oil-bearing lithological system is rather complicated. Hydrocarbon finding is inclined to be in new basin, new formation and in late stage reservoir forming condition[1,2], which is also in line with the exploration practice. However, a good oil generation condition has existed in China since Palaeozoic objectively. Therefore, it should be strengthened to study the hydrocarbon geological regulations from the geo-history evolution point of view.

EARLY PALAEOZOIC CONTINENTAL CRUST CONVERGENCE AND DIVERGENCE ASSOCIATED WITH FORMATION OF HYDROCARBON GENERATION

There are various opinions on the convergence and divergence of the early Chinese continent crust[3-5]. It is considered that three obvious development stages in different nature exist and have common characteristics, even in the separation stage.

Paleo-Chinese Platform Prior to Precambrian (Pt³—C)

The platform was occupied by continental crust[6] with large scale and relatively high consolidated grade[7]. The paleo-Asia ocean in the north was separated by Zhongtianshan-Xilamulun fault into Siberia plate and China plate continental margins. Along approximately 2000 km distance from eastern Liaoning to Inner-Mongolia, Qinghezhen formation with small shelly fossil-bearing or partial Wenduermiao Group (Pt³—C), which is continental marginal and deep sea deposits, are widely distributed. Jiang-Shao Fault and Jinshajiang Fault are considered as the south and west boundaries respectively [13], with mostly deep-metamorphosed marginal sediments.

The major continental crust, except Northern China, is characterized by platform and marginal facies deposits which are the oldest source rock in Yangtze and Tarim area. A sizable gas field in Dengying formation in Weiyuan of Sichuan basin has been found.

China Continental Crust Split in Late Stage of Early Cambrian[7]

During this stage, China continental crust was splitted, while Tarim, north China, Yangtze and "Huaxia" plates were floating in the ocean[15], which expanded the region of continental marginal sediments, and some aulacogen such as Manjiaer, Helan and Back Longmenshan were formed by abyssal basin extending to the intercontinent. With such large thickness and aboundant organic sediments, this is the most brilliant period for hydrocarbon generation formation in China geological history.

On the separated continental crusts, there are carbonate platform dominated deposits with 1-2 km thick and medium aboundant organics and local platform-trench facies deposits. Many medium-small sized and a few large size but low commercial oil and gas fields as Jinbian field in Shanxi were discovered in a wide and relatively stable craton region.

At the platform margin and slope between platform and abyssal basin, there is an advantageous condition for generation, migration and accumulation of the hydrocarbon. This is a quite common model in early Palaeozoic China, but later on, it was destroyed and metamorphosed by the intensive tectonic and overlapped by thrust belt. Fig. 1 shows the partial platform marginal belt, especially the "a thousands of kilometers of paleo-slope belt"[10] in the south margin of Yangtze platform as the most remarkable one, along which paleo-reservoirs, bitumen and oil seeps can be found everywhere[11,12]. The opportunities for finding out the deep remained and reformed hydrocarbon reservoirs in southern China still exist. The preserving condition in south and west margin of Ordos in northern China is better than that in the southern China; while the condition in the northeast and southwest margin of Tarim is even better due to their completely deep buried underground.

Separated Continental Plates Presented as a Compressional Uplift Tendency After Mid-Ordovician

None of the strata above late Ordovician has been seen in north China; no clear deep-sea de-

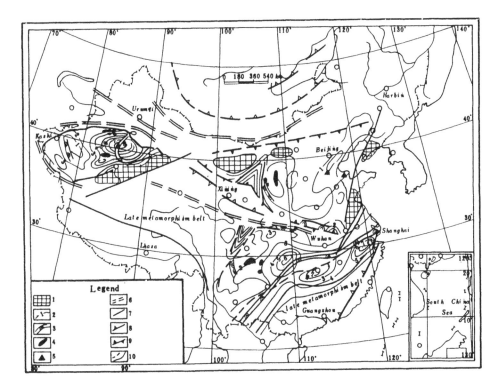

Fig. 1 Sketch map of tectonic and sedimentation of Early Palaeozoic in China and adjacent area

1. uplift in the platform; 2. sedimentary (non-metamorphism) isopack (km); 3. marginal reef facies in the platform; 4. authigenic source; 5. allogenetic source; 6. speculative oceanic crust or ridge; 7. main fault; 8. plate subduction belt; 9. plate suture belt; 10. epimentamorphism belt

posits are found in Tarim and Yangtze continental margin which was transfered to develop clastic depressions toward craton, such as Chuanxiang depression (S) and caohu depression (O₃); the platform at the common border of Jiangsu, Zhejiang and Anhui started to subside, and dominated by clastics since O₂, and the former platform-slope-basin model no longer existed. These continental marginal depressions in the late stage of early Palaeozoic plays the role of source rock trasmission for eastern Sichuan gas areas and Akekula oil and gas areas, and probably plays a similar role for Mid-Yangtze in Jianghan and the Lower Yangtze in southern Jiangsu.

WIDE MARINE-CONTINENTAL FORMATION OF HYDROCARBON GENERATION OF NEW AND OLD PLATFORM IN LATE PALAEOZOIC (might through T₂)

Caledonian movement almost make all the intercontinental subduction zones folded and the plate margin merged, as a result, some of the previous marginal deep sea arose and become a real "paleo-continents" such as "Jiangnan uplift". Even though the plate get closed, each plate increasingly become difference; the new platform on the basis of Caledonia fold is more significant to the petroleum geology. So it is better to discuss it geographically as the following.

Southern China

The southern China has subjected to several times of tectonic deformations in Early Palaeozoic, and the matured continental crust expanded by the end of Caledonia so that the paleotectonic framework varied significantly[13]. Southern China active zone, merged to paleo-Yangtze block, was unified as Hercynian south China plate. Several thousand meters of carbonate and clastic rocks were deposited in the most areas of Guizhou, Guangxi and Hunan (Fig. 2). Oil source rock is very thick with high content of organic carbon. Rift-depression

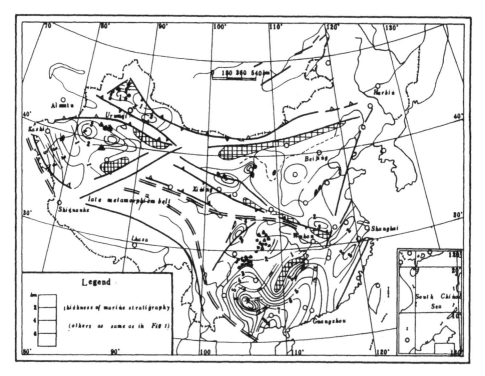

Fig. 2   Sketch map of tectonic and sedimentation of Late Palaeozoic (or with Early-Middle Triassic) in China and adjacent area

zones are also widely developed during each stage. More argillaceous hydrocarbon-generating rocks deposited within the depression and reef belt well-developed on the edge, where many old reef oil field trace and numerous bitumen and oil seeps were discovered, resulted from late intensive compressional folding and crust uplift eroding. The hydrocarbon generating rocks continuously buried more deep in some areas with its vitrinite reflectance reached to the dead limit ($R^\circ > 7\%$). However, this area is a widely distributed multi-type area, so that the potential energy of hydrocarbon is still very large.

Sichuan basin on craton is characterized by uplifting and erosion in the early stage of Late Palaeozoic, sporadic dolomite remained in eastern Sichuan during C_2, and wide area transgressive lasted from P_1 through T_2. Several tens of gas fields have been found in this area.

North China Platform

Devonian and Lower Carboniferous strata are missing at the platform; the trough on the south margin of the platform is primarily closed; and uplift occurred as a belt on the north margin. After Mid-Carboniferous, large scale semi-closed marine-continental transitional and limnetic sedimentary basins were formed, which are the most important coal-bearing regions in China. Particularly, it is very prospective to find out coal-measure related gas field in the gentle deformed Ordos where could become a unique large tight sandstone gas-bearing territory in China. Breakthrough could be achieved from stratigraphic and lithological traps.

Tarim Platform

Tarim platform was in nature of compressional foreland basin[15] in the early stage of late Palaeozoic; tectonic movement affected the whole basin in late Devonian and early Hercynian[16], resulting in row-like fault and fold structure. Transgression from southwest towards northeast in Carboniferous formed a good assembly of source rock, reservoir and seal. As an example, part of the source in some oil and gas fields as north Tarim, middle Tarim and Maigaiti are from Carboniferous itself.

Junger Platform on Early Hercynian Fold Basement

This platform was probably as a unified sedimentary basin with Tuha, Santanghu basins together in late Palaeozoic. Upper Permian is believed to be the most reliable source rock[17].

Qiangtang Area

Qiangtang area, a block with platform nature on the basis of Caledonian folding in north Gandwana continental margin, has a good formation of hydrocarbon generation. The prospects of hydrocarbon depend on the late deformation grade.

PETROLIFEROUS FORMATION AND FORMATION MECHANISM OF CONTINENTAL BASIN IN MESOZOIC AND CENOZOIC

After Hercynian and Indosinian movement, Chinese continents except partial Tibet was sutured into an integrated body basically. In different areas with its internal agent, it was formed that a certain overall tectonic model and sedimentary basin group with their own characteristics under different external action for each stage. There are three major external factors during Mesozoic and Cenozoic as followings: firstly, Gandwana continental block colliding to Chinese plate several times and continuous subducting, forming compressional stress and high gravity potential to the north and the east; secondly, Siberia plate moving southward continuously, remaining activeties between the two Paleo-plates and the associated Paleo-Asia ocean still obvious in Mesozoic; lastly, east of Chinese continent was more and more affected by Pacific ocean plate (include Kula) since Jurassic. In addition, the south margin of Chinese continent was affected by the tensile stress caused by the South China Sea spreading in Cenozoic. The most important zone, new folds, faults and magmatic activities are easy to be developed by the late stage external forces. Moreover, the soft zone inside the crust sphere causes the deposited rocks on the plate subjected to the very intensive deformation.

The three external factors from Tethys, Paleo-Asia ocean and Pacific ocean are not only correlated each other, but also played their major role in different stages and controlled the development of the continental basin and the formation of hydrocarbon generation one by one. So, the characteristics of macro geological regulation and distribution of hydrocarbon in China can be seen from the above mentioned features.

Petroliferous Formation in the Indosinian Foreland Basin $(T_2 - J_2)$

During late Triassic, west Qinling-Kunlun and Jinshajiang zones collided each other, intensively compressing northward and eastward to Chinese plate in a wide area, and resulting in many early folded zones reactivated again. Particularly, in the central, west China adjacent to the collision zone, a series of foreland basins were formed in front of the orogenic belt. Good Upper Triassic to Middle Jurassic dominated oil-generating formation was widely developed. There are deep water lacustrine and marshy facies without magmatic activities. Fig. 3 shows the several foreland basins associated with nappe orogenic belt to the out margin of the collision zone, i.e. Chuxiong basin in front of Ailaoshan; western Sichuan basin in front of Longmenshan; Qingyang-Huachi, Wuqi basins in front of Indosinian uplift in west Ordos; Hexizoulang basin in north side of Qilianshan, including eastern Jiuquan basin with oil discovery[18] and north Caidamu basin and Minhe basin in the south side of Qilanshan; foreland basin as Junger, Tuha and Santang separated from Bogda Indosinian orogenic movement in north Tianshan; the intensified Kuche foreland basin and its expansion to the whole north Tarim on the basis of Hercynian movement in south Tianshan; and southwest Tarim basin to the north side of Kunlun.

The above mentioned foreland basins constitute the first phase cycle of non-marine oil-generating formation after Chinese continent merging, distributed in north and east side of the Jianshajiang, west Qinling-Kunlun collision arc and the affected areas. Many oil and gas fields have been discovered in most of the basins which are the major non-marine territories in central and western China. North of Qiangtang basin inside Jinshajiang collision arc is belong to the foreland basin of Indosinian nappe, Jurassic is a good hydrocarbon-generation formation.

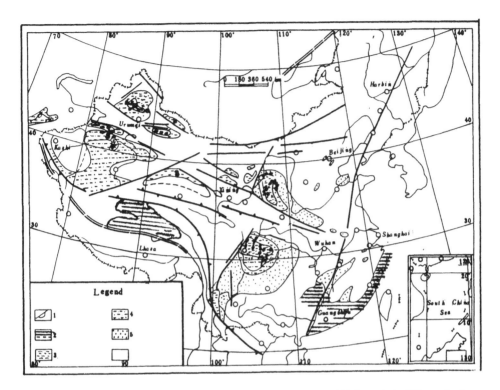

Fig. 3 Sketch map of tectonic and sedimentation of Upper Triassic-Middle Jurassic in China and adjacent area

1. forland basin in orogeny belt; 2. marine deposit (marine and continental facies); 3. deep and semi-deep water lacustrine and swarp; 4. fluvial and lacustrine facies; 5. fluvial facies

This kind of foreland basin is also very important in the middle Asia, such as Karakum basin in front of Mt. Kopet, where Lower-Middle Jurassic coal-bearing hydrocarbon-generation formation became a huge gas source (Fig. 7).

Indosinian fold belt along southeast coast may extend to Vietnam[19] passing north part of Hainan Island before large area of marine sediments were formed but not developed to large size basin any more. The origional sedimentary behavior was obscured by the late tectonic and magmatic activities. Most of the Late Triassic to Middle Jurassic sedimentaries in other region of China are fluvial fillings. Some sizable lacustrine and swamp deposits were subsided down to the Cenozoic depression or without beneficial seals and preserving conditions.

Rift-Depression Basin in North China (J₃−K₁)

Bangonghu-Dingqing collision belt[20] in late Jurassic was far from Chinese continent, with less compressional stress than Indosimian movement. For the foreland basins formed by Indosinian movement in central and west China, mountain region rose, more epicontinental clastic deposited but basin setting velocity decreased, with shallow water so that causes an overall poor oil-gernerating condition. Lianhuakou formation (J_3) and Jianmenguan formation (K_1) in western Sichuan are a thick conglomerate, sandstone and red argillaceous rocks. There is no Upper Jurassic sediments in Ordos, Zhidan formation is dominated by conglomerate and variegated mudstone; Qigu formation (J_3) and the overlain strata in each basin of Xinjiang are mostly red bed. Hydrocarbon reservoirs associated with Upper Jurassic and Lower Cretaceous in western Sichuan, north Tarim and Junger are secondary accumulations migrated from the deep.

However, a very important hydrocarbon generation formation belt associated with rift-depression basin occurred in northern China at this period (Fig. 4). A giant extentional rift-de-

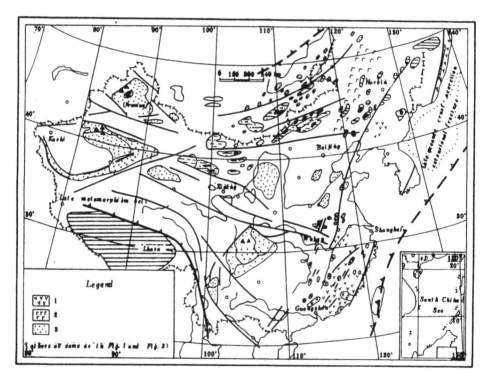

Fig. 4 Sketch map of tectonic and sedimentation of Upper Jurassic-Early Cretaceous in China and adjacent area
1. calc-alkaline volcanic rock in continental margin; 2. weak alkaline volcanic rock in continent; 3. granite

pression belt and more than four hundred basins were formed on the former Hercynian fold belt, from west Jiuquan, Yinger and Badanjilin, passing by Erlian and China-Mongoliao bordor, across Daxinganling and Songliao basin, eastward to the sea. More volcanic activities occurred in Late Jurassic, strong in the east and weak in the west; Early Cretaceous is an important period for source rock deposit, with large thickness, rich organic content and high paleo-geothermal gradient, which formed a giant NEE trend hydrocarbon accumulation zone where many important oil and gas fields have been found with same oil generation period and oil-forming geological conditions. Although this zone is located on the Hercynian fold belt, the crust is rather thin, probably related to the mantle arising. Jurassic geocyncline is very active in north Mongolia-Okhotsk, with some ten thousands of meters thick Middle Jurassic deposits, and an intensive folding occurred subsequently[7]. Meanwhile, rift-depression occurred in the area around Ealian and Daxinganling. Tectonic movement after early Cretaceous tends to be quiet, so that the paleo-magnetic poles of north China plate and Siberia plate could completely meet together[2]. Yanshanian cycle is a typical period when the tectonic system in east, and north China varied from paleo-Asia ocean to Pacific ocean activities, with major tectonic direction changing NEE-NE-NNE, which regularly controlled the oil-bearing formation basin and the deformaiton.

Several thousand kilometers of volcanic eruption and magmatic intrusion in eastern Asia are the most magnificent geological events in the period, when southeast coast was uplifting without back-arc basin sediment[22], some fault-related $J_3 - K_1$ strata are sporadically distributed in a wide area in other places of China. For instance, there is a certain thick dark mudstone in Hehuai area[23], generally dominated by fluvial and lack of stable deep water lacustrine basin.

Rift-Depression Basin $(K_2 - E)$ in Eastern China

Prior to the collision with Indo plate, central and western China was in a relatively quiet period. The former large scale foreland basin formed by Indosinian movement was in shrinking nature, such as Chuxiong, western Sichuan, Ordos, Hexizoulang and basins in Xinjiang, so there are no good oil-generation formations. However, in central and southern China, under the joint reaction of the compressional stress after large scale magmatic activities and the inter-reacted movement caused by the final merge of Yangtze and north China plate, Yanshanian tectonic deformation was intensive and overall, consequently become a tectonic system dominated by Pacific ocean plate subduction.

In eastern China, there is a NNE trend gravity gradient belt (Fig. 5), east of which such rift-depression basins as Songliao, Bohai Bay, Nanxiang, Jianghan and north Jiangsu-south Yellow Sea and East China Sea are the largest oil-bearing zones in China, with high value on Moho depth contour[7]. Seismic sounding data[24] shows that middle and lower crust in northern China basin is only half thick of Shanxi, with clear mantle arising, directly corresponding to the subduction of Pacific ocean plate. The basins in the north of the giant oil-bearing belt is in large area and good oil generation condition, and become poor in the south due to the gradual dispersal. This difference could be the response of deep crust[7,25]. It is probably related to the factors that during 100-25 Ma, subduction zone of Pacific ocean plate was mainly at northeast Asia, while the south end of that was only close to the East China Sea, and transfered fault further to the south[26]; and it also could be related to the spreading oceanic ridge subducting into the continental plate[27] (Fig. 5).

Some small size east-west trend rift-depression basin groups were formed disharmoniously

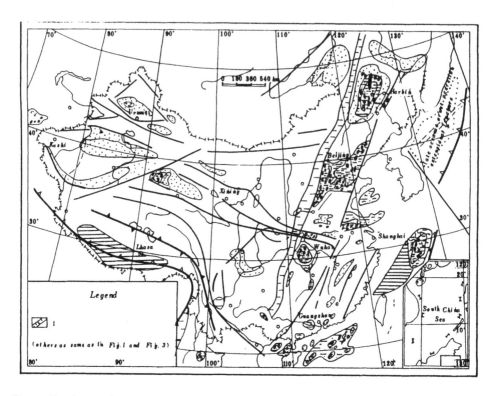

Fig. 5   Sketch map of tectonic and sedimentation of Upper Cretaceous-Palaeogene in China and adjacent area
1. gravity gradient anormaly

with Pacific ocean tectonic belt in southern Guangdong and Guangxi in this period, and the oil generation condition getting better towards the sea area. The crust in Pearl River Mouth basin began thining in middle $K_2 - E_1$ period, forming intercontinental extensional rift-dipression in $E_1 - E_3^1$, so that better source rock[28] deposited, which reflects the stress situation of South China Sea before spreading. NW trend Baise basin sited in the long-term active belt has more typical meanings.

New Passive Continental Margin (N including E_3) in Southern China
There are several important geological events occurred in Chinese continental margin around Pligocene epoch; last intensive collision acted to Chinese continent from Indo plate; while Japanese Sea spreading, subduction zone of Pacific ocean plate far away from the continent; when South China Sea spreading, both side of which became new passive continental margin. The subduction and collision of Indo plate forms the most thick Tibet crust on the earth. Great potential energy [29] and compressional stress continuously acted to the adjacent area which reactivated the paleo-orogenic zone that arose sharply and the foreland subsided with large difference in elevation and subsidence. Many basins were filled rapidly, characterized by shallow water and non-source facies. Such oil and gas fields discovered in Upper Tertiary

red bed as Kekeya, Yaha, Dushanzi and Laojunmiao are all not sourced by itself. However, there is an exception in west Caidamu basin where occurred some local deep water lacustrine source rocks in $E_3 - N_2^1$ and formed several oil fields which are obviously associated with the joint place of two active zones.

Due to the continuous pushing and compressing of Indo plate, China plate itself was compressed out of the east, which some paleo-faults become strike-slip boundary forming a series of transtensional depression, such as Hetao and Fenwei basins in the north and Yunnan and Xikang basins in the south where there are some low matured source rock and light deposits.

The expansion period of South China Sea is in $E_3^2 - N_1^1$, characterized by passive continental margin at the north shelf and transgressive from south to north, $E_3 - N^1$ is the major sedimentary period of source rock in Yingehai and southeastern basin of Hainan. Oil-generating condition in Zhu-2 depression is also good. After Upper Tertiary become thinning towards south, there were 2 km thick deposits [27]distributed in west-east trend under the slope. But the overall Tertiary source rock in north of South China Sea is low matured, so that many experts emphasized the opinion of "non-marine oil-generation (E)" and "marine reservoir (N)"[28,30,31]   (Fig. 6).

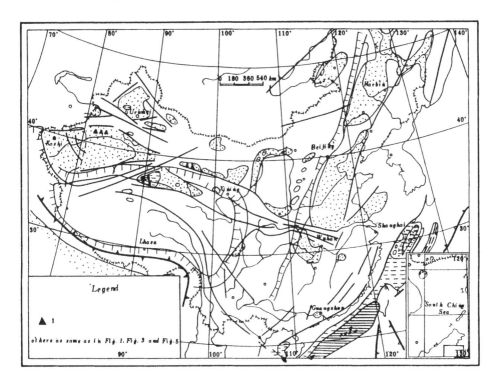

Fig. 6   Sketch map of tectonics and sedimentation of Neogene (or with Upper Palaeogene) in China and adjacent area
1. secondary oil and gas shows

Cenozoic strata in the south shelf of South China Sea are larger in both sedimentary thickness and the area than those strata in the north. Since E_3, marine sediments reached to 3000-9000 m thick, with high geothermal gradient, which became a good formation of oil generation. No doubt, it is a great oil and gas potential area.

During $E_3 - N_1$ period, basins in the north of South China Sea are different from the basins in East China Sea where both Yuquan movement (prior to E_3) and Longjing movement (end of N_1) pressented as compressional feature. Xihu clear material supply from the east [30,32] . Oil-generation condition in this period is not as good as that in $E_2 - K_2$ period, and also much poor than the passive continental margin in South China Sea (Fig. 7).

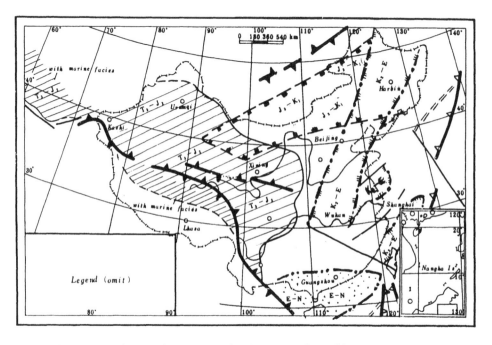

Fig. 7   Schematic map showing the migration of major contiental petroliferous formations in China and adjacent area

CONCLUSIONS

This article is focused on the macro control conditions for the hydrocarbon generation formation in China. Marine-continental transitional facies in passive continental marginal and semi-closed continental platform is very important during the convergence and divergence of Palaeozoic continental crust. After Indosinian movement, China became an integrated continent. The four stages of $T_3 - J_2$, $J_3 - K_1$, $K_2 - E$ and $E_3 - N$ varied with the tectonic framework. The formation of oil-generation has the corresponding regulations as: an order development from west, north, east and south of China; compressional foreland basin features kept in central and western China, while extensional rift-depression features in eastern China; most of the Mesozoic continental basins stacking on Palaeozoic marine platform, constituting the basic petroleum geology of China.

REFERENCES

1. Zhang Yigang. Natural gas generation, accumulation and preservation. Hehai University Press. Nanjing China, 154-156 & 169-174(1991).

2. Deng Zhongfan. Historical review of appraising hydrocarbon-bearing characteristics of Palaeozoic marine carbonate and theoretical basis for later appraisal in South China. Selected papers on Petroleum and Natural gas Geology, Volume 3; Major hydrocarbon-bearing basins geological characteristics & evaluation, Edited by Institute of Petroleum Geology, MGMR. Geological publishing house, Beijing China, 228-238(1991).

3. Wang Hongzhe(Chief editor). The Paleomaps of China. Cartographic Publishing House, Beijing, China, 18-37(1985).

4. Yang Zhende, Xie Mingqian. On the tectonic evolution of East Asia in the late Precambrian. Scientia Geological Sinica, 4,373-382 (1984).

5. Liu Benpei, Xiao Jingdong, Zhou Zhengguo et al.. Contribution to sedimentary geology of paleo-continental margin. China University of Geosciences Press, Wuhan, China, 7(1992).

6. Huang Jiqing. The selected papers of petroleum geology of Huang Jiqing. Scientific Publishing House, Beijing China, 145(1993).

7. Yang Senran, Yang Weiran (Editors). The regional geotectology of China. Geological Publishing House, Beijing, China, 51 & 302-320 (1985).

8. Wang Dongfang et al.. The continental geology in Northern margin of Sino-Korean Platform. Seismological Press, Beijing, China, 3 & 42-51(1992).

9. Wang Jinqi. The problem of hydrocarbon geology of Palaeozoic China. Sun Zhaocai, Zhang Yuchang (Editors); The collected works of discussing X. Zhu Scientific Thought. Petroleum Industry Press, Beijing, China, 145-152 (1993).

10. Yue Wenzhe, Ye Zhizhen. The giant slope lie across a thousand Li. Sun Zhaocai, Zhang Yuchang (Editors); The collected works of discussing X. Zhu Scientific Thought, Petroleum Industry Press, Beijing, China, 1-9 (1993).

11. Han Shiqing, Wang Shoude, Hu Weihuan. The discovery of a paleopool in Majiang and its geological significance. Oil & Gas Geology, 3;4, 316-325 (1982).

12. Xu Keding, Jin Chunmei, Zhou Guangyong. The condition of oil mineralization and characteristic evolution of paleopool in West Zhejiang. Shi Baoheng (Editors); The geology hydrocarbon of marine on Yangtze. Petroleum Industry Press, Beijing, China, 277-285(1993).

13. Liu Benpei. On the tecton-paleogeographical development of South China in the Hercynian-Indosinian stage. Wang Hongzhen, Yang Weiran, Liu Benpei (Editors); Tectonnic history of the ancient continental margins of South China. Geological Institute of Wuhan Press, Wuhan, China, 65(1986).

14. Deng Zonghuo, Cheng Guodong. Research of paleopool in Guizhou and adjacent area. Shi Baoheng (Editors); The geology hydrocarbon of marine on Yangtze, Petroleum Industry Press, Beijing, China, 296-307 (1993).

15. Chen Fajing, Chen Quanmao et al. Tectonic evolution of North Tarim basin and its relation to oil and

gas. Jia Runxu (chief editor): Research of petroleum geology of Northern Tarim basin in China. China University of Geoscience Press, Beijing, China, 32-35 (1993).

16. Wei Guoqi, Jia Chengzhao et al. The characteristic of structure deformation in Early Hercynian and the controllability of hydrocabon accumulation on the Central uplift zone. Tong Xiaoyuan, Liang Digan (Editors): The book of hydrocarbon exploration of Tarim Basin. Xinjiang Scientific and Hygiene Press, Urumqi, China, 229-301(1992).

17. Zhang Guojun, Zhao Bai, Wu Qingfu. The characteristic of hydrocarbon accumulation in Junggar Basin. Advaces in evaluation and research of oil and gas resources No. 5: Oil and gas accumulation and distribution in China. Petroleum Industry Press, Beijing, China, 189-191 (1991).

18. Xu Wang. Oil injection and basin study of the Jurassic in Eastern Jiuquan Basin. Experimental Petroleum Geology. 16:2,103-117 (1994).

19. Wang Hongzhen, Yang Sennan, Li Sitian. Mesozoic and Cenozoic basin formation in East China and adjacent regions and development of the continental margin. Acta Geological Simica 57:3,125 (1983).

20. Wang Naiwen. Qingzang-Indo paleocontinental and its welding to Cathaysia. Li Guangcen, J L Mercier (Editors): Sino-French cooperative investigation in Himalayas, Geological Publishing House, Beijing, China, 51-56 (1984).

21. Zhao Yue, Yang Zhenyu, Ma Xinghua. Geotectonic transition from paleoasian system and paleotethyan system to paleopacific active continental margin in Eastern China. Scientia Geological Sinica. 29:2,105-113 (1994).

22. Weng Shijie, Huang Hai. Plate tectonics of Southeast China and relationship between tectonism and magmatism in Jurassic and Cretaceous time. Acta Geological Sinica. 57:2,119-125 (1983).

23. Han Jingxing. Discussion on structural development and petroleum exploration problem in Hehuai basin. Selected papers on Petroleum and Natural Gas Geology, Volume 3:Major Hydrocarbon-bearing basins geological characteristics & evaluation, Edited by Institute of Petroleum Geology, MGMR. Geological Publishing House, Beijing China, 168-177 (1991).

24. Sun Wucheng, Zhu Zhiping, Zhang Li et al.. Exploration of the crust and uper mantle in North China. Department of Scientific Programming and Earthquake Monitoring (Editor): Developments in the research of deep sturctures of China's continent, Geological Publishing House, Beijing, 33(1988).

25. Yin Xiuhua, Shi Zhihong, Liu Zhanpo et al.. China 1°×1° Bouguer gravity anormaly map. Ma Xingyuan (Editor): The directions of China lithosphere dynamics outline (1 : 300000 China and Adjacent Sea Lithosphere Dynamics Map). Geological Publishing House, Beijing, 18 (1987).

26. Hilde T W, Uyeda C S, Kroenke L. Evolution of the Western Pacific and its margin. Tectonphysics, 38:(1-2), 145-165 (1977).

27. D Jongsma, A J Barber (Editor, and Xu Zhicheng translated to Chinese). Tectonic of East Asia and resource research (SEATAR). Geological Publishing House (Chinese), Beijing, China, 141 (1985).

28. Jin Qinghuan. Rift evolution and hydrocarbons in Pearl River Mouth basin. Selected papers on Petroleum and Natural Gas Geology, Volume 3: Major hydrocarbon-bearing basins geological characteristics & evaluation, Edited by Institute of Petroleum Geology, MGMR. Geological Publishing House, Beijing, China, 29-41 (1991).

29. Zhou Jiu, Huang Xiuwu and Wang Xuechang. Gravity tectonics of the Qinghai-Tibetan plateau. Scientia Geological Sinica, 4, 351 (1983).

30. Wu Chongyun, Xue Shuhao et al.. The sedimentology of petroliferous basins in China. The Petroleum Industry Press, Beijng, China, 367-380 (1992).

31. Cheng Sizhong, Li Zesong, Zhou Yechu. Oil-gas accumulation in Zhu Jiang Kou Basin-An example of tensional basin in continental margin. Advances in evaluation and research of oil and gas resources No. 5: Oil and gas accumulation distribution in China. The Petroleum Industry Press, Beijing, China, 288-301(1991).

32. Zhou Zhiwu, Zhao Jinhai, Yin Peiling, Characteristics and tectonic evolution of the East China Sea. K J Hsu (Series Editor): Sedimentary basins of the world. X. Zhu (Editor): Chinese sedimentary basins. Elserier Amstterdam, 1165-179 (1989).

Proc. 30ᵗʰ Int' l. Geol. Congr. , Vol. 18, pp. 299~311
Sun Z. C. *et al.* (Eds)
© VSP 1997

Petroleum Geology and Exploration Direction of Qingzang Plateau

HE ZHILIANG   LIU JISHUN   JIANG PING

Comprehensive Institute of Petroleum Geology, MGMR, Jingsha, Hubei, CHINA

Abstract

Qingzang Plateau lies on the east part of Tethys tectonic-sedimentary domain. Four major evolutionary phases that have undergone since late Proterozoic, namely, Proterozoic Ocean, paleoTethys, neoTethys, and Indian Ocean, resulted in the formation of polyphasical basin and the composite and superposition of different prototype basins in different tectonic position within the plateau, and thus many potential oil-bearing regions were developed. Intense compression and complex basin deformation initiated from Pliocene complicated the generation and preservation of hydrocarbons. An important exploration question is whether hydrocarbons previously generated are efficiently preserved or not, and hydrocarbon accumulations are valid or not, where to identify the relatively stable tectonic units. Served as the examples, Lunpola Tertiary terrestrial basin and Qiangtang Mesozoic residual marine basin are two important oil-gas exploration regions in Qingzang plateau.

Qingzang plateau is an important part of paleoTethys domain which was located between paleoEurasia and Gondwana. Explortion activities to date have discovered many significant hydrocarbon accumulations in the sedimentary basins of Tethys tectonic belt in varying degree[1]. Compared with the basic geology conditions and oil-gas-forming geology conditions of other basins in Tethys, Qingzang plateau is quite different and exhibits unique styles. However, the success in the exploration of petroleum conducted in the adjacent basins illuminates the prospects of hydrocarbon exploration in Qingzang region, and gives many examples and models for comparative study of basins.

Keywords: Petroleum geology, Paleo-Tethys, Exploration direction, Qingzang plateau

COMPLEX GEOLOGICAL TECTONIC

Qingzang plateau, the higest plateau in the world, is titled as the proof of the world or the third polar of the earth. Its complexities of geological tectonics can be summerized as follows:

Tectonic Framework of Arc Foreland Belt Alternated with Massif
The nearly E-W trending arc faults (Figure 1) are translithospheric fracture which have acti-

vated multiphasically. Their evolutionary history can be divided into
three stage;slipting of microplates or
slivers in the early stage,subduction
in the middle stage,intracontinental
subduction,obduction and strike-slip
in the late stage. However,the intermediate massives, which are al-
ternared with the faults mentioned
above, show respective basal structures, sedimentary covers, volcanic
activities and metamorphism. [2-5]
Henceforth, sedimentary basins and
sedimentary formation of different
types and periods were developed and
persisted , with different petroleum
geology features.

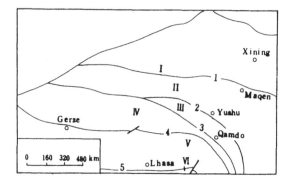

Figure 1   Division of tectonic units of Qingzang Plateau
I . Qimantage-Burhanbudai Terrain; II . Hoh Xil-Songpan Mi-
croplate; III . Qamdo Microplate; IV . Qiangtang Microplate; V .
Lhasa Microplate; VI . Himalayan Composite Terrain
1. Xidatan-Maqen-Xiugou Suture; 2. Xijin Ulan Hu-Jinshajiang
Suture; 3. Gomo Co-Lancangjiang Suture; 4. Bangong-Co-Nujiang
Suture; 5. Yarlungzangbojiang Suture

Complex Crust-Mantle Architecture

Qingzang plateau is characterized by intense uplift of topography and anormally thickening of
crust. The contour map of crustal thickness of this region reveals that the crustal thickness
of northern Tibet plateau(about 72km)is larger than that of the surrounding regions (generally 65km thick). The Lithosphere in this region,with a thickness of 140km,discloses apparent layer-block structure, i. e. several layers or blocks can be divided vertically or laterally,
respectively. The crust is large in thickness but low in density,whereas reverse is the situation for the upper mantle . Data of explosion,earthquake,and magnetotelluric sounding suggest the existence of two widespread low-velocity and low-resistance zones in the interior of
the plateau. The upper one,distributing in the crust ranges from 10 to 30km in depth(dipping northward),and 140km for the lower one in mantle[6,7]. On the basis of gravity data,it
can be concluded that the crustal structure in the interior of the plateau is relatively simple
and in close approach to isostatic equilibrium,whereas the crustal structure in the interior of
the marginal regions are more complex,as evidenced by large isostatic anomalies and the
abrupt change of anomaly gradient and crustal thickness.
Complex crust-mantle structure not only indirectly influence the distribution of sedimentary
package but also granted the interior of the plateau with unique thermal system and thermal
behavior. Limited data for telluric heat-flow measurements[7] conducted on the plateau reflects a regional large heat-flow anomaly,especially for those of Yamzho Yumco (3. 5 ±
0. 4HFU) and Puma Yumco(2. 2±0. 1HFU) where the heat-flow is 2. 3 and 1. 5 times larger
than the average respectively. It is the intracrustal melting and shallow magma bundle,which

are represented by the intracrustal low-velocity and low-resistance zone, that served as the heat sources of local and regional large heat-flow anomaly and resulted in the relatively strong heat background and comparatively high geotemperature gradient, as evidenced by high geotemperature gradient of 6.6℃/100m. Studies suggest that the formation of strong heat background probably coincided with the uplift of the plateau and, in turn, intensified the amplitude and scale of uplift in this region.

Intense Neotectonic Movement
The neotectonic movement occurred in the plateau since Himalayan movement, espcially Pliocene, was very intense in response to northward commpressional forces from India plate and to resistance by Tarim, Sino-Korean and Yangtze plate[6-7]. The effect of intracontinental collision deformation is exhibited as followings three aspects .

A. Contraction and Overlaping
The extent of crust contraction in the plateau is estimated to be more than 1,000km, which was caused by internal folding, structral overlaping of major central faults and major baundary faults, crustal remelting of S-type granite, basin-type acid volcanics, and local alkaline.

B. Strain of Slip Line Field
Deformation caused by plastic flow in the dead band of the plateau interior sheared and cut up the structual zones developed previously, as evidenced by the dextral strike-slip/oblique thrusting of Hoh Xil, Jinshajiang-Honghe deep fracture, and the sinistral strike-ship of Altun deep fracrure, as well as the dextral strike-slip of Karakonrum deep fracture. They, in fact, belong to simple shear deformation related to lateral movement/flow.

C. Strain of Rigid Field
In the shallow crust of the plateau, four arrays of tectonic i. e. E-W trending compressional tectonic belt, N-S trending extensional tectonic belt, N-W trending dextral strike-slip belt and N-E trending sinistral strike-slip belt, were created, which violently deformed and cut up the early tectonic belt. Consequently, some complex block-fault structure and residual faulted-block basins were developed, and internal tectonic of the plateau were more complex.

POLYCYCLIC GEOLOGIC EVOLUTION

Qingzang plateau are composed of two microplates. The southern and the northern corresponded to the microplate of Gondwana and that of Cathysia, respectively. Four phases of evolution are divided: Proterozoic Ocean, paleoTethys, neoTethys, and Indian Ocean (Figure 2). From late Proterozoic to early Paleozoic, Gondwana took shape by pan-nonepeirogeny events, whereas the northern Pangaea was disjointed gradually. Proterzoic Oceam, a open o-

cean then, with features of passive
continental margin on both the south
flank and north flank, exsisted in the
region between the two continents,
approximately along the present
Mazha-Kangxiwa-Maqen-Xiougou
suture, where Himalaya-Gandise region hosts widespread early Paleozoic
sediments of stable environment, locally deposits related to the sedimentation of intracontinental rift.

In Paleozoic, paleoTethys Ocean was
created between Eurasian and Gondwana continent in response to the
gradual close of Proterozoic Ocean
and the piecing together of Eurasian
continent as well as the southward
drift of Gondwana continent. The
opening-close boundary approximately extends westward from Malalak to
Malaysia Peninsula, Luang Prabang,
Changning shuangjiang, Tanianta-
wong, northern Lancangjiang, Gomo
co, Xingdukushi, northern E'erbushi
Mountain, Large Caucasus, northern
Turkey, Crimea, northern Dobrogea.
The subduct of paleoTethys oceanic
crust against Yangtze plate in the late
Paleozoic resulted in the formation of
Jinshajian back-arc marginal sea
basin and the final close of Proterozoic Ocean.

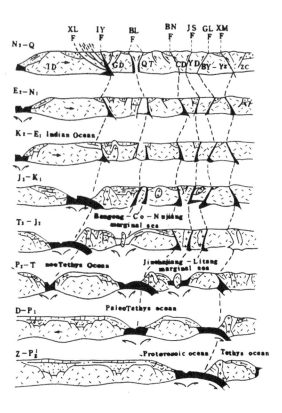

Figure 2   Schematic cross-section showing the evolution
of Qingzang plateau
XLF—Himalayan South Margin Fracture; IYF—Indian River-
Yarlungzangbojiang Fracture; BLF—Bangong-Co-Nujiang Frac-
ture; BNF—Northern Lancanjiang Fracture; JSF—Jinshajiang
Fracture; GLF—Ganze-Litang Fracture; XMF—Xiugou-Maqen
Fracture; ID—Indian Plate; GD—Gandise Terrain; QT—Qiang-
tang Terrain; CD—Qamdo Terrain; YD—Yidun Terrain; BY—
YZ—Bayan Har-Yangtze Plate; ZC—Sino-Korean Plate

During the period of late Paleozoic to Triassic, a large-scale Indo-Sinian folding belt was created in response to the development of neoTethys which resulted in the gradual close of paleo
-Tethys and the following close of Jinshajiang small oceanic basin. In middle Mesozoic, the
northward subduction of neoTethys oceanic crust caused the interoir of Gandise-Qiangtang
plate, which had been pieced up with Eurasian Continent, to split up and spread along Lute,
Bangon-Co-Nujiang and even the whole back-arc region to the south, and further developed
into a small Jurassic-Cretaceous oceanic basin covering Dajiwong, Angren, Naidong areas.

With the formation and spread of Indian Ocean, neoTethys Oceanic crust rapidily subducted nouthward, and, in the form of scissors, converged towards the two flanks in late Cretaceous and Paleogene. Indian River-Yarlungzangbojiang suture was created due to the final piecing together of Eurasian continent with Indian plate and Africa-Arab plate which belong to Gondwana.

Influenced by compression caused by the rapid migrating northward of Indian plate, a series of intracontinental compressional effect was initiated after Eocene. The early effect was demonstrated by the reactivating of early fractures trending in NW-SE such as Bangon-Co-Nujiang, Lancangjiang, Jinshajiang fractures etc, which resulted in the formation of a series of strike-slip-stretch basins locating in the regions adjacent to the fractures and regions sandwiched by faults.

From Pleistocene, the direction of tectonic line in the plateau was changed and dominated, as a whole, by E-W direction due to the further compression, of which the east part was nearly N-S trending arc fault. The previouly developed basins were deformed, and many new types of compressional structures and basins were initiated, such as NE and NW trending strike-slip-stretch basins, N-S trending collisive rift basins, EW trending intermontane, intramontane, foreland basins.

COMPOSITE AND SUPERPOSITION

The evolution of plate tectonics of Qingzang plateau plate has resulted in the composite and superposition of protobasin developed in the same area but in the different stages and by different basin-forming mechanism.

During Ordovician and late Paleozoic, Himalaya region to south of Indian River-Yarlungzang-bojiang was a cratonic depression basin, with the exception of the area west of Zhongba where a early Paleozoic aulogen or intracontinental basin probably existed . In the end of late Paleozoic the separation of Gandise from Gondwana, accompanied with the creation of neoTethys, generated a composite basin composited of by intracontinental rift-intercontinental rift-passive continental margin. From Triassic, apparent differentia between the southern and northern Himalaya emerged. Approximately divided by Laguiganger continental swell, the south part was a shelf covering the north flank of Indian continent, but a continental slope and abyssal plain with huge thickness of deposits for the north part. From late Cretaceous to Paleogene, with the convergence of Indian plate with Gandise microplate, basin was changed into residual oceanic basin and convergent marginal foreland basin.

From Eocene on, owing to the intense compression of Indian plate, northern Himalaya was folded and uplifted, the basins previously developed were deformed and cut into residual blocks by faults. Some small intramontane mollasse basins, small strike-slip basins, and collission-type marginal foreland basins on local area were developed.

Himalayan orogeny violently deformed the pre-Neogene tectonic protobasins, andich sedimentary packages were deformed and metamorphosed. Divided by Laguiganger as the boundary, the south part and the north part show apparent difference. In the south subsidence zone, or Devonian (even the Cambrian of Sipitidi area in the west)to Tertiary System is unmetamorphosed. Sedimentary units of craton-passive continental margin-marginal foreland basin were locally preserved, and which contain several potential regimes for oil-gas exploration. In the north subsidence zone, by contrast, pre-Triassic and Triassic rocks were comparatively intensively deformed and metamorphosed. Only were the Jurassic and Cretaceous rocks developed in passive continental basin-foreland basin preserved, in which several potential frontiers for exploration can be expected(Figure 3).

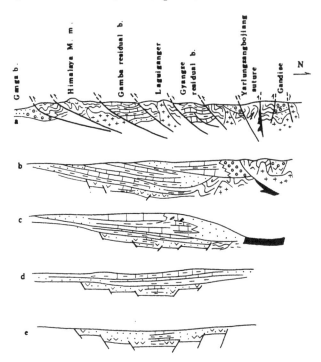

Figure 3   Schematic cross-section showing the superposition of protobasin in northern Himalayan region
a. Collisive Foreland (N-Q) ; b. Peripheral Foreland(K$_2$—E); c. Passive Continental Margin(T—K$_1$); d. Cratonic Depression(O—P); e. Intracontinental Rift(P$_t$~∈)

Three subsidence zones are divided in Gandise-Nyainqentanglha region from south to north. The south zone, represented by Rikaze residual basin, is dominatly composed of coarse clastics and athrogenic flysch, and has undergone violent deformation and metamorphism in the late time. Its exploration significance is doubtful. In Lhasa-Gandise subsidence area, the middle zone, deposits of cratonic depression and para-passive continental margin developed before Triassic are intensively deformed and metamorphosed, and sediments of Jurassic-Cretaceous interarc basin are also in the same situation. Therefore, whether this zone is a potential exploration area remains to be studied, although some thick Tertiary limnetic deposits and volcanic activities which were related to the subduction and collision are recorded, and some of them are considered to be the possible source rocks. In the north zone, which covers the whole region to north of Gandise, most of the pre-Jurassic deposits of cratonic depression were subjected to violent deformation and metamorphism. Initiating from middle Jurassic, the evolution of basin in this area, i. e. from back-arc rift basin (horst alternated with graben) to para-passive marginl basin to residual back-arc basin, was rather complex in late Cretaceous, a turning point of the new tectonic regime, and stepped into the evolution stage of intracontinent . A series of intracontinental strike-slip-stretch basins, which were generated in Paleogene-Miocone and exampled with Lunpola basin, generally distributes along Bangong-Co-Nujiang suture. Strong compression effect in Pliostocene-Quaternary is the main cause of intramontane compression setting basins, longitudinal extensional rift basins, and strike-slip basins, such as Nam Co basin, Yangbajain basin etc. N-E trending Dang-Xiong strike-slip fault zone, which had been probably a transform fault before, was important. Within the area east of this fault zone, a majority of the Mesozoic strata were slightly metamorphosed, and there is little possibility of petroleum derived for the Mesozoic. However, the situation is contrary to the west part, where the Mesozoic is unmetamorphosed and several potential oil-gas-bearing domains (Figure 4) associated with marine and terrestrial basins are present.

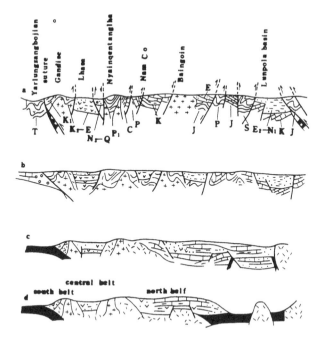

Figure 4   Schematic cross-section showing the superposition of protobasin in Gandise region

a. Residual Collisive Basin (Q—N); b. Strike-slip-stretch Basin, Intraarc Basin (K_2—E); c. Intraarc, Residual back-arc Basin (K_1); d. Intraarc, Back-arc Basin (J)

Qiangtang region is also divided into two zones. Three type of basins are recognized as follows, cratonic rift and depression basin in Permian to Triassic, passive continental margin basin in Jurassic, and intramontane molasse or strike-slip basin in Cretaceous to Tertiary. Of which the hydrocarbon potentials of Triassic to Jurassic rocks deserve to pay attention. The north zone was a cratonic depression-passive continental margin basin from Carboniferous to Triassic, but a peripheral foreland basin in Jurassic, and a intracontinental/strike-slip basin from Cretaceous to Tertiary. Several potential exploration fields can be identified (Figure 5).

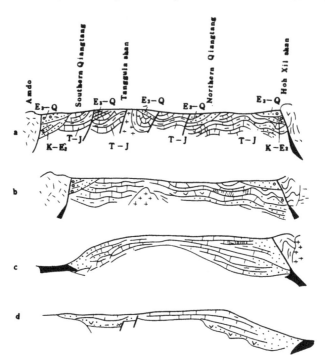

Figure 5   Schematic cross-section showing the superposition of protobasin in Qiangtang region

a. Collision Effect(E_3-Q); b. Strike-slip-stretch Basin, Foreland Basin(K-E_2); c. Peripheral Foreland Basin, Para-passive continental Margin Basin(J); d. Passive Continental Margin, Cratonic Rift Depression(P_2-T)

Qamdo microplate is composed of late Triassic-Jurassic residual interarc basin and numerous Tertiary stike-slip basins which are inlaid and superposited, of which the Tertiary is the major possible target of exploration. However, both Triassic and the underlaid units present in Hoh Xil and Yanghu area are metamorphosed, so the strike-slip basin formed in Cretaceous-Tertiary is the possibe exploration field.

APPLICATIONS OF BASIN AND APPROACHES TO THE APPRECIATING OF HYDRO-CARBON POTENTIALS

Basin analysis of Qingzang plateau should be based on the principle that can bring to light the unique applications of basins existed in the interior of the plateau. Three types of basins which have different implications are divided. They are tectonic protobasin(in the subsidence region), stratigraphic body basin (in the distributing zone of stratigraphy), and residual block or basin(independent oil-gas-water system). Combined with climate and hydrogeologic conditions, these protobasins determine the formation and distribution of sedimentary bodies in basin. Residual stratigraphic body basin supply the material base of petroleum generation. Fluid basin is the site of generation and accumulation of oil and gas, as well as the independent oil-gas-water system, and which serves as the elementary unit of oil-gas appreciating.

It has been generally acknowledged that favorable protobasin composite and spatial superposition style for hydrocarbon have been developed in the interior of the plateau. Data of samples collected in recent years show that pretty source rocks and reservoir rocks are present in the residual stratigraphic units, in which some fair source-reservoir-caprock associations were formed. However, good congenital geological conditions don't mean the good petroleum prospects. Within the whole Qingzang plateau, except for the Pliocene-Quaternary basins which remain the integral basin configuration, each of the early tectonic basins have lost the sense of basin.

Apparently, the cardinal task on the oil-gas appreciating and exploration strategy to identify the relatively stable residual blocks(Figure 6) that present materials for oil-gas formation.

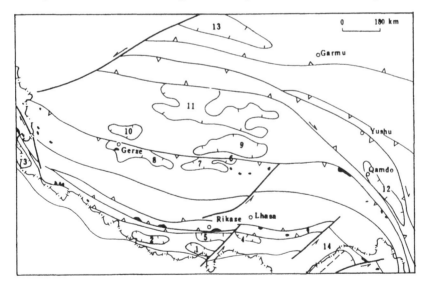

Figure 6   Location of residual basins of Qingzang Plateau

1. Gamba Basin; 2. Tingri Basin; 3. Xiangquanhe Basin; 4. Yamzho Yumco Basin; 5. Gyangze Basin; 6. Lunpola Basin; 7. Baingoin Basin; 8. Gerze Basin; 9. Seqi Basin; 10. Rela Basin; 11. Northern Qiangtang Basin; 12. Qamdo Basin; 13. Kumukuli Basin; 14. Ashamu Basin

Based the information of topographical, morphological, tectonic features revealed by satellite images as well as geological survey, some relatively stable blocks can be recognized in the different tectonic zones of the plateau. These blocks, characterized by relatively gentle mor-

He, Liu and Jiang

phology, light deformation, simple structures, and closed-faults serving as the boundaries, as well as superposition of Cenozoic basins on the older strata supply room for the formation of independent fluid system.

Northern Himalayan subsidence zone, one of violent deformation region in later stage in the plateau, locally preserved relatively stable rhomboid blocks. These blocks are confined by nearly E-W trending thrust faults and NE trending strik-slip or oblique thrust faults, with comparatively simple and gentle interior structure. Because of the lack of understanding on the detached depth of thrust fault and strike-slip faults extending along the basin boundaries, it is inferred that the residual blocks are umrooted. Five relatively stable residual blocks are divided in the whole northern Himalayan subsidence zone. They are Gamba, Dingri, Xiangquanhe, Gyanze and Yangzho Yumco.

Northern Gandise subsidence zone is also one of violent tectonic activity areas in later period. The persistence of stratigraphic units is limited due to the broken basement of block which was caused by spread and splitting. Further more, it is one of the most intense Cenozoic heat activity areas in the whole plateau, and which resulted in the metamorphism of pre-Cenozoic rocks in local area and accelerated the generation of hydrocarbons. Thus, It is not favorable to the generation of hydrocarbon in late period and oil-gas preservation. Gerze-Selincuo area on the north of this zone is the only relatively stable residual blocks. So, the Tertiary and Mesozoic residual tectonic blocks distributing along Bangong-Co-Nujiang are the important areas for oil-gas-bearing evaluating.

Southern and northern Qiangtang subsidence zone are comparatively stable regions in the plateau, of which the latter is more stable than the former. Three relatively stable residual blocks constitute three oil-gas evaluating units, in which Mesozoic rocks serve as the major exploration beds.

Qamdo subsidence zone, a active continental margin in the history of paleoTethys evolution and being voilently deformed in the late period by SE trending Sanjiang strike-slip belt, reveals the features of violently deformed tectonic association. Qamdo-Mangkang residual block, with strong tectonic activities and poor preservation conditions, is minor in the prospects of hydrocarbon exploration.

ANALOGY AND DISCUSSION OF OIL-GAS-BEARING PROSPECTS

Based on the data of regional geology, it is inferred that each of the blocks has undergone the process of generation-accumulation-lose not less than one times. The intense tectonic disturbance occured in the late period destroyed the early petroleum system to different degrees and made the early pools be invalid. Therefore, the key point in oil-gas evaluating is to study the time of trap-forming, the influence of tectonic disturbance in late period, as well as the process of petroleum generation, including the time of secondary generation, marching between the time of trap-forming and the time of hydrocarbon migration. The higher geotemperature background related to the uplift of the plateau in Cenozoic, a little later than the formation of structures, probably resulted in the secondary generation of hydrocarbons and late pool-forming. The higher maturity of organic matters contained in the rocks of the whole plateau means that the products of the secondary generation of hydrocarbons are dominated by dry gases, but with limited amount.

Gas seeps found in the Jurassic strata of Musitang area on northern Himalayan subsidence zone suggest that petroleum pools formed in this area have been destroyed on one hand, and predict the petroleum prospects of other regions that were more stable than Musitang area in the geological history on the other hand.

Petroleum prospects of Gandise region are demonstrated by numerous primary and secondary oil-gas shows that are detected from marine Jurassic-Cretaceous and Tertiary terrestrial rocks. The higher maturity of organic matters (average $T_{max} = 471.5$)[7] derived from marine Jurassic-Cretaceous limits the amount of secondary hydrocarbons. Commercial petroleum produced from the Tertiary of Lunpola basin, which was controlled by tectonics developed since Pliostocene, is an example of late pool-forming [10]. Moreover, numerous oil-gas shows present in the distributing area of Tertiary demonstrate the existence of petroliferous zones or basins in the Tertiary. The significant petroleum prospects in the Tertiary terrestrial basins of the plateau are also suggested by several commercial oil accumulations and the exposed oil-shales (with extracted oil resources of 12.5×10^8 tons) in Thailand's Nae Sot Basin and Fung Basin which, comparing with the Qingzang plateau, are located in the same tectonic zone and are similar in basin size and sedimentary type. In China, the success of petroleum exploration in Baise Basin, Jinggu Basin, which were also influenced by the tectonic evolution of Tethys, also serves as the examples that can be used for reference.

The suitable conditions for petroleum generation in Qiangtang region are illustrated by numerous oil-gas shows recorded in the Jurassic and Tertiary. Oil-gas shows discovered in the north flank of Daheipao anticiline, locating on the area adjacent to No. 144 highway maitenance squard of Qing-Zang (Tibet) road, and in the axial region of Terigawa chine structure are inferred to be the products of Paleozoic oil pools that have been destroyed, and suggest the integrate process of oil-gas-forming that once occurred in this region. Data of examples from Najinqu, Terigawa, southwestern Amuhu, Tumen show that organic matters (type I) are low in maturity. However, samples of Yanshiping area in the east reveal a high maturity of dry gas, which is probably related to the intrusion of granite in late Yanshan stage[9]. Some evidence indicates the existence of cover decollement folds in the hinterland of Qiangtang were probably created in late Jurassic. If the suitable cap rocks were well developed, some valid pool-forming associations of early period would be formed in Qiangtang region, of which the Jurassic and Tertiary are of importance. Middle Jurassic gypsum and rock salt beds, with some thickness, which were developed in the closed bay environment under dry climate and covered certain area, contribute to the source-reservior-caprock association with good sealing capability. The early fluid system of oil-gas-water was probably preserved with some certainty in some areas. Some fair source-reservoir-caprock associations contained in Tertiary basin, accompanied with gypsum and rock salt, are also the suitable sites for petroleum accumulation. Combination of multiperiodical hydrocarbon generation and good source-reservior-caprock properties with folds and faulted-blocks created in different periods may ensure the validity of early and late oil-gas pools. Three types of oil-gas traps can be anticipated, such as traps of generating from older rocks but reserving in the younger rocks, generating from the younger but reserving in the older, and both generating and reserving all in the younger.

The commercial oil-gas accumulations of Central basn in Iran, which is located in the same tectonic zone as Qingtang and reveals the similar geological conditions to Qiangtang's, lighten the significant prospects of hydrocarbons in Qiangtang region. Furthermore, The oil-gas exploration activities in Qiangtang basin can also be guided, to some extent, by the successful experience in Sichuan basin which is also comparable.

Some oil-gas shows recorded in Qamdo and Hoh Xil area indicate that the process of hydrocarbon generating-migrating-accumulating once occured in these two areas and predict some prospecs of oil-gas exploration in these two areas.

A elongated Cenozoic distributing area, with a area of about 8,000km², a part of Assam basin, covers the foreland region of Himalayan M, geographically along Liga-Bangdila in the southeastern frontier of Tibet . The Tertiary deposits in Assam basin, one of the old preductive fields in India, exceed 7,000m in thickness. Eight oil fields have been discovered in the anticlines locating on the frontier of Nagar foreland thrust belt to southeast of Brahmaputra valley, the southeastern flank of Assam basin. It is inferred that some good oil-forming conditions and exploration prospects can be expected in the Tertiary within the boundary of China, especially in the frontier of Himalaya thrust belt frontier and the lower wall of thrust belt where the oil-gas discoveries and exploration breakthrough are most possibly made.

Kumukuli basin, with a area of 30,000km², locating on the southeast corner of Xinjiang Autonomous Region, is a Tertiary basin similar to Mangya depression of western Qaidam basin. The thickness of Tertiary strata in Kumukuli basin reaches up to 5,700m. Some Jurassic rocks are exposed. Controlled by fault extending along the south margin of Qimantage M. and the north margin of Kunlun M. and Altun strike-slip fault, this basin is probably a strike-slip-stretch basin or a terrestrial basin originated from strike-slip transform deformation. The west flank is burried deeper and consists of two depressions and one uplift. Thirteen structural traps, identified by surface exploration methods, reveal suitable geological conditions neccessary for oil-gas-forming, as demonstrated by good oil-gas shows.

STRATEGY AND DIRECTION OF EXPLORATION FOR OIL AND GAS

Breakthrough in Terrestrial Petroteum.

The study on the petroleum prospects of Lunpola basin should adopt the means of working along both lines, that is to say, on one hand detailed studies and intense exploration should be carried out, on the other hand the oil-gas potential of other terrestrial basins and that of marine basal rocks in the corresponding basins which are located on the surrounding region or situated in the same tectonic zone should be inquired into. As a terrestrial composite basin of early strike-slip-stretch basin and late compressive deformation basin, Lunpola basin has the same tectonic and sedimentary features as other stretch basins that are small in size but high in production. So, it can be concluded that the breakthrough in oil and gas is only a matter of time. Of the important targets that are to be studied are lower wall of thrust fault (probably a nappe), structures in the central depression zone, sand bodies of subaqueous fans, and syngenetic strike-slip uplift zones.

Probing into Marine Oil-Gas

Although the marine regime of Qingzang as well as those of other basins in Tethys all are commonly of shattled basin, complex structure, exposed target beds, high maturity of organic matters, poor preservation, low ratio of generation-migration-accumulation, it can't be served as the bases that neglect the potentials of petroleum generated from marine rocks in the plateau. Complex petroleum geology conditions and lack of basic geology survey weaken the reliability of inferences and conclusions, and disilluminate the prospects of hydrocarbons.

REFERENCES

1. Gan Kewen, Li Guoyu, Zhang Liangcheng et al. Atlas of petroliferous basins of the world. Beijing, Petroleum Industry Publishing House, 1978, 2-19

2. Chinese Academy of Science . Tectonic evolution of lithosphere in Himalaya. Beijing, Geological Publishing House, 1986, 31-55

3. Liu Zengqian, Jiao Shupei, ZhanYifu et al. Geological map of Qingzang plateau and adjacent regions (1: 5000000). Beijing, Geglogical Publishing House, 1988, 8-30

4. Liu Zengqian, Xuxian, Pan Guitang et al. Tectonic framework and evolution of Qingzang plateau. Beijing, Geological Publishing House, 1988, 7-50

5. Zhou Xiang, Cao Yiugong, Zhu Mingyu et al. Map of tectonics and architecture of Tibet plate (1: 1500000). Beijing, Geological Publishing House, 1986, 5-38

6. Pan Guitang, Wang Peisheng, Xu Yaorong et al. Cenozoic tectonic evolution of Qingzang plateau. Beijing, Geological Publishing House , 1990, 135-152

7. Ma Xingyuan, Ding Guoyu, Gao Wengxue et al . Atlas of lithosphere dynamics of China. Beijing, China Cartographic Publishing House, 1989, 68

8. Yu Guangming, Wang Chengshan . Sedimentary geology of Tibet Tethys. Beijing, Geological Publishing House, 1990, 8-140

9. Editorial committee of "Petroleum Geology of China", vol. 14, Beijing, Petroleum Industry Publishing House, 8-140

10. Xu Zhengyu . Tertiary and petroleum potentials of Lunpola Basin, Tibet. Oil & Gas Geology, 1980, 1(2): 153-158

Proc. 30ᵗʰ Int'l. Geol. Congr. , Vol.18, pp.313~331
Sun Z.C. et al. (Eds)
© VSP 1997

The Regularity of Oil and Gas Abundant Accumulation in North Jiangsu Basin, China

WEI ZILI   ZHOU LIGING

East China Bureau of Petroleum Geology, MGMR, 182. New Road Pukou District, Nanjing, Jiansu, 210031, China

Abstract

The oil and gas accumulated abundantly in North Jiangsu Cenozoic basin. The basin consists of eleven downwarps seperated by twelve upwarps, but 98% of total proved reserves concentrated on three downwarps— Gaoyou, Jinhu and Qingtong downwarp. These three downwarps have good conditions in petroleum geology: (1)they all were in the position of the area with continuous rapidly depositing during the Cenozoic depocenter moved from west to east; (2)they underwent more intense volcanic activities than other downwarps, so their paleogeothermal gradient was 5-10 °C/km higher than that of the other downwarps and their threshold depeth of petroleum generation is 400-600 meters shallower than that of other downwarps. It was controlled by the volcanic activities during the deposition of source rock, Jinhu downwarp deposited high quality source rock with the depth of the oil window from 1200 to 2000 meters, which is 400-800 meters shallower than the other source rocks; (3)they distributed in the good reservoir facies belts. Fifty oil and gas fields have been explored in the basin, each oil field is small-middle in size, but more than 90% of total proved reserves has been found in fourteen larger oil fields, which distributed in the intersect area of the structural high zones and fault terrace zones of Gaoyou downwarp and Qingtong downwarp, and in the structural high zones of shallow buried marginal facies belts of Jinhu downwarp.

Keywords: North Jiangsu basin, oil and gas, regularity, abundant accumulation.

INTRODUCTION

North Jiangsu Basin is the onshore part of the North Jiangsu—Yellow Sea Basin, on the north of Jiangsu Province with an area of 35000 km². The basin is bounded to the west by TanLu Fault, to the north by the sature zone separating Yangtze and North China blocks, and to the south by the low hills of Yangtze folded belt(Fig. 1).

The basin is the largest Cenozoic fault subsidence petroliferous basin onshore in Southeast China. Now total oil production from this basin is about one million tons oil a year. Fifty oil and gas fields and eight-five million tons proved reserves have been explored. Each oil field is small-middle in sized with proved reserves from 1500×10^4 tons to 10×10^4 tons(Fig. 2).

Since the Cenozoic, North Jiangsu Basin entered an extensional tectonic development stage. A transformation from downfault to downwarp was evoluted on the basin. During Paleocene epoch, controlled by the favorable conditions of paleoclimate, intensity of block-faulting activities and source of material, the fluvial–delta–lacustrine systems was developed by a dominantly graydark muddy sediment with the thickness of 2500 meters, which constitute sevearl assemblages of source rocks, reservoir and seal rocks (Fig. 3). After the Wubao Movement, with the block-faulting activities varying from

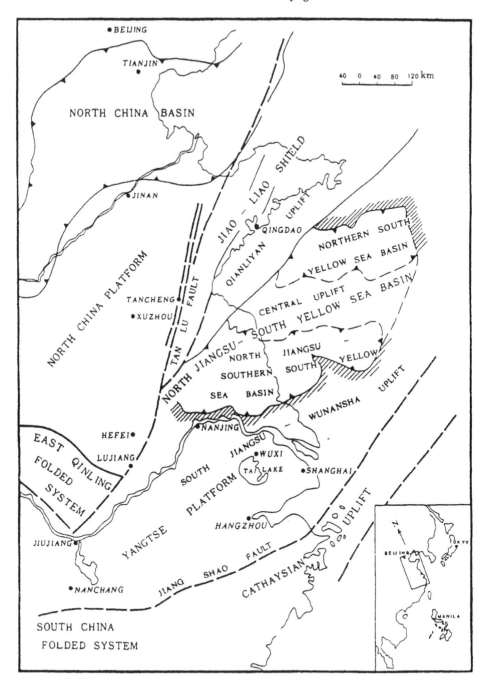

Fig. 1. Regional tectonic setting of the North Jiangsu-South Yellow sea Basin

intensive to weak, the paleoclimate changing gradually from warm and humid to hot and dry, the fluvial fan-delta-floodplain-shallow lacustrine system was developed by dominantly sandstone and gray-yellow and red muddy sediments with the thickness of 3500 meters,

The Regularity of Oil and Gas Abundant Accumulation in North Jiangsu Basin, China

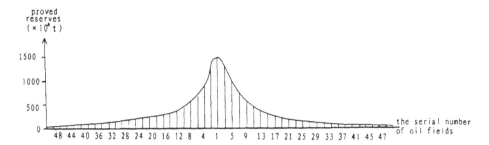

Fig. 2. The proved reserves of the oil fields in the North Jiangsu Basin

which also constitute several assemblages of reservoir and seal rocks. As the volcanic activities became strong accompanied with the strong block-faulting activities, all the continuous deep lacustrine environment, including the depositional environments of the Second member and the Fourth member of the Funing Formation, and the Second member of the Taizhou Formation were developed after the volcanic activities with the level of lake rising and humid paleoclimate. From the Eocene to the Quaternary with the dry and hot paleoclimate, the volcanic activities affected the depostional enviorment strongly, so a favorable oil bearing assemblage of thick reservoir rock with high porisity and thick seal rock were formed, such as the assemblage of the upper mudstone and lower sandstone of the First member of the Sanduo Formation. During other stage, the depositional environment changed gradually and the basin deposited the assemblages of sandstone interbedded with mudstone, such as the First and the Third member of the Funing Formation reservoir rocks with middle-low porosity and permeability.

The oil and gas accumulated abundantly. More than 90% of the total proved reserves has been found in fourteen larger oil fields, 98% of the total proved reserves distributes in Gaoyou downwarp, Jinhu downwarp and Qingtong downwarp. The middle sized oil fields distribute around the fault terrace zones. The distribution of the oil and gas abundant accumulation zones is controlled by tectonization, deposition and volcanic activities.

The relationship between the tectonic framework and the distribution of oil and gas

The architecture of the basin is composed of separated Paleogene half-grabens in its lower part and a unified Neogene panlike depression in its upper part. Detailed data from petroleum exploration demonstrate that North Jiangsu Basin consists of a southern and a northern depression, seperated by a central uplift(Fig. 4). These two depressions consist of eleven downwarps and twelve upwarps, and each downwarp is composed of 2 to 5 sags and structural high zones.

The basin was cut apart so strongly by the reactivation of the Mesozoic widely distributed thrust faults(Fig. 5) and the block-faulting that only small and middle size oil and gas fields were developed.

As upwarps are not able to generate oil and gas themselves, and the distance of the oil and gas lateral migration is less than 10 kilometers, each downwarp has to be an independent oil and gas accumulating unit.

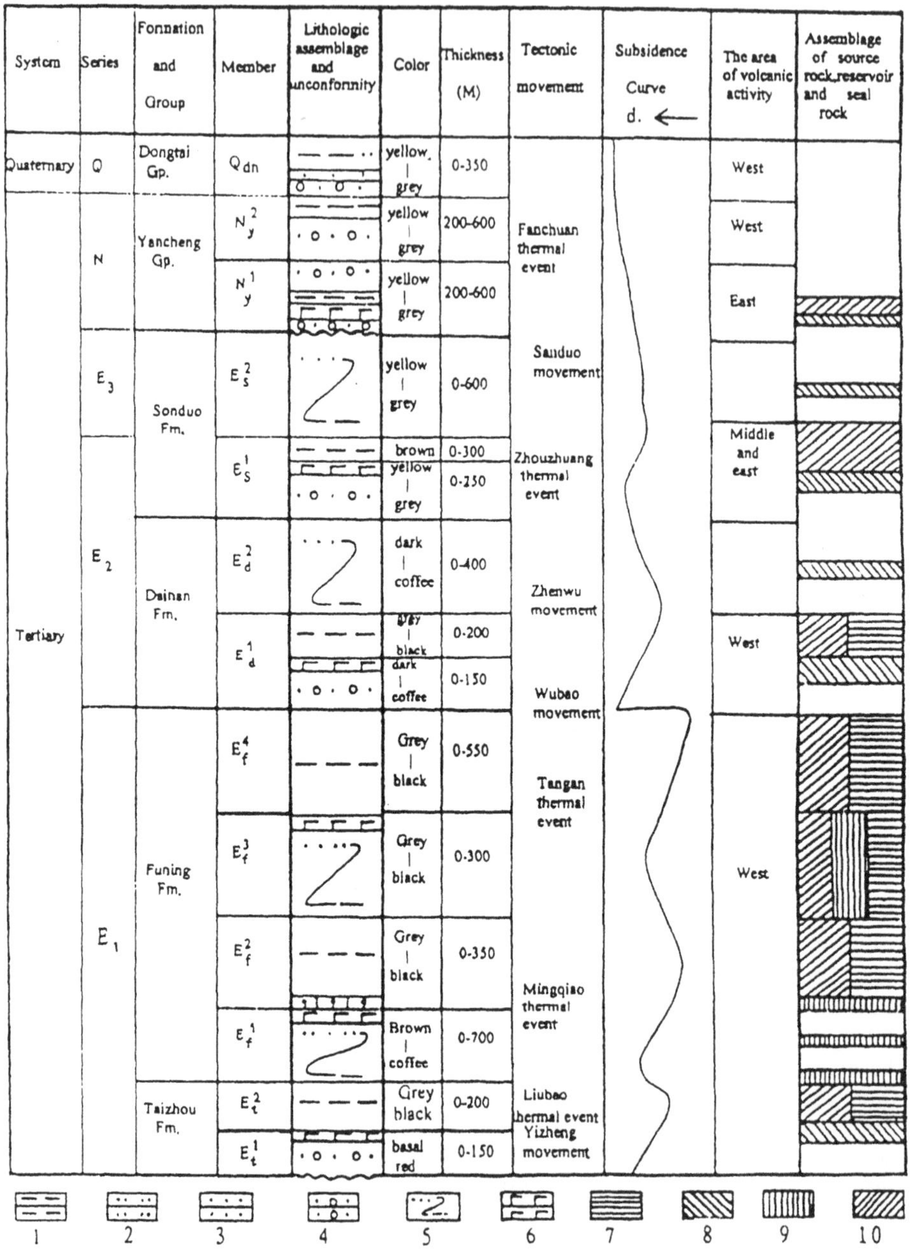

System	Series	Formation and Group	Member	Lithologic assemblage and unconformity	Color	Thickness (M)	Tectonic movement	Subsidence Curve d. ←	The area of volcanic activity	Assemblage of source rock reservoir and seal rock
Quaternary	Q	Dongtai Gp.	Q_{dn}		yellow | grey	0-350			West	
	N	Yancheng Gp.	N_y^2		yellow | grey	200-600	Fanchuan thermal event		West	
			N_y^1		yellow | grey	200-600			East	
Tertiary	E_3	Sonduo Fm.	E_s^2		yellow | grey	0-600	Sanduo movement		Middle and east	
	E_2		E_s^1		brown | yellow | grey	0-300 0-250	Zhouzhuang thermal event			
		Dainan Fm.	E_d^2		dark | coffee	0-400	Zhenwu movement			
			E_d^1		grey | black dark | coffee	0-200 0-150	Wubao movement		West	
	E_1	Funing Fm.	E_f^4		Grey | black	0-550	Tangan thermal event			
			E_f^3		Grey | black	0-300			West	
			E_f^2		Grey | black	0-350				
			E_f^1		Brown | coffee	0-700	Mingqiao thermal event			
		Taizhou Fm.	E_t^2		Grey black	0-200	Liubao thermal event			
			E_t^1		basal red	0-150	Yizheng movement			

Fig. 3 Statigraphic system and tectonic movement in the North Jiangsu Basin (1=mudstone; 2=argillaceous; 3=sandstone; 4=coarse and pebbled sandstone; 5=mudstone interbedded with sandstone; 6=basalt; 7=source rock; 8=reservoir of high and middle porosity porosity; 9=reservoir of low porosity; 10=seal rock)

In the downwarp, the structural high zones divides each downwarp into several oil and gas accumulation units. It is very diffcult for the oil and gas in one sag to migrate

The Regularity of Oil and Gas Abundant Accumulation in North Jiangsu Basin, China

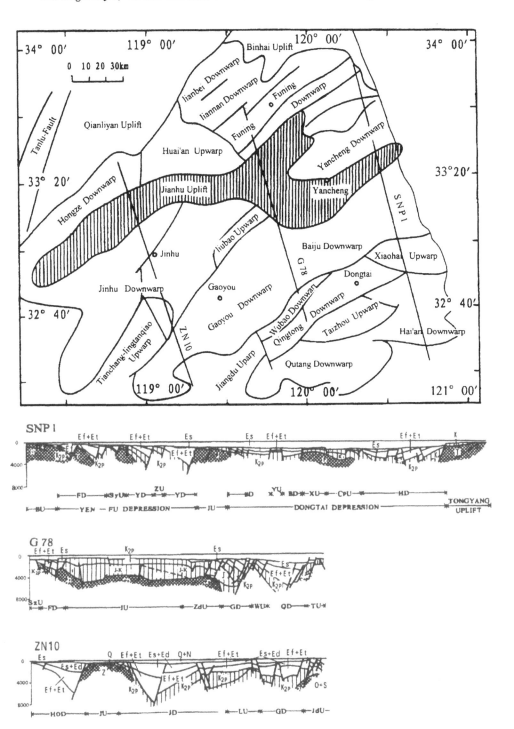

Fig. 4. Cenozoic structural framework of North Jiangsu Basin

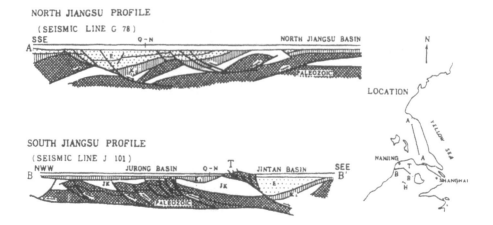

Fig. 5. Mesozoic basement decoupling in Jiangsu Province

through the adjacent structural high zone into another sag or structural high zone. The
oil and gas of the sag just accumulated in the adjacent structural high zones generally.
With a good condition of petroleum geology, including abundant source of oil and gas and
many large traps, the structural high zones become the most important oil and gas
accumulation zones in North Jiangsu Basin(Fig. 6).

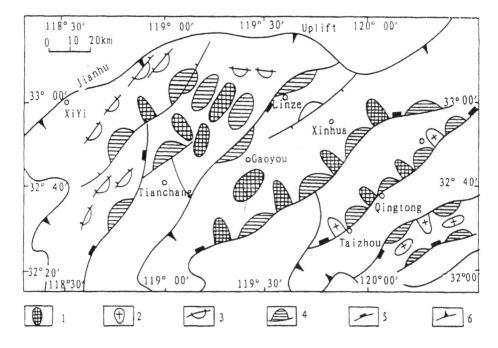

Fig. 6. The oil and gas abundant accumulation zones in the Dongtai Depression of the North Jiangsu Basin
(1=oil and gas abundant accumulation zones; 2=structural high zones of oil and gas prospecting well; 3=dispersed
reservoirs in slope zone; 4=main source of oil and gas; 5=boundary of fault; 6=boundary of overlap)

The Regularity of Oil and Gas Abundant Accumulation in North Jiangsu Basin, China

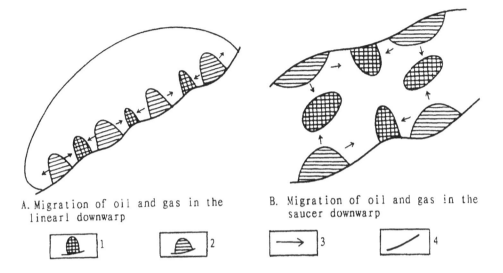

A. Migration of oil and gas in the linearl downwarp

B. Migration of oil and gas in the saucer downwarp

Fig. 7. The structural high zones of the different types of oil and gas migration(1=structural high zones; 2=center of oil and gas source; 3=direction of migration of oil and gas; 4=boundary fault)

There are three types of structural high zones: single, twice, and triple oil source respectively. The types with one or two oil sources were developed in the linear Gaoyou and Qingtong downwarp(Fig. 7A), the types of 2 or 3 oil sources were developed in the saucer Jinhu and Hai'an downwarp(Fig. 7B). In a downwarp, each sag almost provided oil and gas is equal. The amount of oil and gas in each structural high zone varies with the number of the sags of oil source in the same downwarp. For example, in Gaoyou downwarp, 1500–2000 ×10⁴ tons of proved reserves of oil and gas has been found in the structural high zones of two oil sources and 800–1000 ×10⁴ tons in one oil source. Controlled by the difference in petroleum geology, the amount of oil and gas in a structural high zones is different apparently between two downwarps. For example, in Qingtong downwarp, 600–700 ×10⁴ tons of proved reserves distributed in the structural high zones of two sources, while only about 300 ×10⁴ tons has been found in the structural high zones of one oil source.

The relationship between the migration of depocenter of the basin and abundance of oil and gas

Under the affection of a sinstral tenso–shear stress field, which is related to the dextral torsion of Tanlu fault zone and regional tensional rifting, the depocenter of the basin migrated from west to east(Fig. 8, 9, 10), resulting in four types of downwarps: early rapidly depositing(type 1), continuous rapidly depositing (type 2), late rapidly depositing(type 3), continuous slowly depositing(type 4), developping from west to east (Fig. 11).

The difference of the petroleum geological conditions in four types of downwarps are as follows:
(1) The central and western downwarps distrbuting around the depocenter of the depositional stage of source rock deposited better source rocks than other downwarps.

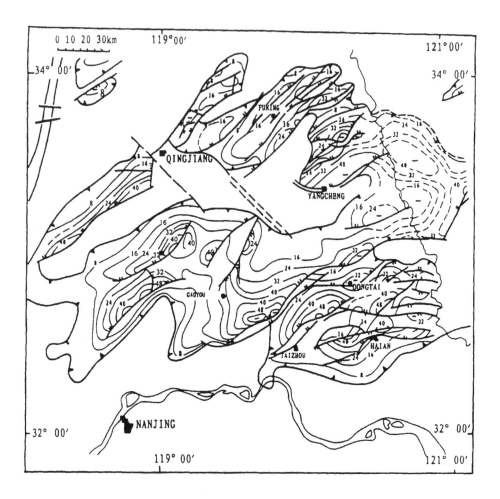

Fig. 8. Contour map of the North Jiangsu Basin, showing the isopachs and features of Paleocene-Eocene downfaulting. Thickness intervals between Paleocene and Eocene are in term of seismic data.

(2) The downwarps distributing around the depocenter after the deposition of the source rocks have a good condition for thermal evolution of source rocks.

(3) A large amount of oil and gas migrated upward along the growth fault zones in the continuous rapidly depositing downwarps.

Controlled by the conditions of different petroleum geology, abundance of oil and gas varies as follows: $4-5 \times 10^4$ tons/km^2 of type 2, $3-4 \times 10^4$ tons/km^2 of type 1, 2×10^4 tons/km^2 of type 3 and 1×10^4 tons/km^2 of type 4. The depth of oil pools varies as follows: 1000 – 6000 meters of type 2, 1200 – 3000 meters of type 1, 2400-4000 meters of type 3 and type 4. Type 2 is very abundant in secondary oil and gas, the ratio of original oil and gas to secondary oil and gas is 1:1 to 1:2. The other types are poor in secondary oil and gas. The bearing oil-stratum varies: the First member of the Yancheng Group to the First member of the Taizhou Formation in type 2, the First member of the Dainan Fourmation to the First member of the Funing Formation in type 1, the Third member of the Funing Formation to the First member of the Taizhou Formation in type 3 and 4.

The Regularity of Oil and Gas Abundant Accumulation in North Jiangsu Basin, China

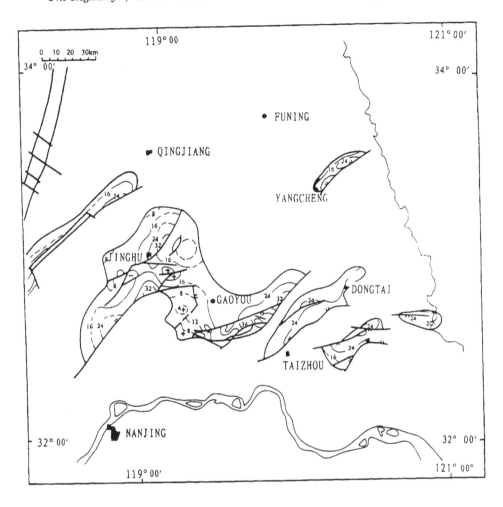

Fig. 9. Contour map of the North Jiangsu Basin, showing the feature of early development of oilgocene downfaulting. Thickness intervals are in terms of seismic data.

The distribution of oil and gas accumulatin zones in the half-grabens

After paleocene epoch, all the downwarps became half-grabens consisting of the fault terrace zones, deep sinks and the slopes(Fig. 12). The geological conditions of oil and gas accumulation are differed in each zone.

Characteristics of oil and gas accumulation in the fault terrace zones
Each fault terrace zone consists of low, middle and high fault terrace, which are divided by 2–3 growth faults with faultthrow from 300 to 3000 meters. The characteristics of petroleum geology in these zones are as follows:
(1) Under concentration of tectonic stress in the zones, there are a great number of traps including early-stage anticlines, fault-block traps, small-scaled reversed drag traps, upheaval interior traps, and lithologic traps etc.
(2) The streams originating from upwarps carried clastic sediments toward the fault terrace

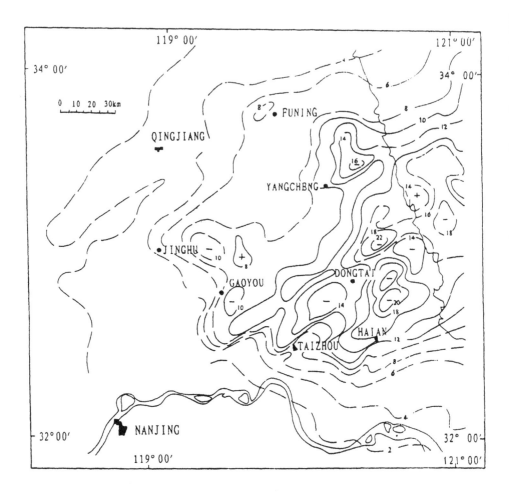

Fig.10. Isopach map of the Miocene to Quaternary, showing the feature of downwarp.

zones, so the zones distributed at more favorable reservoir facies area than the other zones in same downwarp.

(3) Under the threshold depth of petroleum generation, the traps, which distributed in the downthrows of the fault terrace zones, could be laterally sealed by the uplifted walls consisting of the Funing Formation and the Pukou Formation with a sequence of thick mudstone.

(4) The zones distributed around the center of oil sources.

(5) Oil and gas can migrate vertically along the growth faults for a long time.

As a result, these zones are the most important oil and gas accumulation zones with the characteristics of multi-layer and multiple genesis.

Characteristics of oil and gas accumulation in the slope

The dip angles of the slopes are from 5° to 20°. There are a lot of the third and fourth level faults with faultthrow from 20 to 200 meters. The characteristics of petroleum

The Regularity of Oil and Gas Abundant Accumulation in North Jiangsu Basin, China

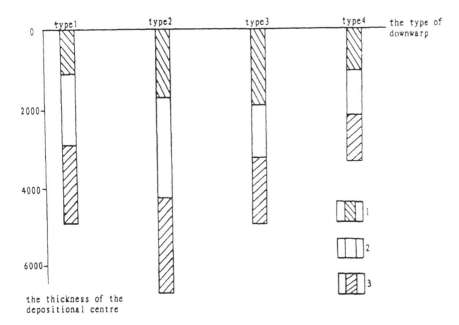

Fig. 11. Four types of downwarp in the North Jiangsu basin (1=the thickness of N-Q; 2=the thickness of E_w-E_d; 3=the thickness of E_t-Ef; type 1=early rapidly subsiding downwarp; type 2=continuous rapidly subsiding downwarp; type 3=late rapidly subsiding downwarp; type 4=continuous slowly subsiding downwarp)

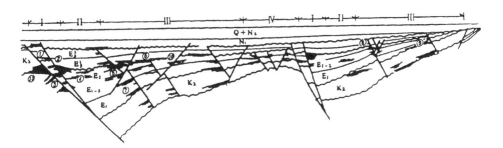

Fig. 12. A model for the prediction of oil and gas traps in the half-grabens of the North Jiangsu Basin (I). Traps formed under control of the joint zone of the downfaults and upheavals along major fault belts: 1=early-stage anticlines of fault barrier traps in updip positions; 2=small-scale reversed drag traps formed by growth faults; 3=upheaval interior traps formed by lateral migration. II. Structural-stratigraphic traps related to deep sink; 4=in-situ traps in sandstone lenses of turbidite sediments; 5=structural traps formed by folding in an early subsidence stage. III. Structural-stratigraphic traps formed on the slope and the hinge zone between the slope and deep sink; 6=unconformity traps formed in overlapping and underlying wedge sandstones; 7=lithologic traps caused by reversed drag; 8=fault-barrier traps at high locations of antithetic fault; 9=traps in beach clastics and reef-like bioclastic bodies; 10=traps in fissured volcanic (basalt) rocks surrounded by source beds)

geology in the slopes are as follows:

(1) Without intense activities of growth faults, the oil and gas mainly migrated laterally in the slope.

(2) The traps of the Dainan Formation and the Sundou Formatin can not be sealed laterally by thick mudstone well.

(3) The traps of the Taizhou Formation and the Funing Formation can get a large amount of oil and gas from themselves and oil and gas can be sealed by several layers of thick mudstone, so that many original oil and gas pools have been formed in the slope.

(4) The third level faults controlled lateral migration of oil and gas. The abundant or dispersed migration of oil and gas was controlled by the combination between the variation of dip of the faults and several assemblages of source rocks, reservoir and seal rocks. The dispersed migration of oil and gas distributed so widely that there are a great number of oil fields, but all in small size.

According to the characteristics of the third level faults, the slope is divided into the interior slope zone, the hinge zone and the exterior slope zone. Because of the barrier of the faults, abundance of oil and gas varies: the interior slope zone is the bset, the hinge zone better, the exterior slope zone still   good. The oil pools distributed from the First member of the Dainan Formation to the First member of the Taizhou Formation in the interior slope zone and the hinge zone. The oil   pools distributed only from the First member of the Funing Formation to the First member of the Taizhou Formation in the exterior zones.

Many types of pools were developed in the slope, such as unicomformity pools, lithologic pools, fault-block pools, fissured volcanic(basalt) rock pools and intrusive mass-barrier pools.

Only small scondary pools were developed in the upheavals. Small sandstone lense-pools and small anticlinal pools were developed in deep sink.

The relationship between the depositional systems and oil and gas accumulation

The evolution of the depositional systems
During the Cenozoic, controlled by tectonic evolution, the depositional systems of the basin changed greatly. During the doposition of source rocks, the western large axial stream had an important effect on the distribution of the most favorable reservoir facies belts including deltaic plain, subaqueous alluvium, delta front, oolite terrace etc. Only western downwarps distributed in the most favorable reservoir facies belts. The other downwarps were affected by small streams originating from upwarps, and the favorable reservoir facies distributed in the fault terrace zones and the exterior slope zones(Fig. 13). After the Wubao Movement, two large axial streams from the west and the east controlled depositional system of the basin. Associated with the western, eastern and central part, there are three areas: undeveloped seal rocks, well developed seal and reservoir rocks, non-reservoir rocks respectively(Fig. 14).

The assemblages of reservoir and seal rocks in different areas
Affected by the transformation of depositional systems and the difference of conditions in petroleum geology, the assemblages of source rocks , reservoir and seal rocks varied in each area.

There are three important assemblages in Jinhu downwarp and Hongze downwarp: the Fourth member and the Third member of the Funing Formation, the upper part and the lower part of the Second member of the Funing Formation, and the Second member and the First member of the Funing Formation. With favorable reservoir facies belts and high quality

The Regularity of Oil and Gas Abundant Accumulation in North Jiangsu Basin, China

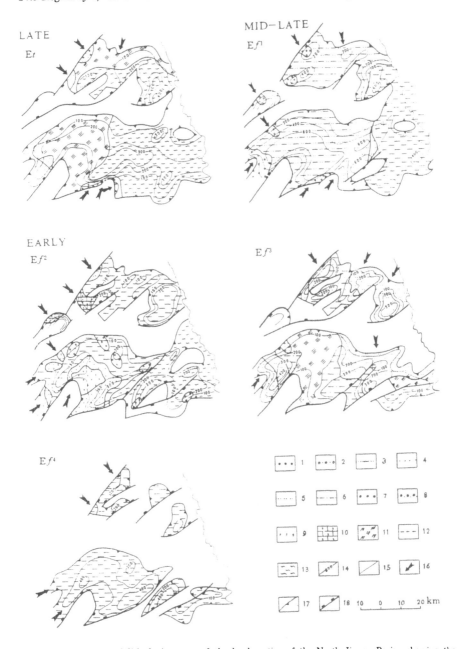

Fig. 13. Paleogeographic and lithofacies map of the land portion of the North JiangsuBasin, showing the depositional developments of earlier downfaults during the period from the Paleocene to Eocene (1=diluvial facies; 2=sub-aqueous alluvium; 3=floodplain; 4=deltaic plain; 5=delta front; 6=prodelta; 7=oolite terrace; 8=biodetritus oolite beach; 9=biodetritus beach; 10=distribution of biological limestones; 11=shallow lacustrine facies; 12=lacustrine facies; 13=deeper lacustrine facies; 14=isopach; 15=facies boundary; 16=direction of supply; 17=basin boundary; 18=fault line)

source rock of the Funing Formation in the west area, there are many shallow pools of the Funing Formation. Because of poor secondary oil source, the assemblage of the upper part

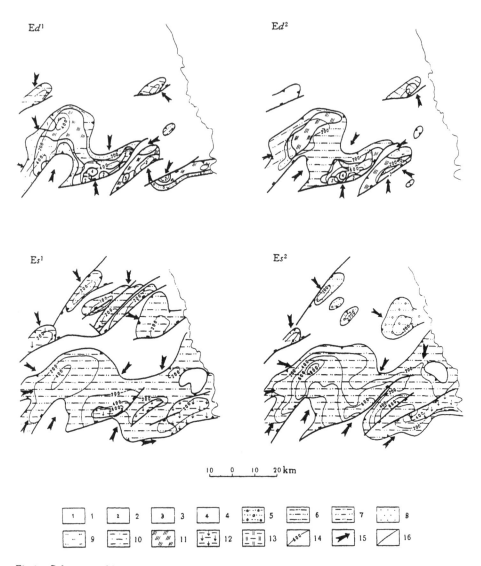

Fig. 14. Paleogeographic and lithofacies map of the land portion of Jiangsu Basin, showing the depositional developments of later downfaults during the Oligocene (1=diluvial facies; 2=sub-aqueous alluvium; 3=turbidites; 4=deep lacustrine facies; 5=alluvial facies; 6=alluvial plain; 7=floodplain; 8= deltaic plain; 9=delta front; 10=prodelta; 11=shallow lacustrine facies; 12=seasonal catchment; 13=lake bay; 14=isopach; 15=supplying direction of debris; 16=facies boundary)

and the lower part of the First member of the Dainan Formation is a less important one.

There are two oil and gas bearing system in Gaoyou downwarp and Qingtong downwarp: middle oil and gas bearing system consisting of the Sanduo Formation and Dainan Foramtion; lower oil and gas bearing system consisting of the Funing Formation and the Taizhou Formation. The main reservoir rocks of the middle system are deltaic sandstone and alluvial sandstone with 15%-30% in porosities and 50-2000 MD in permeabilities. The

The Regularity of Oil and Gas Abundant Accumulation in North Jiangsu Basin, China

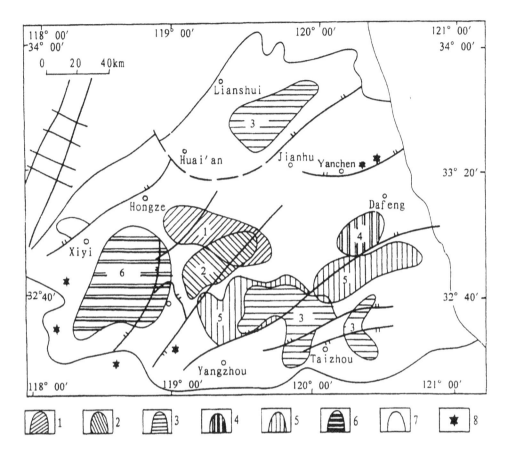

Fig. 15. Distribution of the volcanic rocks of different stages in the North Jiangsu Basin (1=E₁-E₁; 2=E₁²-E₁²; 3=E₂¹; 4=E₂²; 5=N₂¹; 6=N₂²; 7=Q; 8=the scattered volcanic rock)

main reservoir rocks of the lower system are deltaic plain siltstone and delta front sandstone with 10%–18% in porosities and 1–100 MD in permeabilities. In the central area, the most favorable lower oil and gas bearing system distributed in the fault terrace zones, the exterior slope zones and the hinge zones with a favorable buried depth and a good reservoir facies belt.

In Hai'an downwarp and Yancheng downwarp etc., without sufficient oil source in the middle–upper system, the lower system, which consists of the Funing Formation and the Taizhou Formation, is the only important oil and gas bearing system. Because the reservoir facies of the Funing Formation are unfavorable, there is only one favorable exploration layer—the First member of the Taizhou Formation in these area.

The relationship between volcanic activities and the oil and gas accumulation

There are four stages of basaltic volcanic activities (Fig. 15). The first stage was developed between Jinhu downwarp and Gaoyou downwarp during the deposition of the source rocks of the Taizhou and the Funing Formation. The second stage was developed

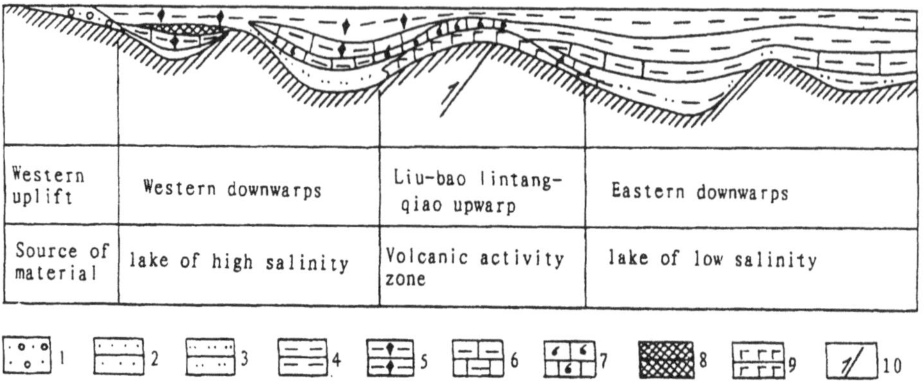

Western uplift	Western downwarps	Liu-bao lintang-qiao upwarp	Eastern downwarps
Source of material	lake of high salinity	Volcanic activity zone	lake of low salinity

1   2   3   4   5   6   7   8   9   10

Fig. 16. The effect of volcanic activity on the quality of source rock during the deposition of oil source rock in the North Jiangsu Basin (1=coarse and pebbled sandstone; 2=sandstone; 3=argillaceous siltstone; 4= mudstone; 5=gypseous mudstone (high quality source rock); 6=argillaceous limestone; 7=bioclastic limestone; 8=halogen salt; 9=basalt; 10=magmatic hearth)

in a vast area of the basin during the deposition of the First member of the Sandou Formation, and volcanic activity was especially strong in the middle and east area. The third stage was developed in the middle and east area during the deposition of the First member of the Yancheng Group. The fourth stage was developed in the west area of Jinhu downwarp around the Tanlu fault during the deposition of the second member of the Yancheng Group.

The volcanic activities of the basin were very useful in enhancing the abundance of oil and gas of the downwarps.

The effect of the volcanic activities during the deposition of source rocks on oil and gas accumulation

During the deposition of the Second and Fourth member of the Funing Formation, a great quantity of terrigenous organic matter was brought into the lake by the western drainage system, increasing the amount of nutrients in the water in western downwarps where aquatic organism grew up and reproduced rapidly. At the same time, the basaltic volcanic activities took place continuously along the east margin of Jinhu downwarp, the lake basin was separated by the project topography and thermal barrier of volcanic activities, and the lake water in western downwarps gradually evaporated and became concentrated to a high salinty(Fig. 16). As a result, the organic matter was preserved very well. Therefore, the western downwarps deposited high quality source rock with the content of organic carbon 2%-5%, the hydrocarbon content 350-1000 ppm and the value of potential 10-20 kgt^{-1}. The kerogens pertain to types I and II. The depth of the oil window with the high quality source rock is 1200-2000 meters, 400-800 meters shallower than the other source rocks.

Therefore, abundance of oil and gas of Jinhu downwarp increased, and a large amount of oil fields distributed at the depth from 1200 to 2000 meters.

A good oil and gas bearing assemblage of source rock, reservoir and seal rock deposited on the basalt plateau during the deposition of the Second member of the Funing

The Regularity of Oil and Gas Abundant Accumulation in North Jiangsu Basin, China

Column	Thickness (m)	Lithologic Characteristics
	40–60	grey black mudstone (good oil source rock)
	30–60	argillaceous limestone
	10–20	bioclastic limestone
	10–20	dolomite
	10–20	basalt volcaniclastic rock
	10–20	top layer of basalt (brocken layer)
	20 –40	basalt

Fig.17. A model showing a depositional cycle in basalt plateau during the deposition of oil source rock

Formation(Fig. 17). The proved reserves in this assemblage is more than 500×10^4 tons.

The effect of the volcanic activities after the deposition of oil source rocks on oil and gas accumulation

The volcanic activities were in the energy-releasing stage of the Earth's crustal movement which caused the basin to have a high heat-flow value(Fig. 18). The paleogeothermal gradient of the downwarps undergoing intense volcanic activities was 5–10 ℃/km higher than the other downwarps (Fig. 19). This condition was very useful in decreasing the

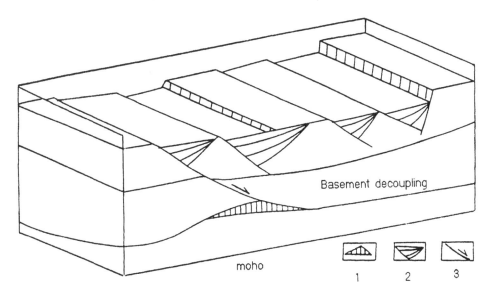

Fig.18. A dynamothermal modulus in the North Jiangsu basin (1=magma cushion; 2=half-graben; 3=shear zone)

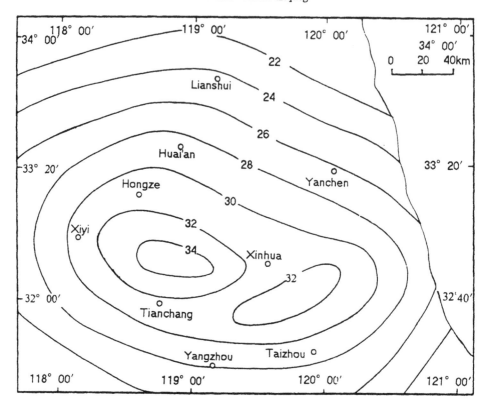

Fig. 19. Distribution of geothermal gradient in the North Jiangsu Basin

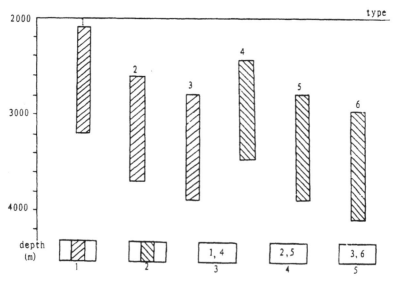

Fig. 20. The depth of petroleum generation of two types of oil source rock in different downwarps (1=the depth of petroleum generation downwarps of E_f^2 and E_f^4 (kerogens pertain to type II and I); 2=the depth of petroleum generation of E_f^3 and E_t^2 (kerogens pertain to type II and III); 3=the downwarp with 2-4 stages of vocanic activity; 4=the downwarp with 1-2 stages of volcanic activity; 5=the downwarp without volcanic activity)

threshold depth of petroleum generation(Fig. 20). The downwarps, undergoing 2-3 intense volcanic activities, have a depth of the oil window from 2000 to 3500 meters (type 1). The downwarps, undergoing one intense volcanic activities, have a depth of the oil window from 2500 to 3800 meters(type 2). The downwarps without intense volcanic activities have a depth of the oil window from 2800 to 4200 meters(type 3).

To the threshold depth of prtroleum generation, the deeper it is, the worse the conditions are. There are the following unfavorable factors: (a) A large amount of shallow-buried source rock are immature so that their petroleum-generating potential decreases correspondingly; (b) When formations are more deeply buried, they are heavily compacted by the overburden so that petroleum generation, migration and accumulation are seriously delayed.

Gaoyou, Qingtong and Jinhu downwarps, which belong to type 1, are the most abundant oil and gas accumulation area with 98% of total proved reserves. In Hai'an downwarp which belongs to type 2, only three small oil fields have been found. In Yancheng downwarp which belongs to type 3, although the thickness of the Cenozoic formation is over 5000 meters, the abundance of oil and gas is so poor that none oil field has been found.

In the shallow buried area of source rocks, including the exterior slope zones and low upwarps, some oil pools were formed by the intrusive masses.

CONCLUSIONS

As a result, most abundant oil and gas accumulation zones distributed in the intersect areas of the structural high zones and the fault terrace zones of Gaoyou downwarp and Qingtong downwarp, and in the structural high zones of shallow buried marginal facies belts of Jinhu downwarp. All the fourteen middle-sized oil and gas fields and more than 90% of the proved reserves of petroleum have been found in these zones. Only small sized oil and gas fields distributed in the other areas.

REFERENCES

1. Zhang Y.C., Wei Z.L., Xu W.L., Tao R.M. and Chen R.G., 1989, The North Jiangsu-South Yellow Sea Basin. In: Zhu X. (Editor), Chinese Sedimentary Basins. People's Republic of China, Tongji University, Shanghai, PP. 107-123.

Proc. 30ᵗʰ Int' l. Geol. Congr., Vol. 18, pp. 333~340
Sun Z. C. *et al.* (Eds)
© VSP 1997

Numerical Simulation of Histories of Hydrocarbon Generation and Primary Migration in Several Typical Basins in China

XU SIHUANG

Department of Petroleum Geology, China University of Geosciences, Wuhan, CHINA

Abstract

The basic geological model of Hydrocarbon (HC) generation and primary migration can be showed as: being buried of source rock-maturing-increasing of pressure-micro fracturing-expulsion and discharging of abnormal pressure. The computer software to simulate the geology model has been written by the author. The histories of generation and primary migration of five typical basins in China are simulated in this paper. These basins are Songliao basin, Liaodongwan area of Bohaiwan basin, Sichuan basin, Lunpola basin in Xizang, and Tarim basin. The geological character and character of histories of generation and expulsion of these five basins are different from each other. Among these five basin, Lunpola basin, which has the highest geothermal gradient, is a typical Cenozoic "hot basin". Tarim basin, which has the lowest geothermal gradient, is a typical Palaeozoic "cold basin".

Keywords: Hydrocarbon Generation, Primary Migration, Numerical Simulation

INTRODUCTION

The thermal evolution, hydrocarbon (HC) generation and expulsion are the basic geological condition for forming oil and gas reservoirs. Traditionally, they were studied by geological and geochemical methods, and the main study contents were the geochemical properties and generation amount. But recently, with the development of petroleum geology theory and wide application of computer technology, it is believed that to study the process of thermal evolution, HC generation and expulsion is very important. In other words, petroleum geologists pay attention to the study of histories of HC generation and primary migration. Various models were put forwards (Nakayama K. ,1987; Ungerer P. et al. 1990; Ozkaya I. , 1991; Hunt J M. , et al 1991).

Based on these models, according to the study of several typical basins in China, a modeling system of histories of thermal evolution, HC generation and expulsion from source rock is put forward in this paper. Its application in these basins is effectual.

MODELING SYSTEM USED IN THE STUDY

Model Blocks in the System

To study the petroleum geologic process of source rock in petroleum basins is the main function of our modeling system. In order to study conveniently, the modeling system is divided into 5 model blocks: the burial history block, geothermal history block, maturity history block, HC generation history block and HC expulsion history block. Each block has its input conditions, output results and processing function (Fig. 1).

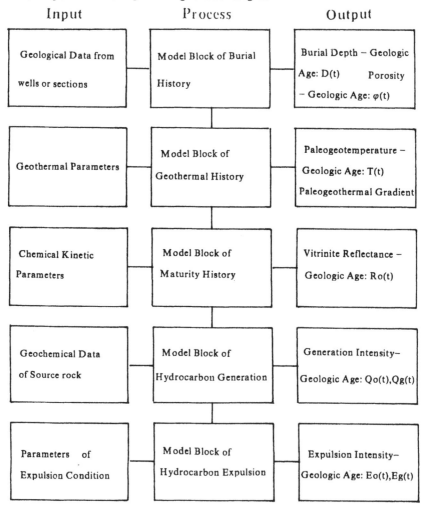

Fig. 1   The flowchart of the modeling system

Ceneral Geological Model

The general geological model of thermal evolution, generation and expulsion of HC of
source rocks can be summed up in deposi-
tion-burial-being heated-maturity-HC gene-
ration-causing high pressure-microfrac-
turing-HC expulsion-discharge of abnormal
pressure. After expulsion and discharge, as
generation continues, the process of causing
high pressure-microfracturing expulsion and
discharge may take place agian and agian.
Anyway, we think that the HC expulsion
caused by microfracture is intermittent of
pulse process.

Calculation of abnormal Pressure Caused by
HC Generation

It is believed that the expulsion is caused by
microfracturing of source rock, and the microfracturing of source rock is caused by
high abnormal fluid pressure. But the high
abnormal fluid pressure may not be caused
by abnormal compaction, the HC generation
can also cause abnormal pressure. We de-
signed the volume model for calculating this
kind of abnormal pressure (Fig. 2).

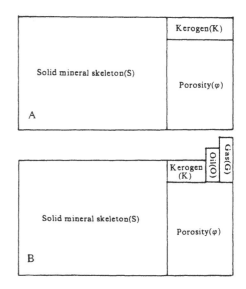

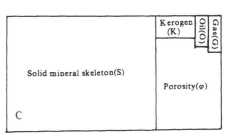

Fig. 2 The volume model for calculating the
pressure hydrocarbon generation

The source rock is composed of solid mineral skeleton (S), pore (Φ) and solid kerogen pow-
der (K) (Fig. 2a). If the volume of source rock is 1 unit, then $S+\Phi+K=1$. When the source
rock became mature, a part of kerogen cracked into oil and gas. Because the density of oil
and gas is lighter than that of kerogen, if pressure were constant, the volume of oil and gas
would be larger than the volume caused by kerogen cracking, the total volume of source rock
would be larger than 1. But in fact it is imposible to source rock that the volume will be increased, unless the microfracturing take place. Before microfracture take place, oil and gas
must be compacted in the volume caused by kerogen cracking (Fig. 2c). So the pore pressure
will became abnormal and high.

According to the relationship of fluid volume and pressure, following formula can be de-
duced.

$$\Delta p = \frac{M_g/\rho_g + M_o/\rho_o - M_k/\rho_k}{C_g \cdot M_g/\rho_\pi + C_o \cdot M_o/\rho_o + C_w\varphi}$$

In the formala, Δp is abnormal pressure caused by generation. M_g, M_o are the mass of free gas and oil generated. M_g is the volume that kerogen lost. ρ_g, ρ_o, ρ_k are the density of free gas, oil and kerogen under ground. C_g, C_o, C_w are the compressiility factors of free gas, oil and pore water, and Φ is porosity of source rock. If no free gas, then M_g is zero. So the total fluid pressure in pore of source rock can be shown as:

$$p = p_n + \Delta p + \Delta p'$$

Where, p_n is hydrostatic pressure, Δp is abnormal pressure caused by generation, $\Delta p'$ is the abnormal pressure caused by other factors such as abnormal compaction.

Expulsion Model

with the capulsion of HC and/or pore water, the pore pressure discharged, and the microfrature will close when the pressure discharged to the critical pressure which is the lowest pressure keeping microfraturing. The expulsion amount can be calculated by following:

$$E_{o/g} = C_{o/g} \cdot \rho_{o/g} \cdot (p - p_c) \cdot V_{o/g}$$

Where, $E_{o/g}$ is expulsion amount of oil/gas from 1 unit of source rock, called as expulsion intensity of oil/gas, $C_{o/g}$ is the compressibility factor of oil/gas , $\rho_{o/g}$ is the density of oil/gas under ground. $V_{o/g}$ is the oil/gas volume before expulsion. p is the pore pressure. p_c is the critical pressure of keeping microfracturing.

The Flowchart of the Modeling System

According to above principle, the application software is written used FORTRAN language, its flowchart was shown as figure 3.

APPLICATION IN SEVERAL BASINS IN CHINA

Geological Setting

Five typical petroleum basin in China have been studied applied this modeling system.

Songliao basin: Songliao basin lied in northeast of China, it is a large Mesozoic-Cenozoic continental basin. The dark mudstone of lower Cretaceous system is the main source rock. Organic matter is type **I** , organic carbon contents is 1. 2%. The temperature of surface and constant layer is 12°C, mean geothermal gradient is 33°C/km.

Bohaiwan basin: It lied in north of China. It is a large Mesozoic-Cenozoic fault basin. the dark mudstone of Eogene is the main source rock. Organic matter is type **I** $_1$, organic carbon contents is 0. 4%-4. 4%. The temperature of surface and constant layere is 10°C, mean geothermal gradient is 30°C/km.

Sichuan basin: It is a middle marine carbonate basin in sourhwest of China. One of the source layer is Da'anzhai formation of lower Jurassic system. Both dark mudstone and shell

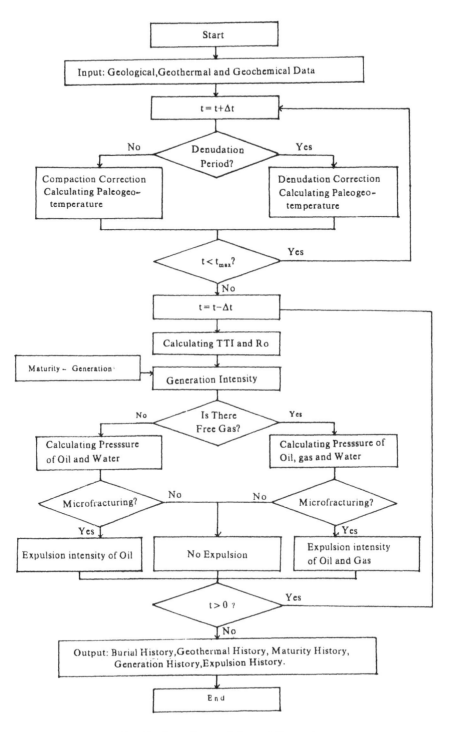

Fig. 3 The flowchart of application software of the modeling system

limestone are source rocks. The temperature of surface and constant layer is 20℃ , geothermal flux 0. 047-0. 056 W/m², so geothermal gradient is only 18. 8-27. 6℃/km.

Tarim basin: It is a macroscopic Palaeozoic marine carbonate basin in northwest of China.
The Cambrian-Ordovician carbonate is its main source rock. The tempreature of surface and
constant layer is 13℃ , geothermal gradient is 25℃/km. Organic matter is type I , but organic carbon contents is less than 0. 2%. The vitrinite reflectance is 0. 8%-1. 7%.

Lunpola basin: It is a small Tertiary intermountain fault basin in Qingzang highland(Tibet).
Its main source rock is the dark mudstone of Eogene. The temperature of surface and constant layer is −8℃ , but geothermal flux is 0. 105W/m², geothermal gradient is very high,
50-60℃/km.

Modeling Results

The geohistory plots of well Kun4 in Lunpola basin and well Sha32 in Tarim basin were
shown in figure 4. It is clear that the rate of sedimentation in Lunpola basin is higher, while
that in Tarim basin is lower. The paleogeothermal gradient in Lunpola basin is about 70-
90℃/km, while that in Tarim basin is only about 26-33℃/km. In other words, Lunpola
basin is a typical HOT basin, but Tarim basin is a typical COLD basin.

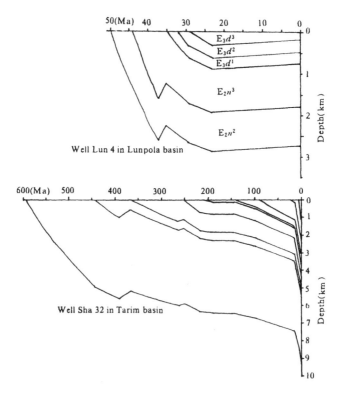

Fig. 4   The geohistory plot of wells

In Lunpola basin, the generation intensity of HC of main source rock, Niu-2 of Eogene, usually is 10-60kg/m³, the expulsion intensity of HC of the same source rock usually is 5-10 kg/m³. In Tarim basin, the generation intensity of HC of main source rock, the Cambrian-Ordovician carbonate, usually is 2.8-7.4kg/m³, the expulsion intensity of HC of the same source rock usually is only 0.05-0.5kg/m³. The expulsion history plots of well Lun4 in Lunpola basin and well Sha32 in Tarim basin were shown in figure 5.

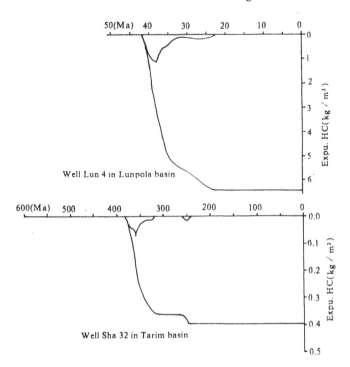

Fig. 5   The expulsion history plot of wells

The general modeling results, such as geothermal gradient, generation intensity and expulsion intensity, of Bohaiwan, Songliao and Sichuan basins usually are between that of Tarim basin and Lunpola basin. In Bohaiwan basin, the generation intensity of HC of main source rock, Shahejie-3 of Eogene, usually is 4-25kg/m³, the expulsion intensity of HC of the same source rock usually is 0.3-3.5kg/m³. In Songliao basin, the generation intensity of HC of main source rock usually is 4.4-13 kg/m³, the expulsion intensity of HC of the same source rock usually is 1-6 kg/m³. In Sichuan basin, the generation intensity of HC of the mudstone source rock of Da'anzhai formation of lower Jurassic is 4-18kg/m³, and the shell limestone of Da'anzhai formation only 2-8kg/m³. The expulsion intensity of HC of the mudstone source rock of the same source bed usually is 0.1-6kg/m³, and that of shell limestone is only 0-1.8 kg/m³.

DISCUSSION

1. The geotectonic framework controlled the properties of basins. The basin which located in stable area of plate, such as Tarim basin, had lower rate of sedimentation and geothermal gradient. The basin which located in active area of plate, such as Lunpola basin, had higher rate of sedimentation and geothermal gradient.

2. The properties of basins and the geochemical properties of source rocks in the basins controlled the HC generation. The basin in which the geothermal gradient was high and the organic matter was rich usually had high generation intensity.

3. The HC expulsion was controlle by both the properties of basin and the HC generation. The basin in which the geothermal gradient was high and the generation intensity was high usually had high expulsion intensity.

REFERENCES

1. K. Nakayama. Hydrocarbon expulsion model and its application to Nilgata area, Japan. *AAPG Bull*, 1987, 7(7): 810-821

2. P. Ungerer, et al.. Basin evaluation by integrated two-dimensional modeling of heat transfer, fluid flow, hydrocarbon generation and migration. *AAPG Bull*, 1990,74(3): 309-335

3. I. Ozkaya. Computer simulation of primary oil migration in Kuwait. *Journal of Petroleum Geology*, 1991,14(1):37-48

4. J. M. Hunt, et al.. Modeling oil generation with time-temperature index graphs based on the Arrhenius equation. *AAPG Bull*, 1991,75(4): 795-807

Proc. 30th Int' l. Geol. Congr. , Vol. 18, pp. 341~345
Sun Z. C. *et al.* (Eds)

Prediction and Assemssment of Natural Gas Resources in China

ZHANG HONGNIAN, YANG DENGWEI

Institute of Petroleum Geology, MGMR, Beijing 100083

YANG RUIZHAO

China University of Geosciences, Beijing 100083

INTRODUCTION

There are approximately 335 sedimentary basins in the onshore and offshore areas in China. It is reported that China has about 1. 67 trillion cubic meters (TCM) of natural gas reserves in place with current production of 17 billion m^3/a (in 1995). The proven reserves of natural gas are mainly distributed in the Sichuan basin and Bohaiwan basin. At a median level of probability, it is estimated that the natural gas resources in China is 43. 3 trillion m^3. China' s ratio of reserves-to-resources is only 0. 03 for natural gas. In addition, coalbed methane resources are estimated to be 16. 44 trillion m^3. A vast natural gas potential still remains in the sedimentary basins in China.

DISTRIBUTION. OF NATURAL GAS RESOURCES

1. Distribution Pattern of Conventional Natural Gas Resources
The conventional natural gas resources in the 30 Meso-Cenozoic basins and 12 marine sedimentary regions in China are estimated to be 30. 4-60. 7 trillion m^3, with a median value of 43. 3 trillion m^3.
(1) Proportional distribution in different strata: Sinian-Ordovician 26. 1% (11. 31 TCM), Permian and Carboniferous 25. 5% (11. 02 TCM), Jurassic and Triassic 6. 6% (2. 86 TCM), Cretaceous 4. 5% (1. 95 TCM), Palaeogene 32. 8% (14. 21 TCM), and Neogene and Quaternary 4. 5% (1. 95 TCM).
(2) Geographic distribution of natural gas: Offshore province 32. 6%, Bohaiwan-Songliao province 12 %, Central province 29. 1%, Northwestern province 26. 3%. In the offshore province, the natural gas resources is 14. 1 TCM, which is mainly distributed in the Palaeogene (10. 58 TCM or 75%) and Quaternary-Neogene (3. 53 TCM or 25%). In the Bohaiwan-Songliao province, the natural gas resources is 5. 2 TCM, which is mainly distributed in

Meso-Cenozoic (4. 16 TCM or 80%) and Permian-Carboniferous (0. 85 TCM or 17%) strata. In the Central province, the natural gas resources is 12. 6 TCM, which is distributed in Permian-Carboniferous (7. 81 TCM or 62%) and Ordovician-Sinian (4. 28 TCM or 34%). In the Northwestern province, the natural gas resources is 11. 4 TCM, which is distributed in Ordovician-Sinian (6. 73 TCM or 59%), Permian-Carboniferous (2. 17 TCM or 19%) and Jurassic-Triassic (2. 17 TCM or 19%).

(3) Distribution of natural gas resources in sedimentary basins

The ten richest basins with gas resources of more than 1 TCM each account for 87% of China's total gas resources (Table 1).

Table 1   The sedimentary basins with the natural gas resources of more than 1 TCM

Basin (Region)	Natural gas resources $(10^8 m^3)$	% of China's total
Tarim	8128. 3	18. 77
Sichuan	5771. 4	13. 33
East China Sea	5106. 8	11. 79
Southeast Hainan	3970. 0	9. 17
Ordos	3536. 2	8. 16
South China Carbonate rock	2938. 8	6. 78
Junggar	2750. 1	6. 35
Bohaiwan	2496. 2	5. 76
Songliao	1872. 7	4. 32
Southwest Taiwan	1139. 2	2. 63
Total	37709. 4	87. 07

(4)Proportional distribution in genetic-type gases: biogenic gas 6. 5% (2. 82 TCM) and thermogenic gas 93. 5% (40. 45 TCM). The latter can be subdivided into three types: gas sourced from lacustrine shales 26% (11. 25 TCM), gas generated in carbonate rocks 39. 1% (16. 92 TCM) and coal-derived gas 28. 4% (12. 32 TCM). More than half of biogenic gas resources is in the sedimentary basins of China's offshore area (Figure 1). Gas resources sourced from lacustrine shales are mostly distributed in the Bohaiwan, Junggar, Southwestern Hainan basins (Figure 2). Coal-derived gas resources are mainly distributed in the East China sea, Ordos, Junggar, Yingehai and Songliao basins (Figure 3). The gas resources generated in carbonate rocks are concentrated in Tarim basin, Sichuan basin and carbonate rock area in South China (Figure 4).

2. Coalbed Methane

The coalbed methane resources in China are estimated to be 10. 6-25. 23 trillion m³, with a median value of 16. 46 TCM (Table 2), of which about 6. 4 TCM is at depth shallower than 1000m. They are mainly distributed in Northern and Northwestern China. We can expect the coalbed methane will play an important role in China's future gas supplies.

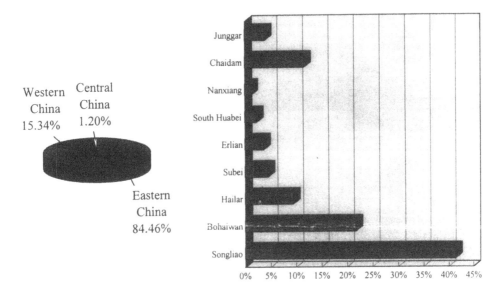

Figure 1   Distribution of Biogenic Gas Resources in Onshore Region

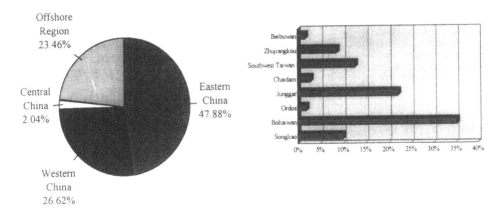

Figure 2   Distribution of Thermogenic Gas Resources

Table 2   Distribution of Coalbed Methane Resources in China

Coal Province	Coalbed Methane Resources ($10^8 m^3$)	%	Depth<1000m Coalbed Methane Resources	%
Northeastern	272. 8	1. 6	35. 47	0. 6
Huabei	9049. 4	55. 0	2510. 03	39. 2
Northwestern	5579. 7	33. 9	3129. 64	48. 9
South China	1560. 3	9. 5	727. 03	11. 3
Total	16462. 2	100	6402. 17	100

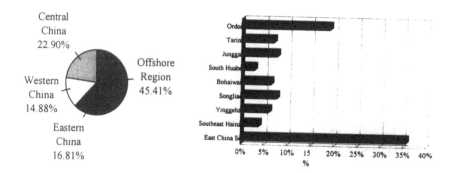

Figure 3   Distribution of Coal-derived Gas Resources

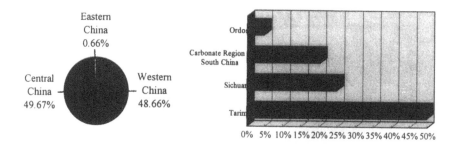

Figure 4   Distribution of Gas Resources Generated in Carbonate Rocks

EXPLORATION PROSPECT OF NATURAL GAS

Based on the natural gas assessment in China, the 27 regions are selected for decision analyses.

1. There are a total of 9 prospective areas with the gross profits more than 50 billion RMB yuan (the Tabei uplift, Tazhong uplift and Southwestern depression in Tarim basin; the central Tianhuan-Yishan slope in Ordos basin; the Xihu depression, East China Sea basin, Yinggehai-Hainan basin; the southern and eastern regions, Sichuan basin; the Dongpu and Jiyang depressions, Bohaiwan basin) and 5 perspective areas with the return on investment of more than 6 (the Tabei uplift, Tarim basin; Sanzhao region, Songliao basin; Bohaiwan offshore area, Liaohe depression, Jizhong depression).

2. The main target regions for near term natural gas exploration and development plans in China are suggested to be:

(1) Tabei uplift area, Tarim basin.

(2) Xihu depression, East China Sea Continental basin.

(3) Yinggehai-Hainan basin.

(4) the eastern and western regions in Sichuan basin.

(5) Sanzhao depression in Songliao basin.

(6) Bohaiwan basin.

(7) the central Tianhuan-Yishan slope in Ordos basin (table 3).

Table 3   Economic Assessment and Decision Analyses of Prospective Areas

Prospective Regions	Gross Profits (10^8 RMB Yuan)	Returnon Investment	Comprehensive Discreminant Scores
Tabei Uplift, Tarim Basin	2989. 05	17. 89	770
Xihu Depression, East China Sea Basin	616. 36	3. 82	580
Yinggehai-Southeastern Hainan Basin	969. 00	1. 40	570
Eastern Region, Sichuan Basin	582. 50	2. 86	580
Western Region, Sichuan Basin	117. 42	1. 93	550
Sanzhao Depression, Songliao Basin	143. 14	5. 47	560
Bohaiwan Basin (Offshore Region)	339. 18	6. 88	560
Tianhuan-Yishan Slope, Ordos basin	1661. 58	2. 60	560

3. 8 prospective regions, such as the Taikang area in Songliao basin, the northeastern and northwestern areas in Sichuan basin, and the southwestern depression in Tarim basin, can be considered as natural gas strategic backup exploration areas.

CONCLUSION

1. A vast natural gas potential still remains in the sedimentary basins in China.

2. The main exploration target of natural gases in China are the Palaeogene play, Permian-Carboniferous play and Ordovician-Sinian play.

3. The favorable position for searching the medium-large gas fields are considered to be large coal-bearing basins, porous carbonate reservoirs, palaeouplift and regional unconformity.

Proc. 30ᵗʰ Int' l. Geol. Congr. , Vol. 18, pp. 347～360
Sun Z. C. *et al.* (Eds)
© VSP 1997

GEOLOGY AND HYDROCARBON OCCURRENCE OF TERTIARY STRIKE-SLIP BASINS IN EASTERN TIBET AND WESTERN SICHUAN

Tian Zaiyi, Zhang Qingchun

Research Institute of Petroleum Exploration and Development, China National Petroleum Corporation, Beijing, PRC 100083

Abstract

The eastern Tibet and western Sichuan experienced a new tectonic development stage during the Cenozoic. Sedimentary characteristics, volcanic and magmatic activities and structural deformation styles changed, so did the stress field. Therefore, the whole development stage can be clearly divided into two sub-stages, that is, the Early Tertiary strike-slip and extension and the Late Tertiary-Quaternary interior continental convergence.

During the Early Tertiary, the crust below the whole Qinghai-Tibet Plateau was in a compressional environment due to nearly N-S collision between the Indian plate and Asian Continent. The Indian plate moved northward and the Yangtze plate moved slowly southeastward, which generated a unique strike-slip background in the eastern Tibet and western Sichuan. A series of large scale and deep strike-slip faults and associated pull-apart basins were thus formed, striking N-S and NWN about 100 km in length and about 5-20 km in width. The bounding faults often occurred along pre-existing N-S or NWN trending shear faults. The strikes of faulted basins were obviously controlled by pre-existing basement faults.

The sedimentary environments in the basins south of the Dingqing-Changdu-Litang line are clearly different from those in the basins north of it. The southern part is dominated by dark coaly fine clastics about 4,000 m thick representing a warm and humid climate. The sediments may contain source materials. The northern part is dominated by complex-colored clastics interbedded with anhydrite and salt rocks under a relatively hot and arid climate. Volcanic rocks mainly consists of basalt and rhyolite association of calc-alkaline and alkaline series were generated in pull-apart basins due to strike-slip extension. Volcanic activities were more intensive in the western part than in the eastern part.

This region has mainly undergone nearly E-W compressional contraction since the Late Tertiary due to a differential movement between the Indian plate and the Yangtze plate, accompanied by a dextral force pair. Bounding faults of the Early Tertiary pull-apart basins were changed into oblique thrusting faults under this background. Folds in the basins were often developed on one side of major oblique thrusting faults. These folds have a plane sinistral pattern, which indicates that the folds were generated by strike-slip thrusting, that is, folding was caused by thrusting. This kind of 3-D structural framework indicates a united thrusting and strike-slip transformation type structural displacement field.

Keywords: hydrocarbon occurrence, strike-slip basins, Tertiary, Eastern Tibet and western sichuan

GENERAL DESCRIPTION

The pull-apart basins in the Hengduan mountain region of eastern Tibet and western Sichuan were formed in the Tertiary as an important tectonic unit linked to a strike-slip fault system. Such a tectonic unit was formed by the pull-apart action caused by i(slip of a bent strike-slip fault, ii(branched sections of a strike-slip fault, or iii(the overlapped sections of an en echelon strike-slip fault.

In the Tertiary, because of continuing thrusting of the Indian plate below the Asian continental crust and intra-continental convergence, the entire Qinghai-Tibet plateau was under a compressive tectonic setting and spread out in the East West direction (Molnar and Tapponnier, 1975; Li Chunyu et al. , 1979; Tapponnier et al. , 1986). However, in eastern Tibet, the Hengduan mountain belt appears as upheaval and rift valleys in a nearly North South direction. This is closely related to the tectonic location of the Hengduan mountains (Pan Guitang et al. , 1990), which is located along a peculiar tectonic belt lying at the boundary of the Yangtze plate on one side and the Indian plate on the other side (Fig. 1).

In the Paleogene, this region experienced a new tectonic stage. In the Eocene-Oligocene right lateral pull-apart motion came into force and a series of pull-apart strike-slip basins were developed, striking N-S and NNW, generally 100 km long and 5-20 km wide. The bounding faults often occurred along pre-existing N-S or NNW trending faults. The opening of pull-a-part basins with graben or half graben was obviously controlled by the pre-existing basement faults. In such a broad strain field of strike-slip faulting, several sedimentary basins developed(Fig. 2). For instance, along the pre-existing fault zone of Nujiang there were developed pull-apart strike-slip basins of Basu (Baxoi), Luolong (Lhorong) and Dingqing (Dengqen). Between the Lancangjiang fault zone and a branched brush-shaped fault there were developed pull-apart strike-slip basins of Zaduo (Zadoi)-Jiqu (Gyiqu), Shisuzhan and Nangqen. Along the Zigasi-Deqin (Deqen) fault zone there were developed the pull-apart basins of Gongjiao (Gonjo), Mangkang (Markam) and Mangcuo. Along the Ganze (Garze)-Litang fault zone there were developed the pull-apart basins of Relu etc (Pan Guitang et al. , 1990).

In these pull-apart strike-slip basins of different sizes, a remarkable sedimentary environment change can be observed from north and south of a boundary line trending from Dingqing (Dengqen) to Changdu (Qamdo) and Litang. In the south, sediments are predominantly dark coaly fine clastics, representing a warm, humid climate, about 4,000 m thick and possibly containing hydrocarbon source rocks. In the north sediments are red coarse anhydrite and salt containing clastics under a relativley dry and hot climate. With strike-slip motion, volcanic rocks were developed in the pull-apart basins, mainly as trachytic basalt-rhyolite association of calc-alkaline, slightly alkaline series. Volcanic activities were generally stronger to the west than to the east. Geochemical studies showed large extension of volcanism, possibly cutting through the whole crust and reaching the deep mantle.

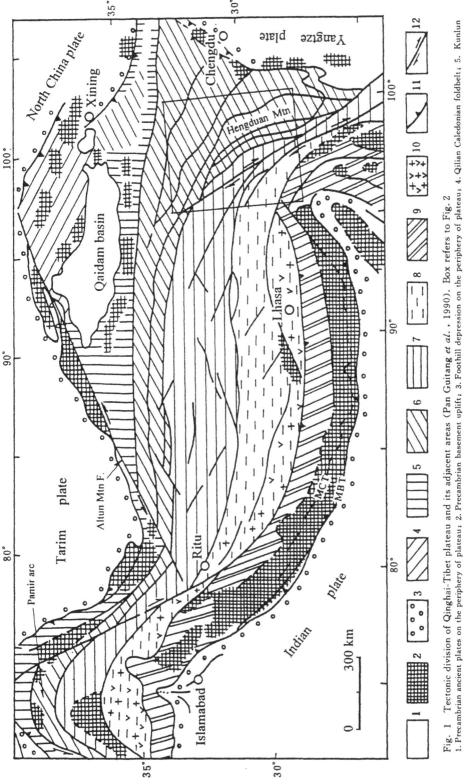

Fig. 1 Tectonic division of Qinghai–Tibet plateau and its adjacent areas (Pan Guitang *et al.*, 1990). Box refers to Fig. 2
1. Precambrian ancient plates on the periphery of plateau; 2. Precambrian basement uplift; 3. Foothill depression on the periphery of plateau; 4. Qilian Caledonian foldbelt; 5. Kunlun Hercynian foldbelt and adjacent areas; 6. Bayan Har Indosinian foldbelt and adjacent areas; 7. Tanggula early Yanshanian foldbelt and adjacent areas; 8. Gangdhis late Yanshanian foldbelt and adjacent areas; 9. Himalayan fold belt; 10. Volcanic magmatic rock; 11. Thrust or overthrust belt; 12. Strike-slip fault

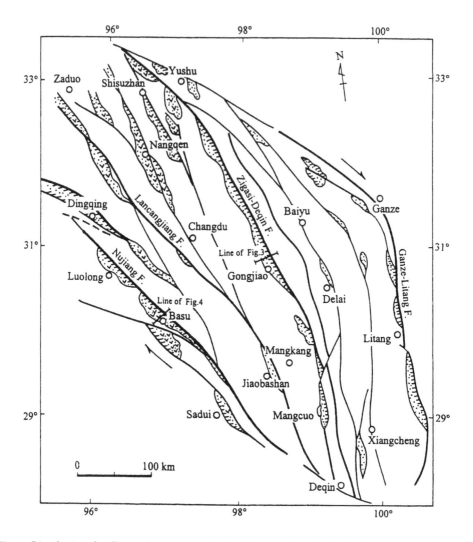

Fig. 2 Distribution of pull-apart basins controlled by right-lateral strike-slip fault system in the Eogene Hengduan mountain region (Pan Guitang *et al.* , 1990)

Since the Neogene, this region has experienced nearly E-W shortening caused by differential movement between the Indian and the Yangtze plate, which was superimposed upon a dextral shear couple. Under this tectonic setting the bounding faults of Palaeogene pull-apart basins were transformed into oblique thrust faults, with intrabasinal fold belts developed on one side of the oblique thrust faults. These folds have a sinistral pattern in horizon, which indicates the causal relationship between thrusting and folding: folding generated by strike-slip thrusting.

Fan-shaped thrust nappes are observed on cross sections. Reverse thrusting occurrs from the Nujiang fault zone to the Ganze (Garze)-Litang fault zone. Secondary asymmetrical fan-shaped thrust elements inside appear as rhombic strike-slip structures on a planar view. Such a spatial structural framework visually reflects the unity of thrusting and strike-slip transformation type structural displacement field in this region.

GEOLOGICAL FEATURES OF STRIKE-SLIP BASINS

Regional tectonic features

In the Hengduan mountains, the change in stress field over time can be inferred from the structural deformation, volcanism and sedimentary features. From the period of the Eocene-Miocene, the Assam tectonic knot of the Indian plate moved nearly perpendicular to the Hengduan mountain belt and generated a series of regional right lateral strike-slip faults along which pull-apart basins are located. Continental fluvial-incustrine clastics were deposited in the basins, which were shallow in the west and deep in the east, with sediments thinner in the west than in the east. Volcanic activities also occurred. After the Miocene only a few minor rift basins were developed, with accompanying weak volcanism. Tectonic activities in the Hengduan mountain region were mainly a series of N-S thrust belts and NW left lateral strike-slip faults in a tectonic setting of rhombic pattern. The combination of fold axis, thrust and strike-slip fault implies a strong intra-continental convergence from late Tertiary to Quaternary that caused nearly E-W regional shortening.

The configuration and size of pull-apart basins are controlled by the tectonic setting demonstrated by the nature of bounding faults and basement rock properties. Each sedimentary-basin has its unique history of growth. The relationship between sedimentation and subsidence indicates synchronism between both phenomena. The strike-slip pull-apart basins in the Hengduan mountain region are distributed in NS or NNW direction. The bounding faults are connected with pre-existing outside major faults. The basins vary in size and linear dimensions. They are mainly irregular shaped. In the western part of the Hengduan mountains larger basins were developed to the west of Jinshajiang, dozens of km long and several km wide. Smaller basins were developed to the east of Jinshajiang in western Sichuan, generally several km long and hundreds of meters wide. Such features reflect a control by basement faulting on basin development.

Distribution of sediments, thickness of sediments and change in sedimentary facies are closely related to the bounding strike-slip faults. Graben type sediments are distributed on both sides of the basin, i.c., near the bounding faults. The sediments are coarse-grained, composed of alluvial fan and slump debris flow. Depocenter often does not coincide with the center of subsidence. For a single faulted asymmetrical half graben depression (Fig. 3), the maximum thickness of deposition is located in the east of the basin, close to the bounding fault. Fluvial alluvial fan of fan deltaic deposits are distributed along the fault. On the other side of the basin braided river deposits dominate, inclining as basement monocline and wedging into the basin center (Pan Gultang *et al.*, 1990).

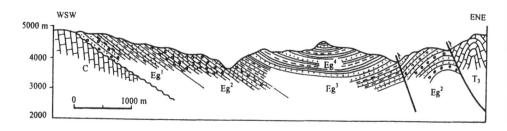

Fig. 3   Eogene Youzha section in Gongjiao basin (Pan Guitang *et al.* , 1990)

In a strike-slip pull-apart basin, the wrenching structure caused by the rotation of a bounding fault gave rise to tilting and folding of sediments. Left lateral en echelon folds on a planar
view with an axial direction N50(W and intersecting the major bounding fault with an angle
of 15(- 20(indicate the features of clockwise wrenching movement on the east side of the
basin. In cross section, the formations are gently dipping NE on the northeast flank of an anticline and steeply dipping SW on the southwest flank; the formations are steeply dipping on
the northeast flank of a syncline and gently dipping on the southwest flank (Fig. 3).

Oligocene-Miocene tectonic movement in the Hengduan mountains consists of converging
overthrust and oblique slippage between blocks. Presently the bounding faults of all basins
are thrust faults generated by strong E-W shortening (Pan Guitang *et al.* , 1990) (Fig. 4).
Accompanying folding indicates the action of shear couple. A series of NWW left lateral con-
jugate shear faults were developed inside the basins, which intersected major fault zones to
dissect the basin into several fault blocks.

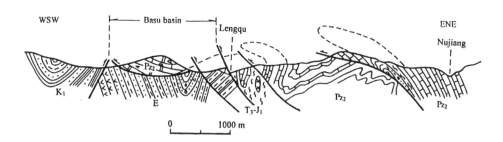

Fig. 4   Cross-section through Nujiang thust zone in Basu basin (Pan Guitang *et al.* , 1990)

Volcanic rock growth features

Volcanic activities are developed in the Hengduan pull-apart basins, they vary with the
downcutting depth and activities of faults. Magmatic formation of a certain type was developed under a certain regime of tectonic movement. Volcanic rocks developed in the Nangqen
basin occurred mainly in two stages. Volcaniclastic rocks were predominant in the early
stage, i.e. , the early Eocene, and were distributed in the east part of the basin as interbeds

of Palaeogene continental clastic rocks. The late stage of volcanic activity occurred at the beginning of the Oligocene, mainly as trachyte and epigenetic volcanism that were widely distributed in the basin. Dykes and volcanic veins cut through sedimentary rocks (Fig. 5), or
penetrated along the beds with a baked metamorphic zone near the contact. The extension of
volcanic rocks basically concides with the direction of the basin, indicating predominant eruption along linear fractures. The type of volcanic rock is mainly calc-alkaline trachyte, and
secondly trachyandesite. Alkaline trachyte, volcanic clastic rock, basalt and rhyolite are less
outcropping. The above rocks are mostly eruptive, and some are subvolcanic. The former is
mainly all kinds of trachyte and trachyandesite, and the latter is mainly alkaline-slightly alkaline rocks (Pan Guitang *et al.*, 1990). Strong and frequent alkaline-neutral-acidic volcanic
eruptions occurred in the Basu (Baxoi) basin with relatively long duration and a complete assemblage of rock types. The contact of volcanic rocks with surrounding rocks occurr generally as fault dated, from late Yanshanian to early Himalayan. The rock assemblage is mainly o-
livine basalt-basaltic andesite-dacite-ignimbrite-breccia lava.

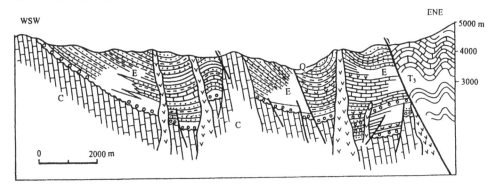

Fig. 5   Cross-section of Nangqen basin (Pan Guitang *et al.*, 1990)

Sedimentary rock features

Tertiary pull-apart basins in the Hengduan mountains received sediments of metamorphosed
Mesozoic or even older rocks in provenance from the surrounding uplifted areas. Lacustrine-
fluvial sandy mudstone, clastic rock sediments are about 1,000-4,000 m thick, of the
Eocene-Miocene. Fossil evidence indicate the absence of Palaeocene formations. Pliocene uplifting caused widespread lack of sediments or erosion. Quaternary sedimentation was
widespread. According to the records of "Cenozoic tectonic evolution of the Qinghai-Tibet
plateau" (Pan Guitang *et al.*, 1990), the sediments in the pull-apart basins are typical continental fluvial-lacustrine: alluvial fan, braided river, meandering river, fluvial deltaic, lacustrine sediments (Fig. 6).

In the foothill of a fault zone with a big topograhic difference, mountain rivers carry a lot of
clastics in a torrential season, which are quickly laid out and piled up to form alluvial fan
when the rivers flow into the plain area of a gentle slope to reduce the flow rate and carrying
capacity. Braided river deposits are characterized by shallow and wide river channel,

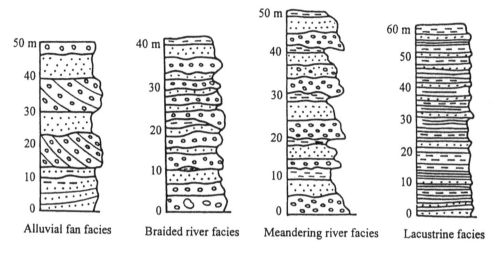

Fig. 6   Model section of sedimentary facies of pull-apart basin in Hengduan mountain region (Pan Gui-tang *et al.* , 1990)

large reduction in the slope of river bed, varying river channel in branched or braided pattern. Because of large drop in slope, fast flow and significant silt load, the sediments are
coarse-grained and featured by downcut erosion and channel bar growth. Point bar and flood
plain are hardly or not developed. Braided river sediments are most developed between alluvial fan and deltaic plain facies. Meandering river facies is featured by narrow and deep river
channel, small drop in slope of river bed, and stable river channel. Sidewise erosion is dominant, making the river channel migrate in a markedly curved shape. Capturing occurs when
curving of river channel goes beyond its limit to form an oxbow lake. Point bar and flood
plain are developed with very few channel bar. The carrying capacity of a meandering river is
rather stable and the sediments are relativley fine-grained. A delta is deposited when a river
carrying a large amount of terrestrial clasitcs flows past a flood plain to reach a lowland and
the shallow waters near the lake basin. Longitudinally, there are deltaic plain, deltaic front,
and prodelta facies. Vertically there are bottomset, foreset and topset beds. Lacustrine deposits of deep water and shallow water origins were developed in pull-apart basins. Fresh
water and brackish water lakes exist due to different weather conditions. The salinity also
changed in the process of lake basin evolution. Rain mark and mud crack are common in shallow water lake deposits. Deposits in deeper lakes, such as the Nangqen basin, showed bibli-
olite and thin carbonaceous mudstone.

Sedimentary rock assemblage differs depending on the different stages of basin growth and
different places within one basin. Gongjiao (Gonjo) basin is 150 km long in NNW direction
and is 5-8 km wide. Four lithological sections are identified from top to bottom. The first
section is fluvial sandy mudstone interbedded with sandy conglomerate, thickening from
south to north, in a range of 600-1,800 m. The second section is an assemblage of meander-

ing river sandstone and muddy siltstone, thinning from south to north, in a range of 400-1, 400 m. The third section is composed of lacustrine thin-bedded sandstone and siltstone, interbedded with shale and muddy dolomite and containing gypsum-salt deposits, 400-1,400 m thick. The fourth section is fluvial lacustrine feldspathic quartzose sandstone, containing conglomeratic sandstone and mudstone, interbedded with rhyolitic tuff and quartzose albitophyre, about 140-380 m thick,

MECHANISM OF BASIN FORMATION

The tectonic activities and setting of the Hengduan mountains is part of the tectonic activities of the Qinghai-Tibet plateau. Crustal deformation was generated from strong intra-continental convergence after the late Tertiary. The deformation zones of various patterns vary in size, spatial distribution and degree of generation, but reflect as a whole the general stress field of the crustal material in the plateau. Therefore the analysis of the tectonic features and regional stress field of Hengduan mountains should be linked with the tectonic deformation of the plateau itself.

After the disappearance of the Tethys, the Qinghai-Tibet plateau was surrounded by rigid blocks of Indian plate, Yangtze plate, North China plate and Tarim plate. The surrounding blocks existed not only as the boundary condition for the movement of plateau material, but also in the form of horizontal differential movement between the blocks. Northward push of the Indian plate, NWW push of the Yangtze plate and southward push of the North China-Tarim plates caused intra-continental convergence that put the whole plateau under a special compressive displacement field (Fig. 7). The collision between the Indian plate and the Asian continent could not be the product of a single movement, and could hardly happen at the same time on the contact surface, but happened many times from point contact to line contact and further to surface contact at several stages. For the collision between the Indian plate and the Asian continent, the Pamir arc top in the west first collided in the Palaeocene (about 70 million years ago), and the Assam arc top in the east collided subsequently in the Oligocene (about 50 million years ago), marking the contact between two plates in full length at the end of the Oligocene (about 30 million years ago). Further intense compression caused serious crustal deformation. The Himalaya kept rising and strong horizontal contraction occurred in the lithosphere (about 2,000 km) (Dewey and Bird, 1970; Tapponnier and Molnar, 1977), to form a massive earth crust.

The Hengduan mountain region and the three rivers area to the south, located tectonically between the Yangtze plate and the Indian plate, correspond to the Assam tectonic knot. The Yangtze plate is bounded by Ganze (Garze)-Litang and Songpan, Ganze (Garze), Yanyuan and Lijiang to the east of the Ailao mountain. The two sedimentary basins of Songpan-Ganze (Garze) and Yanyuan-Lijiang inherit the basement of the Yangtze landmass and the stable

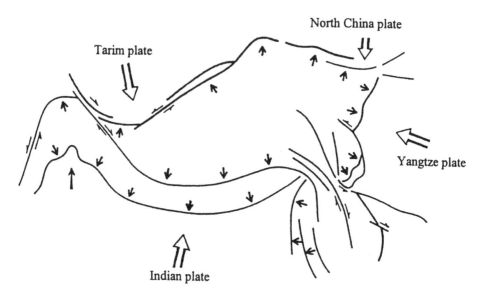

Fig. 7   Planar distribution of crustal horizontal stress of Qinghai-Tibet plateau

cover of the Palaeozoic and Mesozoic (Fig. 8). The Indian plate is bounded by western Yunnan and southeastern Tibet to the west of two boundaries of Menglian-Cangyuan ophiolite melange belt and north Lancangjiang belt. In the Sinian-Cambrian the sedimentary environment was mainly sea trough. In the Palaeozoic after the Ordovician it was stable platform sedimentation. The Hengduan mountains and the three rivers area are located at the junction of these two plates, where several ophiolite belts of the Tethys oceanic crust, Hercynian trench-arc tectonic and Indosinian trench-arc- basin tectonic and corresponding sedimentary zones were developed. A number of stratigraphic and metamorphic bodies were left from various ages and experienced structural compression and shearing to form elongated or rhombic block masses that vary in size (Chen Bingwei *et al.*, 1991). In terms of crustal structure, it is a plastic landmass in comparison with the Yangtze plate and the Indian plate.

Hengduan mountains-three rivers area is plastic with non-uniform mechanical properties of the crust and low degree of hardening. The NNE push of the Indian plate was even stronger at the location to the east of the Assam tectonic knot. The Yangtze plate crept relatively to SSE and multiple associations of complicated transcurrent faults appeared in a pinnate pattern (Fig. 2), and sometimes in a reticulated pattern. These transcurrent faults were generated by the shearing between the Yangtze plate and the Indian plate, as the result of simple shear. It also had something to do with the pure shear caused by N-S compression. From the properties of the plane of fracture of transcurrent fault, there existed on the plane of fracture not only shear stress but also normal stress when the crust was fractured. The plane of fracture can be tenso-shear or compresso-shear. From the direction of transcurrent fault, its dis-

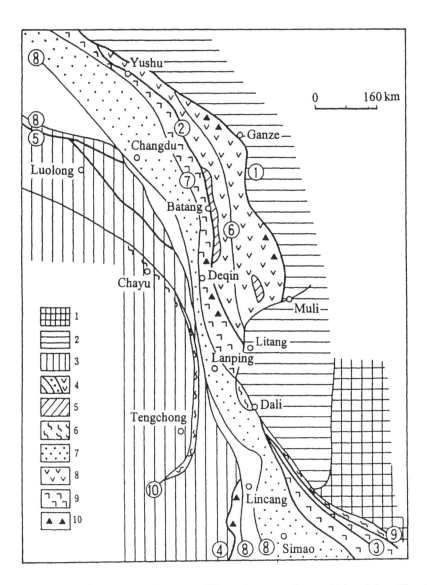

Fig. 8   Simplified map of tectonic units in eastern Tibet and western Yunnan (Adapted from Chen Bing-wei *et al.* ,1991).

1. Yangtze plate; 2. Yangtze plate margin; 3. Indian plate margin; 4. Transform tectonic zone; 5. Terrane in fold belts; 6. Metamorphic zone; 7. Interior Massif; 8. Trench-arc-basin fold belt; 9. Trench-arc fold belt; 10. Ophiolite-melange; ① Ganze(Garze)-Litang ophiolite belt; ② Jinshajiang ophiolite belt; ③ Tengtiaojiang ophiolite belt; ④ Cangyuan-Menglian ophiolite melange belt; ⑤ Dingqing(Dengqen) ophiolite belt; ⑥ Keludong-Xiangcheng fault; ⑦ Zigasi-Deqin(Deqen) fault; ⑧ Lancangjiang dual fault zone; ⑨ Honghe fault; ⑩ Tengchong fault.

placement could not be strictly horizontal because of gravity, and often appeared olique slid-
ing with a certain vertical component to form normal-transcurrent or reversed-transcurrent

fault. Considering the horizontal stress field, the former is tenso-shear, while the latter is compresso-shear. The result was a variety of strike-slip pull-apart basins of all sizes.

PALAEOCLIMATE AS CONTROL OF HYDROCARBON OCCURRENCE

Pull-apart basins in the Hengduan mountains are Tertiary continental lake basins. In the lake basin sediments the richness of organic matter is determined by the palaeoclimate and the ancient aqueous medium conditions. Variation of climate influences the pH value, salinity and redox potential of the aqueous medium and also the growth and multiplication of various organisms (Lanzhou Institute of Geology, 1981; Tian Zaiyi and Zhang Qingchun, 1993). Exploration practice in China indicates that organic content is higher in sedimentary formations under a humid, semi-humid climate, where excellent source rocks can be found. Under an arid, semi-arid climate, the lake basin is salinized and organic content decreases considerably.

The palaeoclimate in the Tertiary is featured by zonation from north to south (Fig. 9). Generally three zones of botanical palaeogeography from north to south are pan-Arctic Tertiary floral zone, palaeo-tropical, subtropical Tertiary floral zone and pan-Antarctic floral zone. In the Palaeogene the boundary between palaeo-tropical floral zone and pan- Arctic floral zone is approximately lattitude 42(north (Han Dexin *et al.* , 1980).

The northern zone covered northeast China, north China, northern Jiangsu, Nanyang and Jianghan, with a climate generally warm and humid. The vegetation was mainly deciduous and evergreen. Organisms flourished. Organic matter could be preserved in deep lake basins under a reducing environment to form dark-colored or variegated rock formations mainly as lignite and oil shale. In the west part of the northern zone, farther away from the sea, the climate tended to be dry due to lack of influence of oceanic climate.

The intermediate zone covered a vast area to the south of Tianshan mountain, Liupanshan mountain and Dabieshan mountain and to the north of Gangdise-Nanling. The sediments were red beds and contained anhydrite and salt, which indicated significant evaporation. The sediments were formed under an oxidizing or semi-oxidizing environment. The vegetation was featured by xerophytes of Ephedraceae. The climate was arid, sub-tropical.

The southern zone was southern Tibet and southern Guangxi and Guangdong to the south of Gangdise-Nanling. The climate was arid in the early stage and became humid in the middle and late stages due to the influence of the seasonal winds from the Indian Ocean and Pacific Ocean. The vegetation was mainly evergreen of the tropical and sub-tropical climate. A certain amount of coaly beds were formed. The sediments were dark-colored and variegated sedimentary formations, favorable to the formation of source rocks.

Tertiary sediments in the pull-apart basins in the Hengduan mountain region are over 4000 m thick, deposited under a transitional humid-dry climate. It is a prospective area for petroleum exploration. Oil/gas fields or seepages have been found in some basins, such as

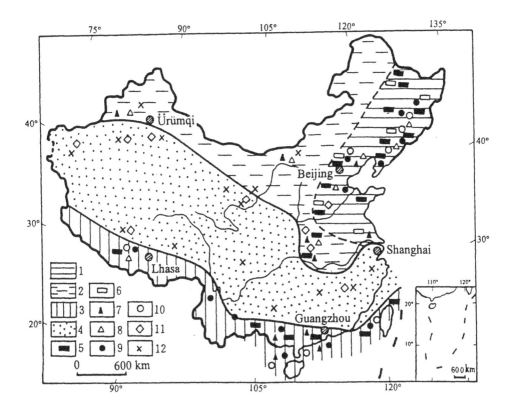

Fig. 9   Sketch map of early Tertiary palaeoclimate in China (Tian Zaiyi and Wan Lunkun, 1994)

1. Humid climate zone; 2. Semi-humid climate zone; 3. Climate zone, dry earlier and humid later; 4. Dry climate zone; 5. Evergreen,deciduous mixed forest; 6. Deciduous forest; 7. Dark mudstone; 8. Dark shale; 9. Coal; 10. Oil shale; 11. Anhydrite and salt; 12. Red beds.

Lunpola basin in Tibet, Jinggu basin , Longchuan basin in Yunnan. More than 30 oil/gas bearing basins have been identified in Thailand. More than 100 basins of different sizes remain unexplored. Good prospect can be predicted in such basins.

REFERENCE

1. Chen Bingwei, Li Yongsen, Qu Jingchuan, Wang Kaiyuan, Ai Changxing, and Zhu Zhizhi, 1991. Major tectonic issues in the three rivers area and their relations with mineral formation. Geological Publishing House, Beijing. (in Chinese with English abstract).

2. Dewey, J. F. and Bird J. M. Mountain belts and the new global tectonics. J. Geophys. Res. , 1970,75 (14): 2625-2647.

3. Han Dexin and Yang Qi (eds). Geology of Coal Fields in China. Coal Industry Press, Beijing, 1980, 415 pp. (in Chinese).

4. Lanzhou Institute of Geology, Chinese Academia Sinica. Formation, evolution and migration of non-ma-

rine hydrocarbons in China. Gansu People's Publishing House, Lanzhou, 1981, 269 pp. (in Chinese).

5. Li Chunyu, Liu Xueya, Wang Quan and Zhang Zhimeng. A tentative contribution to plate tectonics of China. Chin. Acad. Geol. Sci. (unpubl.). ,1979.

6. Molnar, P. and Tapponnier, P. Cenozoic tectonics of Asia: effects of continental collision. Science, 1975,189: 418-426.

7. Pan Guitang, Wang Peisheng, Xu Yaorong, Jiao Shupei, and Xiang Tianxiu. Cenozoic tectonic evolution of Qinghai-Tibet plateau. Geological Publishing House, Beijing, 1990,190 pp. (in Chinese with English abstract).

8. Tapponnier, P. and Molnar, P. Active faulting and tectonics of China. J. Geophys. Res. , 1987,82: 2905-2930.

9. Tapponnier, P. , Peltzer, G. and Armijo, R. On the mechanics of the collsion between India and Asia. In: M. P. Coward and A. C. Ries (eds), Collsion Tectonics. Geol. Soc. Spec. Publ. , 1977,19: 115-157.

10. Tian Zaiyi and Zhang Qingchun. Factors controlling oil and gas occurrence in a sedimentary basin. Acta Petrolei Sinica, 1993,14(4): 1-19. (in Chinese with English abstract).

11. Tian Zaiyi and Wan Lunkun. Tertiary petrofacies paleogeography and oil-gas prospect in China. Henan Petroleum, 1994,8(2): 1-10. (in Chinese with English abstract).